AUTHORS

Shannon McCaw

Beth Armstrong • Matt McCaw • Sarah Schuhl • Michelle Terry • Scott Valway

COVER PHOTOGRAPH

Golden Gate Bridge
A majestic and highly recognizable bridge over San Francisco Bay, the Golden Gate bridge is nearly two miles in length. Over 10 million visitors come to see the famous bridge each year.
©*iStockphoto.com/Juan Silva*

SMc curriculum
improving math.

ISBN: 978-1-935033-71-4

T 530992

1 2 3 4 5 6 7 8 9 10

ABOUT THE AUTHORS

Series Author

Shannon McCaw has been a classroom teacher in the Newberg and Parkrose School Districts. She has been trained in Professional Learning Communities, Differentiated Instruction and Critical Friends. Shannon currently works as a consultant with math teachers from over 100 districts around Oregon. Shannon's areas of expertise include the Common Core State Standards, curriculum alignment, assessment best practices and instructional strategies. She has a degree in Mathematics from George Fox University and a Masters of Arts in Secondary Math Education from Colorado College.

Contributing Authors

Beth Armstrong has been a classroom teacher in the Beaverton School District in Oregon. She has received training in Talented and Gifted Instruction. She has a Masters in Curriculum and Instruction from Washington State University.

Matt McCaw has been a classroom teacher, math/science TOSA and special education case-manager in several Oregon school districts. Matt has most recently worked as a curriculum developer and math coach for grades 6-8. He is trained in Differentiated Instruction, Professional Learning Communities, Critical Friends Groups and Understanding Poverty. Matt has a Masters of Special Education from Western Oregon University.

Sarah Schuhl has worked as an educator teaching seventh grade math through calculus and currently serves Centennial School District as the high school instructional coach and district-wide K-6 math instructional specialist. Sarah is also a Solution Tree associate and a consultant with On Target working to support teachers in the areas of math instruction and alignment to the Common Core State Standards, common assessments for all subjects and grade levels and professional learning communities. Since 2010, Sarah has been a member and chair of the National Council of Teachers of Mathematics editorial panel for their Mathematics Teacher journal. Sarah has a Masters of Science in Teaching Mathematics from Portland State University.

Michelle Terry has been a classroom teacher in the Estacada and Newberg School Districts in Oregon. Michelle has received training in Professional Learning Communities, Critical Friends, ELL Instructional Strategies, Proficiency-Based Grading and Lesson Design, Power Strategies for Effective Teaching, and Classroom Love and Logic. Michelle has an Interdisciplinary Masters from Western Oregon University. She currently teaches mathematics at Newberg High School.

Scott Valway has been a classroom teacher in the Tigard-Tualatin, Newberg and Parkrose School Districts in Oregon. Scott has been trained in Differentiated Instruction, Professional Learning Communities, Critical Friends, Discovering Algebra, Pre-Advanced Placement, Assessment Writing and Credit by Proficiency. Scott has a Masters of Science in Teaching from Oregon State University. He currently teaches math at Parkrose High School.

COMMON CORE STATE STANDARDS

Grade 8 Overview

The complete set of Common Core State Standards can be found at http://www.corestandards.org/. This book focuses on the highlighted Common Core State Standards shown below.

The Number System

- Know that there are numbers that are not rational, and approximate them by rational numbers.

Expressions and Equations

- Work with radicals and integer exponents.
- Understand the connections between proportional relationships, lines and linear equations.
- Analyze and solve linear equations and pairs of simultaneous linear equations.

Functions

- Define, evaluate and compare functions.
- Use functions to model relationships between quantities.

Geometry

- Understand congruence and similarity using physical models, transparencies or geometry software.
- Understand and apply the Pythagorean Theorem.
- Solve real-world and mathematical problems involving volume of cylinders, cones and spheres.

Statistics and Probability

- Investigate patterns of association in bivariate data.

Mathematical Practices

1. Make sense of problems and persevere in solving them.
2. Reason abstractly and quantitatively.
3. Construct viable arguments and critique the reasoning of others.
4. Model with mathematics.
5. Use appropriate tools strategically.
6. Attend to precision.
7. Look for and make use of structure.
8. Look for and express regularity in repeated reasoning.

CORE FOCUS ON GEOMETRY

CONTENTS IN BRIEF

CORE FOCUS ON GEOMETRY

BLOCK 1 ~ ANGLES AND TRIANGLES

BLOCK 2 ~ THE PYTHAGOREAN THEOREM

BLOCK 3 ~ TRANSFORMATIONS

BLOCK 4 ~ EXPONENTS AND VOLUME

HOW TO USE YOUR MATH BOOK

Your math book has features that will help you be successful in this course. Use this guide to help you understand how to use this book.

Lesson Target

Look in this box at the beginning of every lesson to know what you will be learning about in each lesson.

Vocabulary

Each new vocabulary word is printed in **red**. The definition can be found with the word. You can also find the definition of the word in the glossary which is in the back of this book.

Explore!

Some lessons have **EXPLORE!** activities which allow you to discover mathematical concepts. Look for these activities in the Table of Contents and in lessons next to the purple line.

Examples

Examples are useful because they remind you how to work through different types of problems. Look for the word **EXAMPLE** and the green line.

Helpful Hints

Helpful hints and important things to remember can be found in green callout boxes.

Blue Boxes

A blue box holds important information or a process that will be used in that lesson. Not every lesson has a blue box.

This calculator icon will appear in Lessons and Exercises where a calculator is needed. Your teacher may want you to use your calculator at other times, too. If you are unsure, make sure to ask if it is the right time to use it.

Exercises

The **EXERCISES** are a place for you to find practice problems to determine if you understand the lesson's target. You can find selected answers in the back of this book so you can check your progress.

Review

The **REVIEW** provides a set of problems for you to practice concepts you have already learned in this book. The **REVIEW** follows the **EXERCISES** in each lesson. There is also a **REVIEW** section at the end of each Block.

Tic-Tac-Toe Activities

Each Block has a Tic-Tac-Toe board at the beginning with activities that extend beyond the Common Core State Standards. The Tic-Tac-Toe activities described on the board can be found throughout each Block in yellow boxes.

Career Focus

At the end of each Block, you will find an autobiography of an individual. Each one describes what they like about their job and how math is used in their career.

Core Focus on Math
Stage 3

CORE FOCUS ON GEOMETRY

BLOCK 1 ~ ANGLES AND TRIANGLES

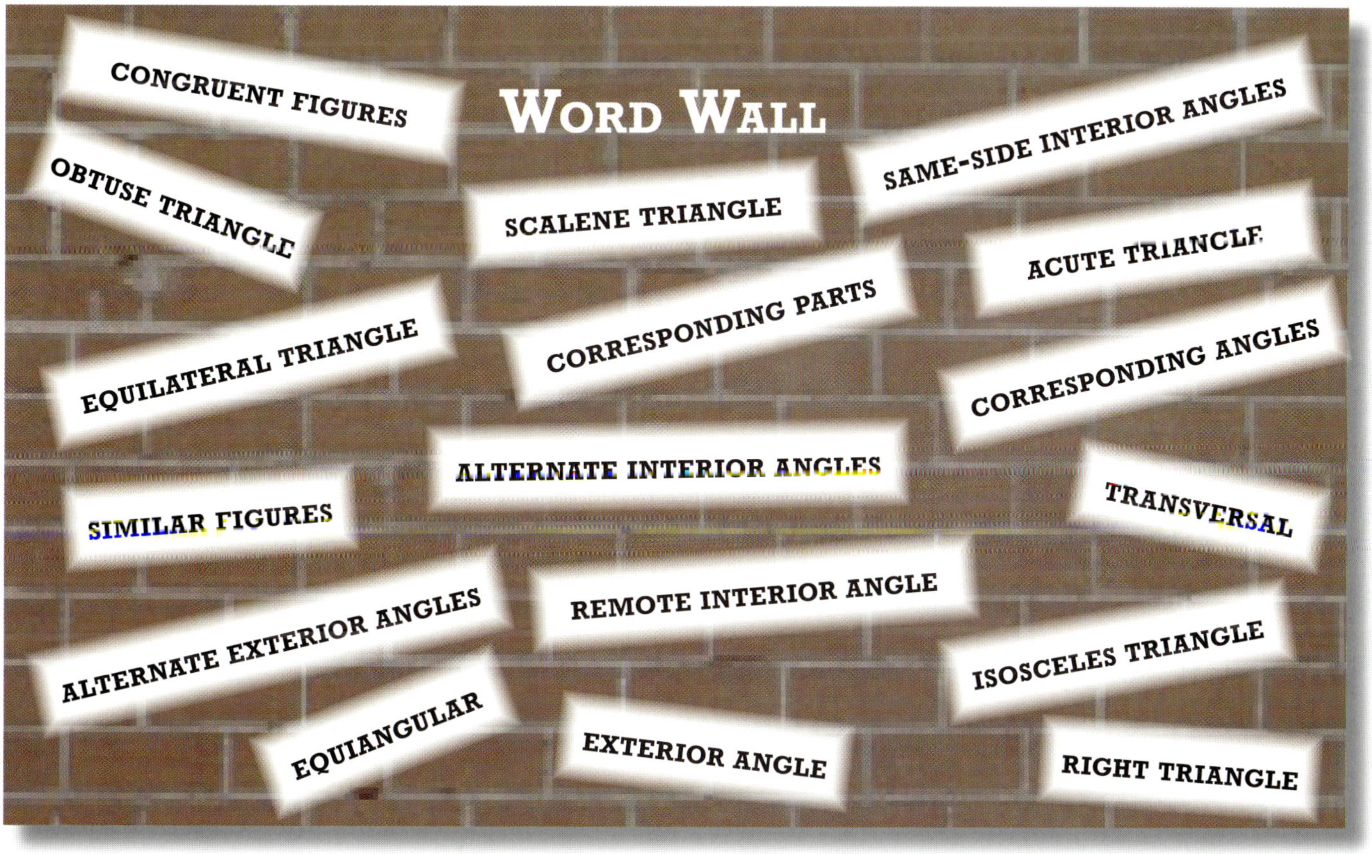

BLOCK 1 ~ ANGLES AND TRIANGLES
TIC-TAC-TOE

High School Choices

Investigate math courses offered at the high school you will attend.

See page 17 for details.

Four Corners

Discover a rule about the angle sum of any quadrilateral.

See page 22 for details.

Triangle Mobile

Create an original piece of art showing the properties of triangles.

See page 26 for details.

Transversal Collage

Find or take pictures of transversals and display them.

See page 43 for details.

Sum of Interior Angles

Discover the rule for the sum of the interior angles in any polygon.

See page 44 for details.

Constructing Segments

Use a compass and straightedge to duplicate and combine segments.

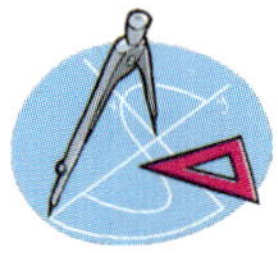

See page 37 for details.

Book of Poetry

Write poems about special angle pairs. Make an illustrated poetry booklet.

See page 12 for details.

Classification Game

Make a memory card game where players classify triangles by sides and angles.

See page 17 for details.

Too Tall To Know

Find the height of objects that are too tall to measure with a measuring tape.

See page 36 for details.

ALTERNATE EXTERIOR AND INTERIOR ANGLES

LESSON 1.1

Recognize and apply properties of alternate exterior and alternate interior angles.

In this lesson you will be introduced to two special angle pairs. There are many special angle pairs. In previous math courses you may have learned about complementary angles, linear pairs, supplementary angles and vertical angles. Below, each of these special angle pairs are revisited with a definition and a diagram.

Complementary Angles
Two angles with a sum of 90°.

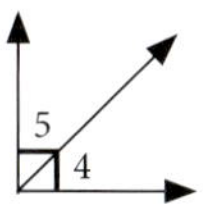

$m\angle 5 + m\angle 4 = 90°$

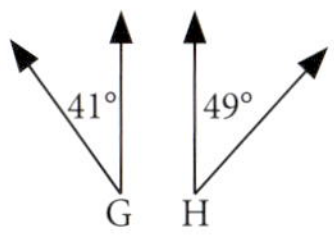

$m\angle G + m\angle H = 90°$

Linear Pair
Two adjacent angles with a sum of 180°.

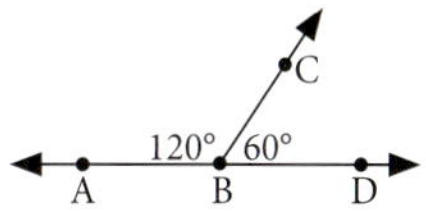

$m\angle ABC + m\angle CBD = 180°$

Supplementary Angles
Two angles with a sum of 180°.

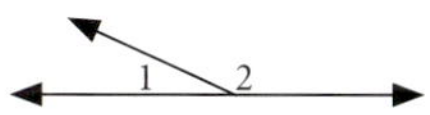

$m\angle 1 + m\angle 2 = 180°$

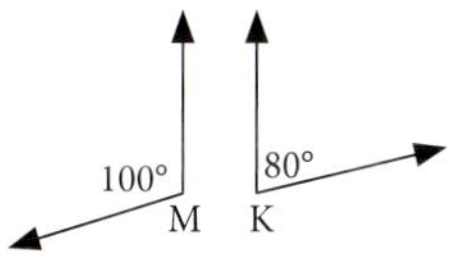

$m\angle M + m\angle K = 180°$

Vertical Angles
Two congruent, non-adjacent angles formed by intersecting lines.

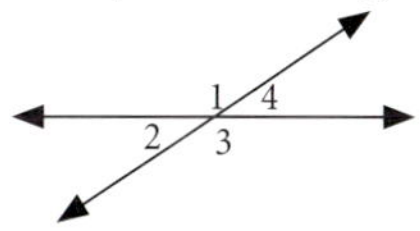

$m\angle 1 = m\angle 3$ $m\angle 2 = m\angle 4$

The words "interior" and "exterior" are used in many non-math situations. The word "interior" is used to describe things that are inside. Things that are "exterior" are on the outside. The two special angle pairs in this lesson are called alternate interior angles and alternate exterior angles. Look at the diagrams in the table below. What does it mean for a pair of angles to be alternate exterior angles? What about alternate interior angles?

Alternate Exterior Angles		Alternate Interior Angles	
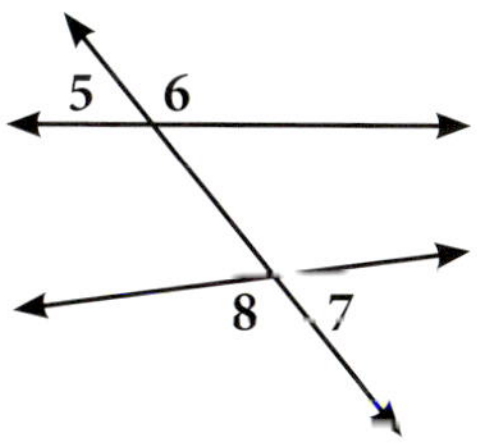	$\angle 6$ and $\angle 8$ are alternate exterior angles. $\angle 5$ and $\angle 7$ are alternate exterior angles.	$\angle 1$ and $\angle 2$ are alternate interior angles. $\angle 3$ and $\angle 4$ are alternate interior angles.	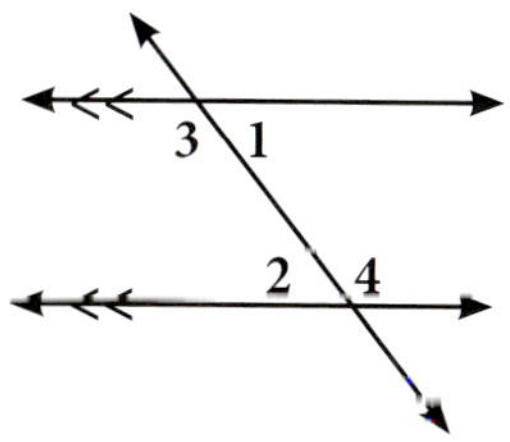
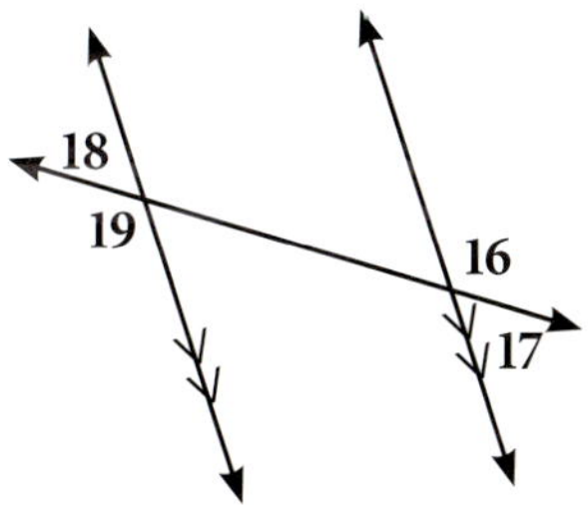	$\angle 17$ and $\angle 18$ are alternate exterior angles. $\angle 16$ and $\angle 19$ are alternate exterior angles.	$\angle 10$ and $\angle 13$ are alternate interior angles. $\angle 12$ and $\angle 14$ are alternate interior angles.	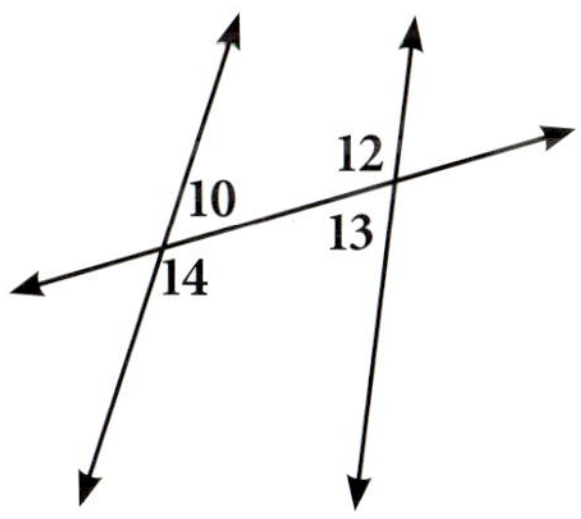

A **transversal** is a line that intersects two or more lines. **Alternate exterior angles** are two angles on the outside of two lines and on opposite sides of a transversal. **Alternate interior angles** are two angles on the inside of two lines and on opposite sides of a transversal.

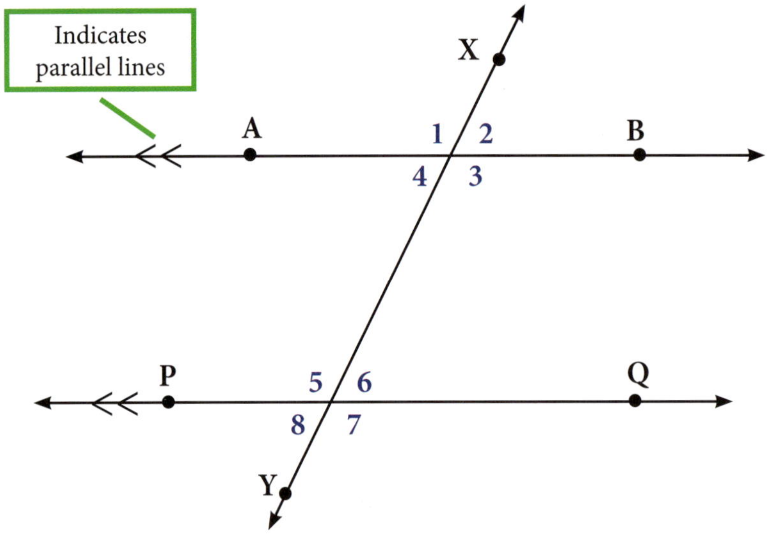

In this diagram, $\overleftrightarrow{XY}$ is the transversal.

$\overleftrightarrow{AB} \parallel \overleftrightarrow{PQ}$ reads, "Line AB is parallel to line PQ."

EXPLORE! **ALTERNATE EXTERIOR AND ALTERNATE INTERIOR**

Step 1: Draw two parallel lines intersected by a transversal. Make the diagram large enough to easily measure the angles with a protractor. Label your angles 1 through 8.

Step 2: Use a protractor to measure the eight angles on the drawing from **Step 1**. Record their measures.

Step 3: Which pairs of angles in your diagram are alternate exterior angles? What do you notice about the degree measure of each pair of alternate exterior angles?

Step 4: Which pairs of angles are alternate interior angles? What do you notice about the degree measure of each pair of alternate interior angles?

Step 5: Trace the figure from the top of this page onto a piece of paper.

Step 6: Slide your tracing until angles 1, 2, 3 and 4 are on top of angles 5, 6, 7 and 8.
- **a.** What do you observe? Explain.
- **b.** Does this confirm your conclusion in **Steps 3 and 4**?

Step 7: Draw two lines that are not parallel. Draw a transversal that intersects both lines.

Step 8: Use a protractor to measure the eight angles created in **Step 7**.

Step 9: Write a conclusion about the alternate interior angles and alternate exterior angles formed by parallel lines compared to those formed by non-parallel lines.

Alternate Interior and Alternate Exterior Angles

If two parallel lines are intersected by a transversal, then the alternate interior angles are congruent.

If two parallel lines are intersected by a transversal, then the alternate exterior angles are congruent.

EXAMPLE 1 **Name the angle relationship between ∠1 and ∠2. Determine whether ∠1 and ∠2 are congruent.**

a.

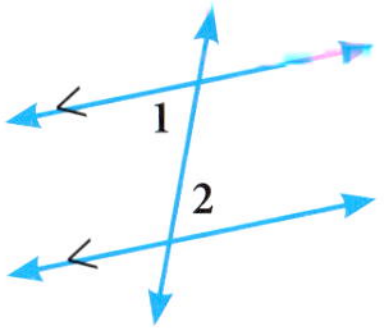

b.

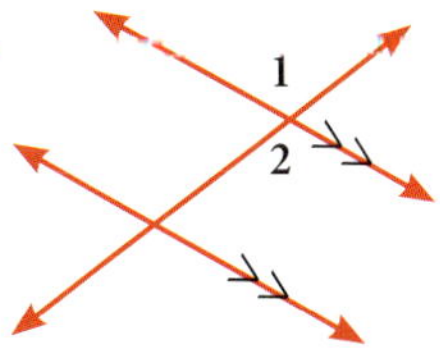

c.

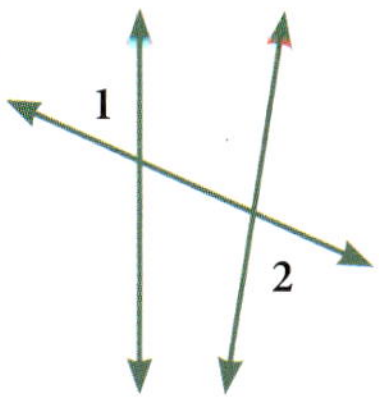

Solutions

a. Alternate interior. The angles are congruent because the lines are parallel.
b. Vertical angles. Vertical angles are congruent to one another.
c. Alternate exterior. The lines are not parallel, so the angles are not congruent.

EXAMPLE 2 **Identify the special angle pair relationship. Solve for x.**

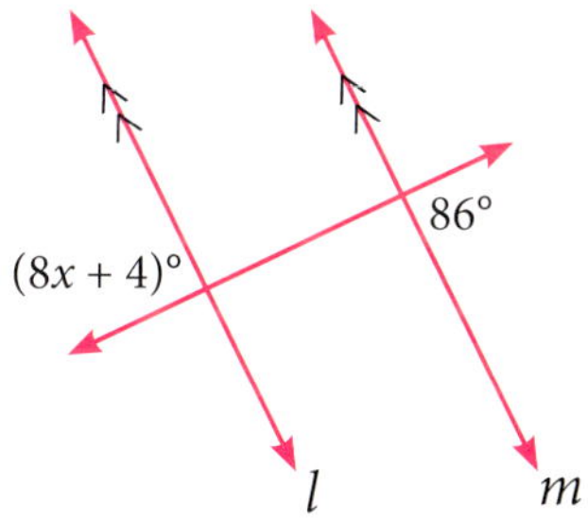

Solution

The angles are alternate exterior angles.

Lines l and m are parallel so the alternate exterior angles are congruent.

Subtract 4 from each side of the equation.

Divide each side of the equation by 8.

$$8x + 4 = 86$$
$$-4 \quad -4$$
$$\frac{8x}{8} = \frac{82}{8}$$
$$x = 10.25$$

☑ Check the solution. $(8x + 4)° = 86°$

Substitute 10.25 for x. $(8(10.25) + 4)° \stackrel{?}{=} 86°$

Multiply. $82 + 4° \stackrel{?}{=} 86°$

Add. $86° = 86°$

The angles are alternate exterior angles and $x = 10.25$

EXAMPLE 3

Use the figure at right.
a. Solve for x.
b. Find the measure of the angles.

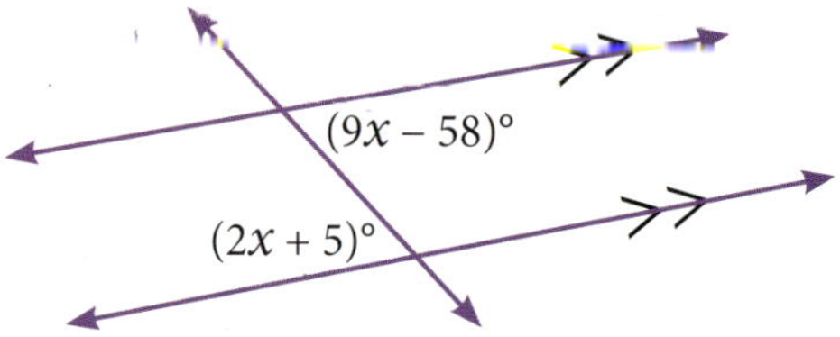

SOLUTIONS

a. The lines are parallel so alternate interior angles are congruent.

$$9x - 58 = 2x + 5$$

Subtract $2x$ from each side of the equation.

$$\begin{aligned} 9x - 58 - 2x &= 2x + 5 - 2x \\ 7x - 58 &= 5 \end{aligned}$$

Add 58 to each side of the equation.

$$7x - 58 + 58 = 5 + 58$$

Divide each side of the equation by 7.

$$\frac{7x}{7} = \frac{63}{7}$$

$$x = 9$$

b. Write the given expression for each angle.

Write the given expression for each angle.	$(9x - 58)°$	$(2x + 5)°$
Substitute 9 for x.	$(9(9) - 58)°$	$(2(9) + 5)°$
Multiply.	$(81 - 58)°$	$(18 + 5)°$
Simplify.	23°	23°

Each angle measures 23°. Alternate interior angles are congruent so **part b** verifies the solution for x.

EXERCISES

Use one of the following special angle pairs to identify the relationship of the angles shown.

Alternate Exterior	Alternate Interior	Vertical	Linear Pair

1.

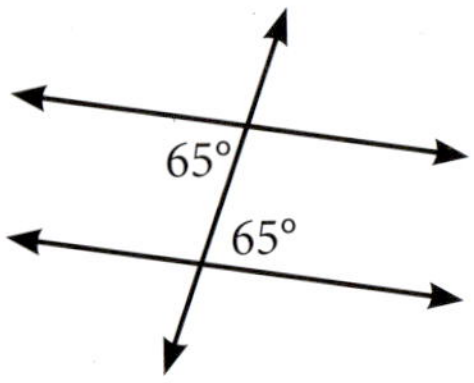

2.

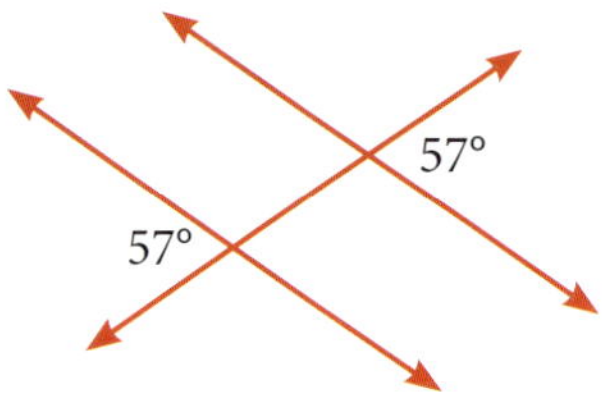

3.

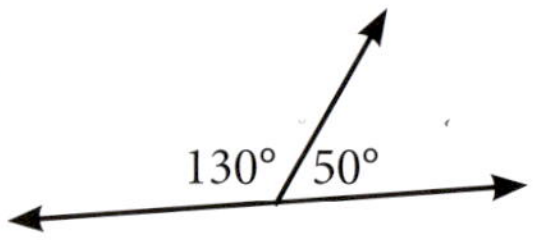

4.

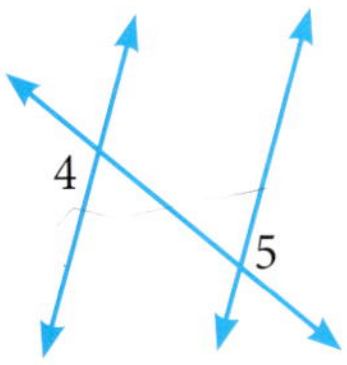

5.

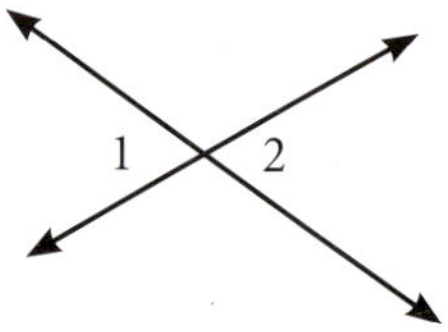

6.

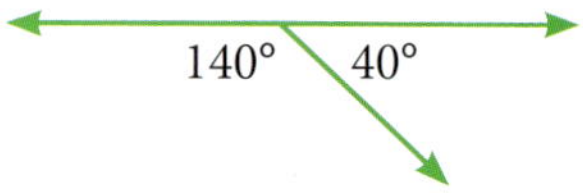

7.

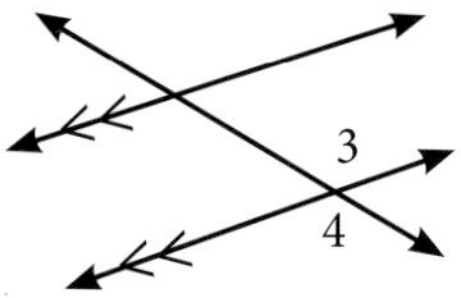

8.

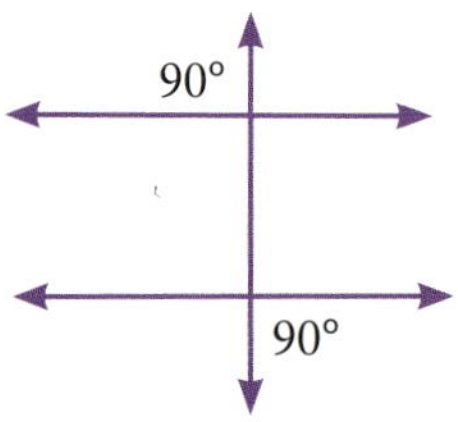

9.

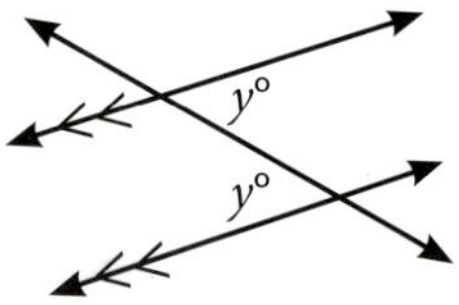

Name the special angle pair relationship between $\angle 1$ and $\angle 2$. Explain whether $\angle 1 \cong \angle 2$.

10.

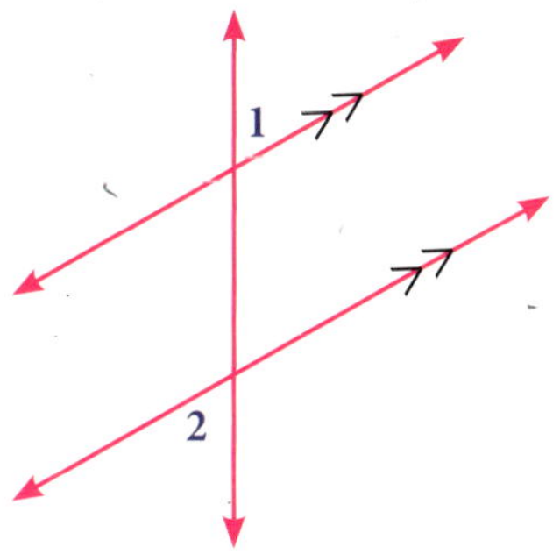

11.

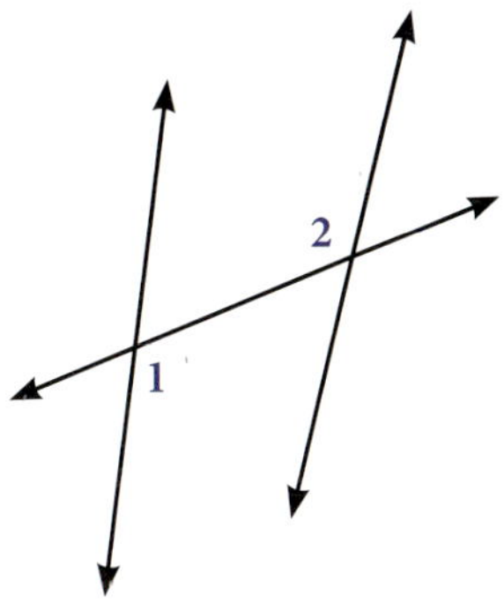

12.

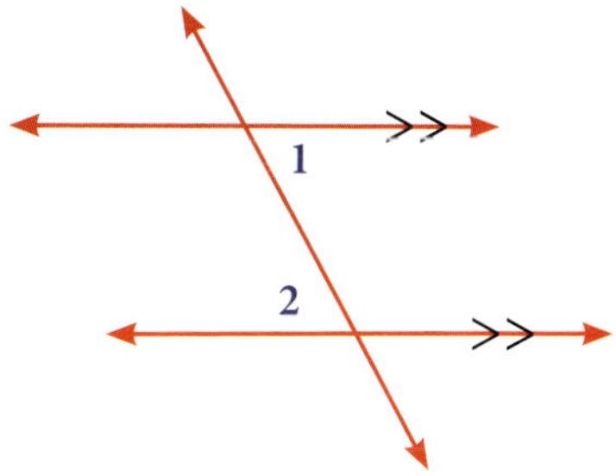

Name the special angle pair relationship. Solve for x.

13.

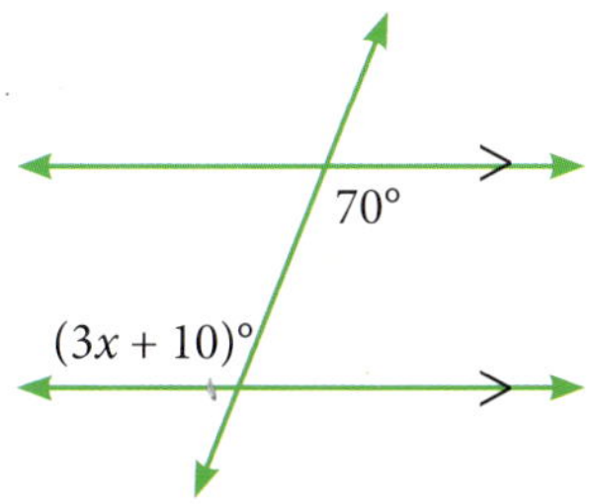

14.

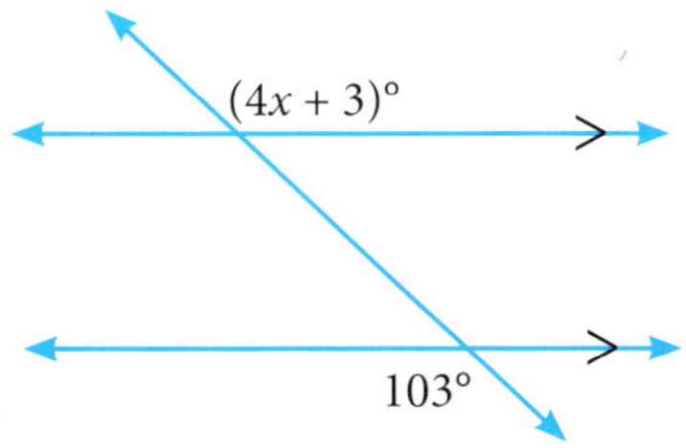

15.

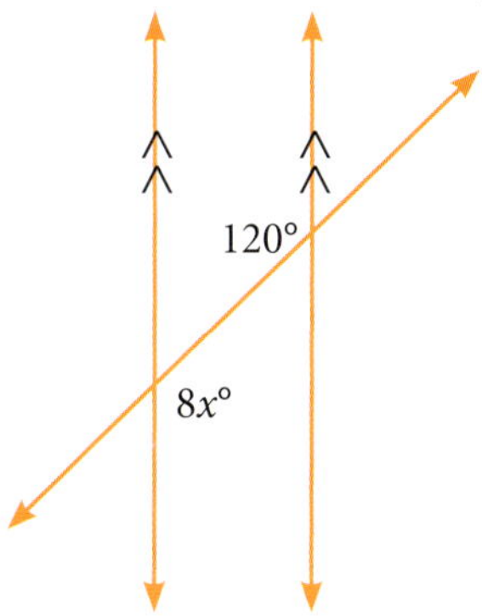

Solve for x. Then find the measure of each identified angle. Check your solution.

16.

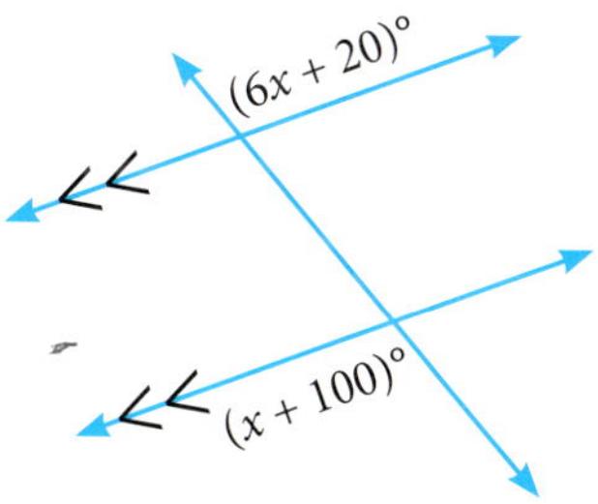

17.

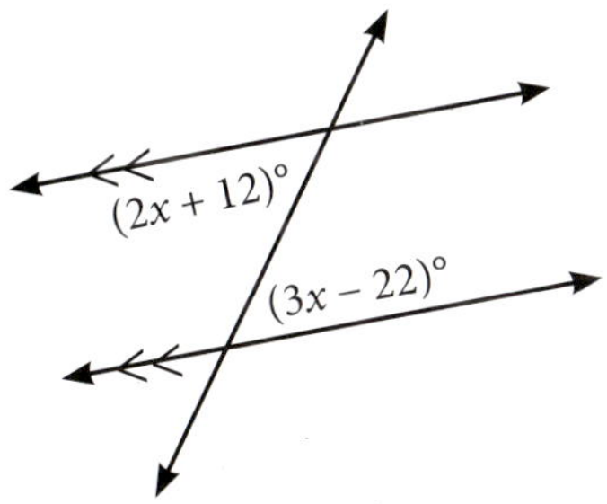

18.

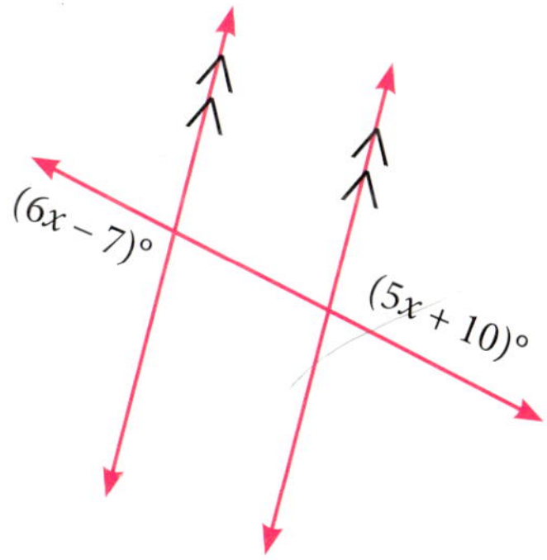

19.

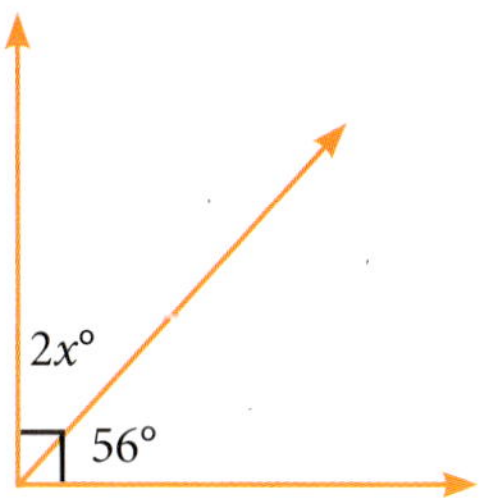

20. $\angle 1$ and $\angle 2$ are vertical angles
$m\angle 1 = (5x + 7)°$
$m\angle 2 = (3x + 15)°$

21. $\angle 5$ and $\angle 8$ are supplementary
$m\angle 5 = (3x - 40)°$
$m\angle 8 = (7x - 120)°$

22. Explain how to distinguish between alternate exterior angles and alternate interior angles.

CORRESPONDING AND SAME-SIDE INTERIOR ANGLES

LESSON 1.2

Recognize and apply properties of corresponding and same-side interior angles.

Two additional special pairs of angles formed by two lines and a transversal are corresponding angles and same-side interior angles.

Corresponding angles are two angles on the same side of a transversal. One angle is an exterior angle and the other angle is an interior angle. They are not adjacent angles.

Same-side interior angles are between the two lines on the same side of a transversal.

Corresponding angles:
$\angle 1$ and $\angle 6$
$\angle 2$ and $\angle 5$
$\angle 3$ and $\angle 8$
$\angle 4$ and $\angle 7$

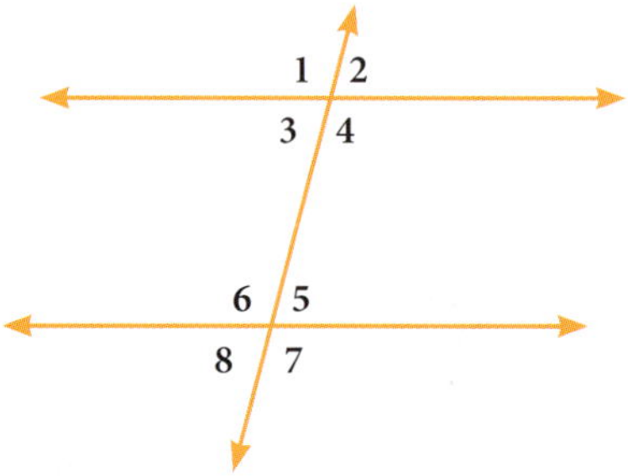

Same-side interior angles:
$\angle 3$ and $\angle 6$
$\angle 4$ and $\angle 5$

EXPLORE! MORE ANGLE PAIRS

Step 1: If two parallel lines are cut by a transversal, predict:
- **a.** What is the relationship between a pair of same-side interior angles?
- **b.** What is the relationship between a pair of corresponding angles?

Step 2: Use a straight edge to draw a pair of parallel lines with a transversal. Make the figure large enough that you can measure the angles with a protractor.

Step 3: Use a protractor to measure each of the eight angles. Record each measure.

Step 4: What do you observe about the corresponding angle measures?

Step 5: Look at the measures of the same-side interior angles. What do you observe about their measures?

Step 6: Must the lines be parallel for these relationships to occur? Draw a figure without parallel lines to test your answer.

EXAMPLE 1 **Name the special angle pair relationship between ∠1 and ∠2.**

a.

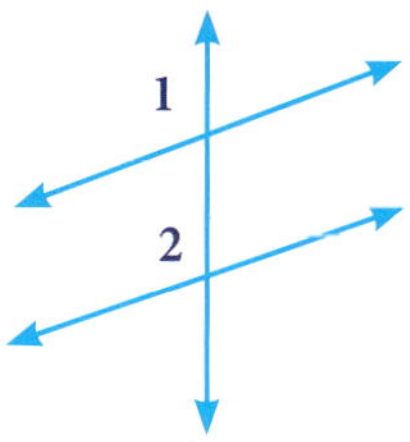

b.

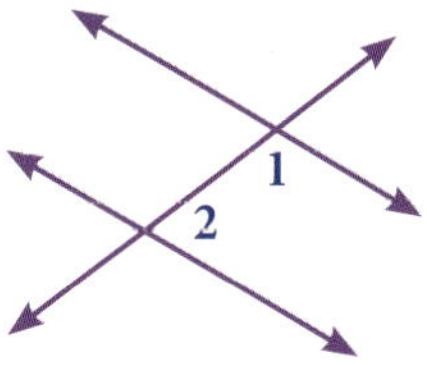

c.

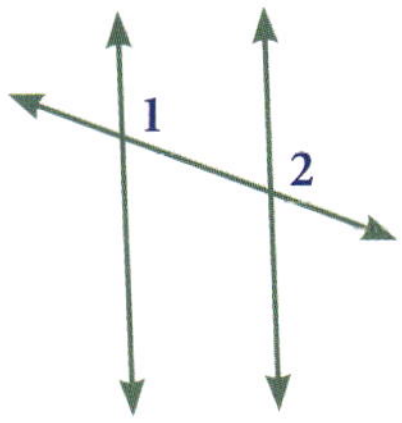

SOLUTIONS

a. corresponding angles
b. same-side interior angles
c. corresponding angles

ANGLES FORMED BY TWO PARALLEL LINES INTERSECTED BY A TRANSVERSAL

Alternate exterior angles are congruent.
$\angle 1 \cong \angle 7$
$\angle 2 \cong \angle 8$

Alternate interior angles are congruent.
$\angle 4 \cong \angle 6$
$\angle 3 \cong \angle 5$

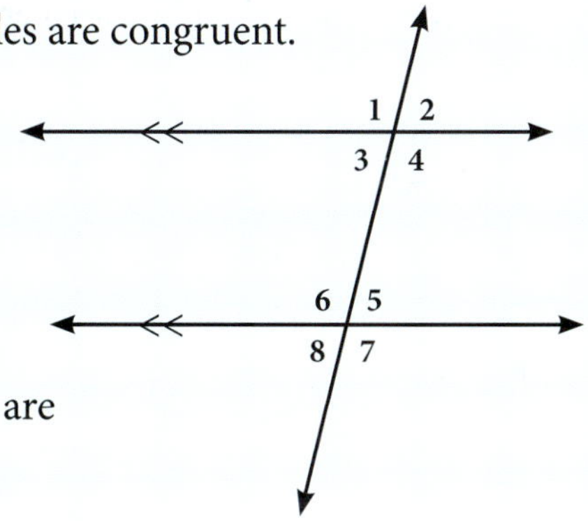

Corresponding angles are congruent.
$\angle 1 \cong \angle 6$ $\angle 2 \cong \angle 5$
$\angle 3 \cong \angle 8$ $\angle 4 \cong \angle 7$

Same-side interior angles are supplementary.
$m\angle 4 + m\angle 5 = 180°$
$m\angle 3 + m\angle 6 = 180°$

Claudia and Ping come to different conclusions when finding the missing angles in the diagram below. Determine who is correct and why.

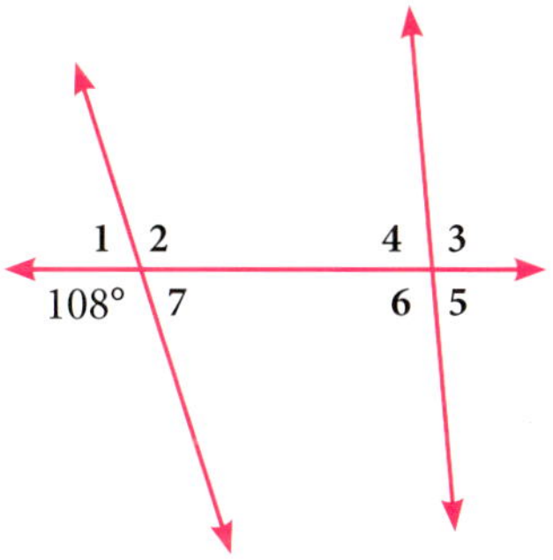

Claudia's Work	Ping's Work
$m\angle 1 = 72°$	$m\angle 1 = 72°$
$m\angle 2 = 108°$	$m\angle 2 = 108°$
$m\angle 3 = 108°$	$m\angle 3$ = cannot determine
$m\angle 4 = 72°$	$m\angle 4$ = cannot determine
$m\angle 5 = 72°$	$m\angle 5$ = cannot determine
$m\angle 6 = 108°$	$m\angle 6$ = cannot determine
$m\angle 7 = 72°$	$m\angle 7 = 72°$

Two lines in the diagram are intersected by a transversal. It is not known whether the lines are parallel. Unless lines are marked parallel it cannot be assumed they are parallel. Ping is correct. The measures of ∠3, ∠4, ∠5 and ∠6 cannot be determined.

EXAMPLE 2 **Write an equation and solve for x.**

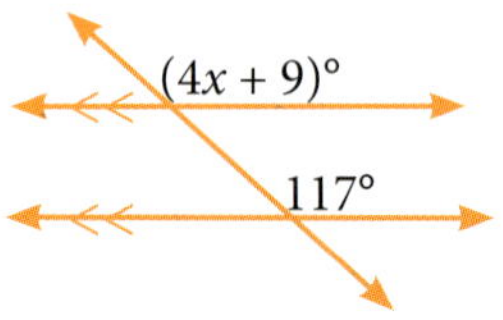

SOLUTION

Corresponding angles have equal measures.

Write an equation. $4x + 9 = 117$

Subtract 9 from each side of the equation. $-9 \quad -9$

Divide each side of the equation by 4. $\frac{4x}{4} = \frac{108}{4}$

$x = 27$

☑ Check the solution.

Substitute 27 for x in the equation.

$4(27) + 9 \overset{?}{=} 117$

$108 + 9 \overset{?}{=} 117$

$117 = 117$

EXAMPLE 3 **Write an equation and solve for x. Then find the measure of each identified angle.**

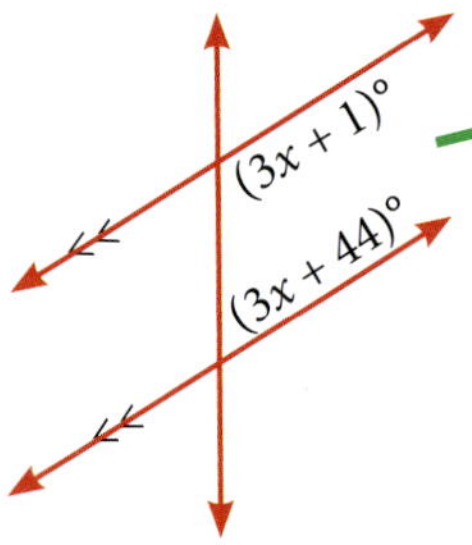

The identified angles are the unknown angles that are marked with algebraic expressions.

SOLUTION

The lines are parallel so the same-side interior angles sum to 180°.

Write an equation. $(3x + 1) + (3x + 44) = 180$

Combine like terms. $6x + 45 = 180$

Subtract 45 from each side of the equation. $-45 \quad -45$

Divide each side of the equation by 6. $\frac{6x}{6} = \frac{135}{6}$

$x = 22.5$

Find the measure of each angle.	$(3x + 1)°$	$(3x + 44)°$
Substitute 22.5 for x.	$(3(22.5) + 1)°$	$(3(22.5) + 44)°$
Multiply.	$(67.5 + 1)°$	$(67.5 + 44)°$
Add.	68.5°	111.5°

☑ Check the solution.

$68.5° + 111.5° \overset{?}{=} 180°$

$180° = 180°$

EXERCISES

Name the special angle pair relationship between ∠1 and ∠2 in each diagram.

1.

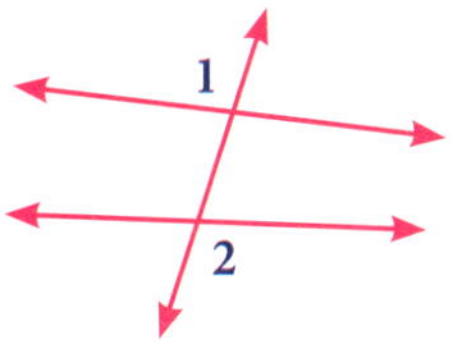

2.

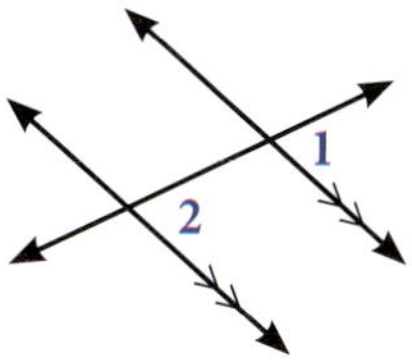

3.

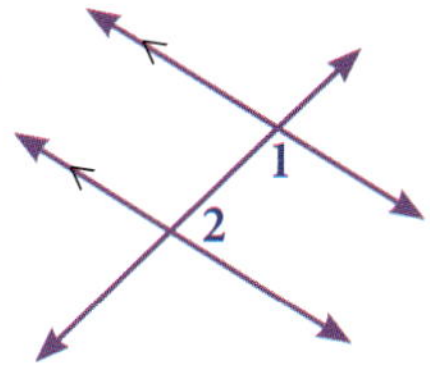

4.

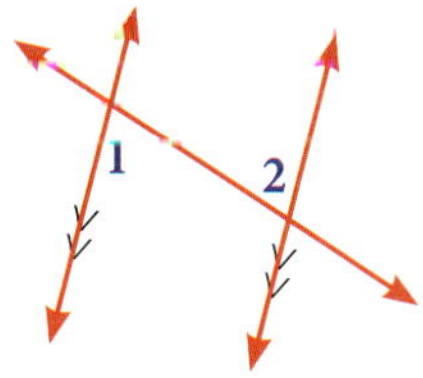

5.

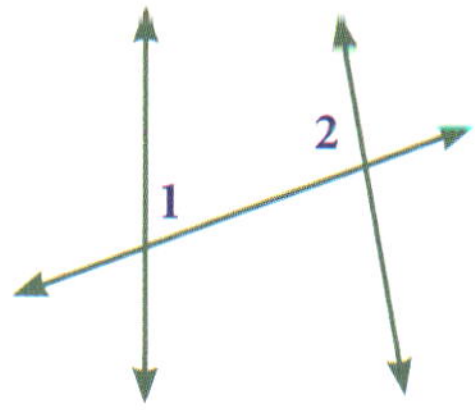

6.

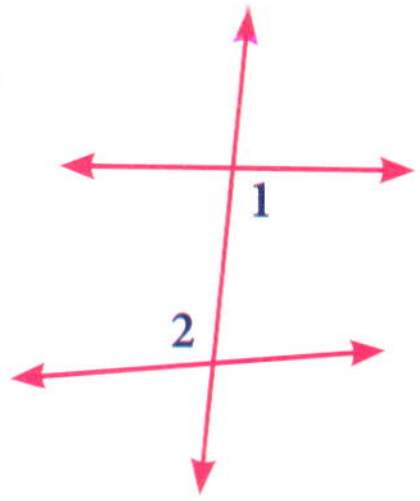

7.

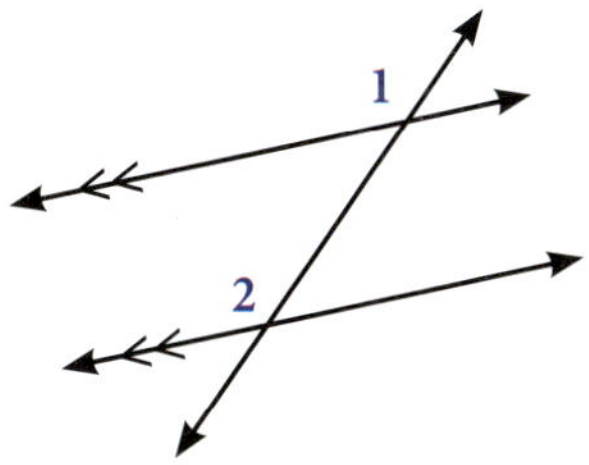

8.

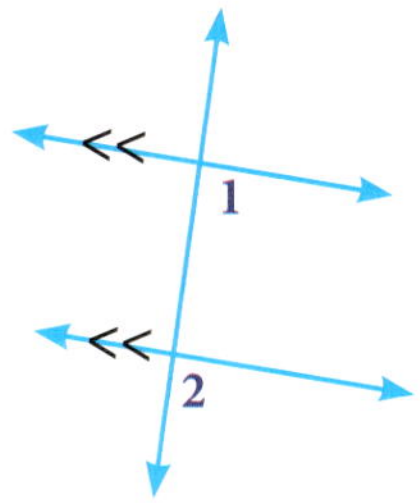

9.

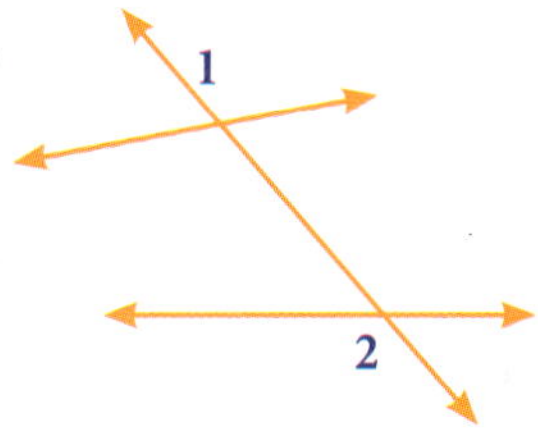

10. Which diagrams in **Exercises 1-9** have parallel lines?

Solve for *x*.

11.

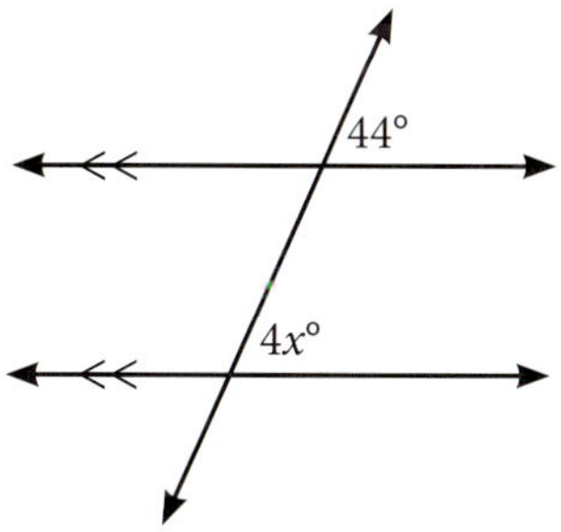

12.

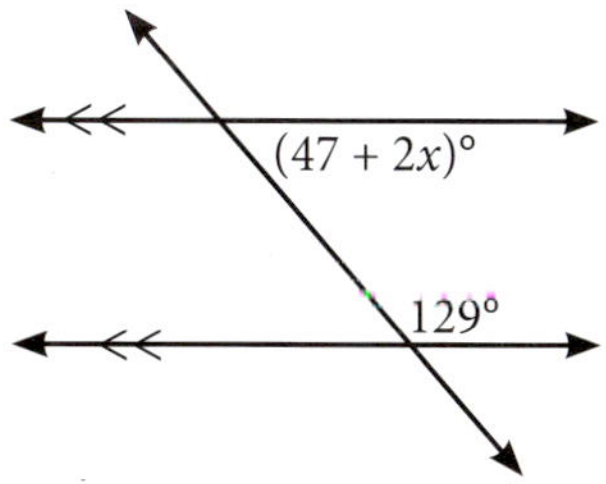

13.

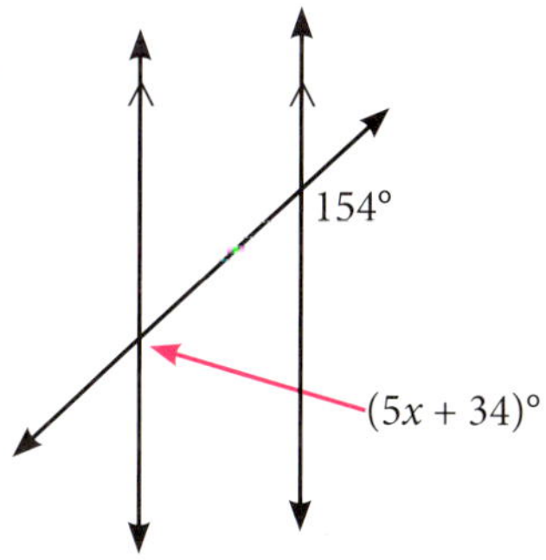

Solve for *x*. Then find the measure of each identified angle. Check your solution.

14.

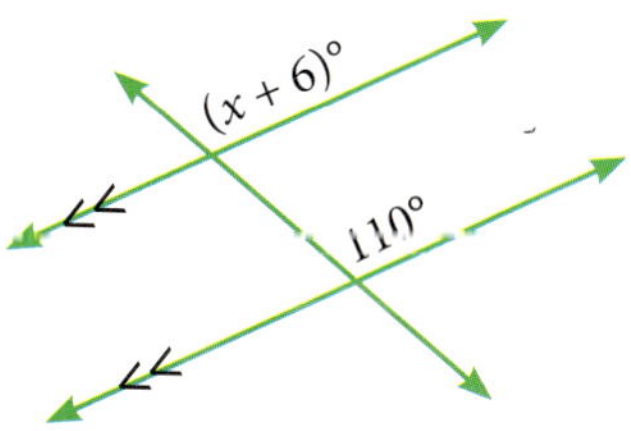

15.

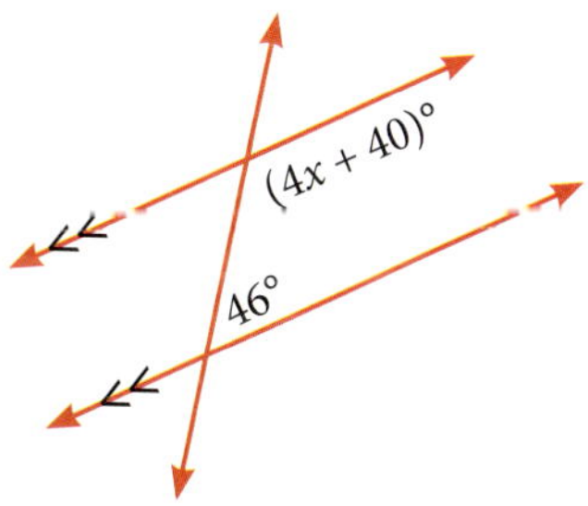

16.

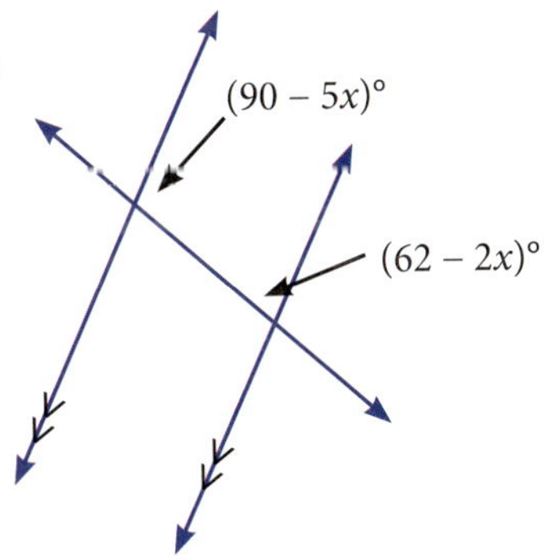

Find the measure of each numbered angle. Check your solution.

17.

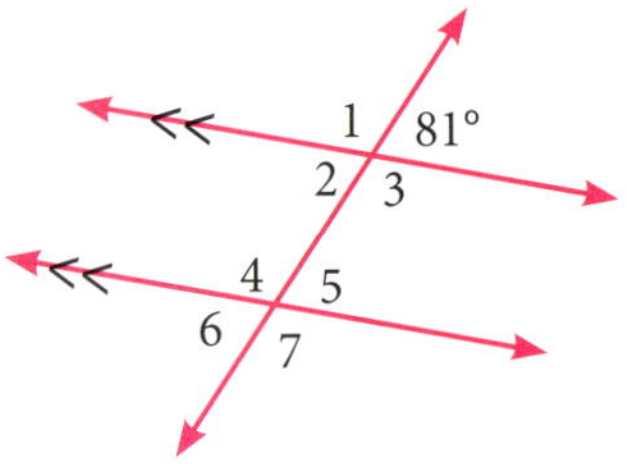

18.

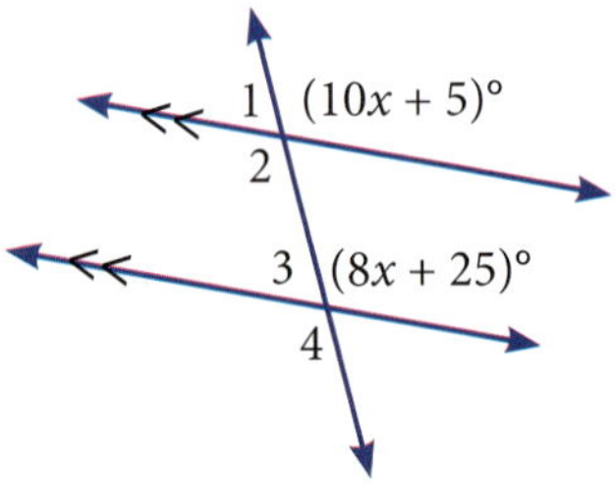

19.

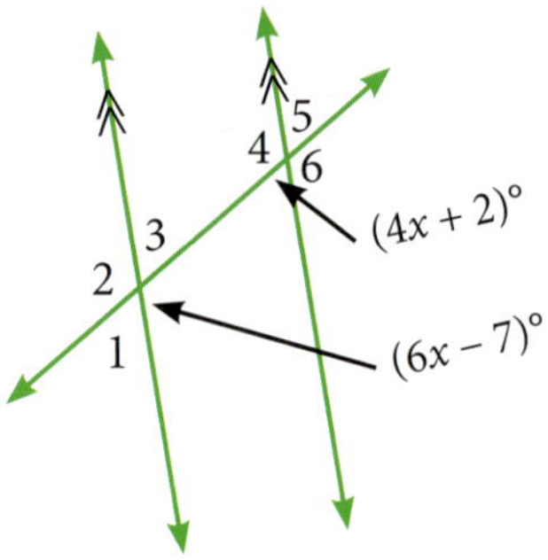

Sketch and label a diagram to represent each pair of angles. Label each pair of angles ∠1 and ∠2.

20. congruent same-side interior angles

21. alternate interior angles that are not congruent

22. acute corresponding angles

23. right alternate exterior angles

24. obtuse same-side interior angles

25. acute vertical angles

REVIEW

Use a protractor and/or straightedge to draw each diagram. Label each diagram.

26. $m\angle GUM = 162°$

27. $m\angle RPQ = 20°$

28. 3 parallel lines with transversal $\overleftrightarrow{AB}$

29. right angle named ∠2

Name each special angle pair. Solve for *x*.

30.

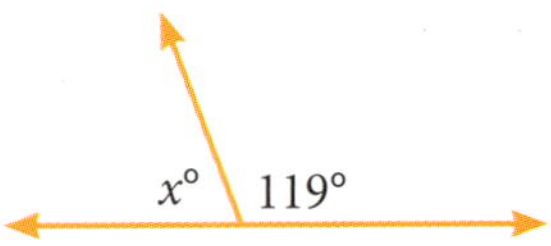

31.

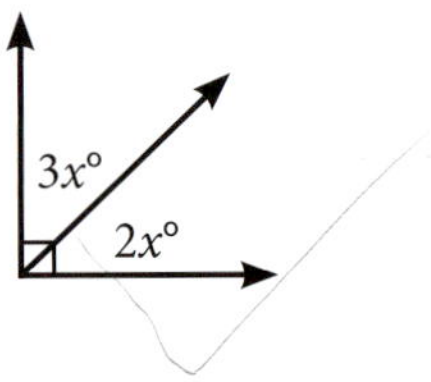

32.

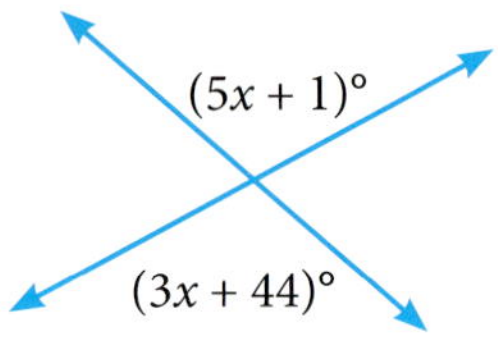

Tic-Tac-Toe ~ Book of Poetry

Step 1: Write a separate poem about each special angle pair: linear pair, supplementary, complementary, vertical, alternate interior, alternate exterior, corresponding and same-side interior angles. Title each poem with the name of the special angle pair.

Step 2: Put the eight poems in a booklet. Include diagrams and illustrations.

CLASSIFYING TRIANGLES

LESSON 1.3

Classify triangles based on sides and angles.

A triangle is a polygon with three sides. A triangle can be classified by its angles and by the lengths of its sides. There are four types of angles: acute, obtuse, right and straight. When a triangle is classified by its angles, three of these same words are used. Since a triangle cannot have a straight angle, there are three classifications of triangles by angles.

Acute triangles **have three acute angles**	Right triangles **have one right angle**	Obtuse triangles **have one obtuse angle**
	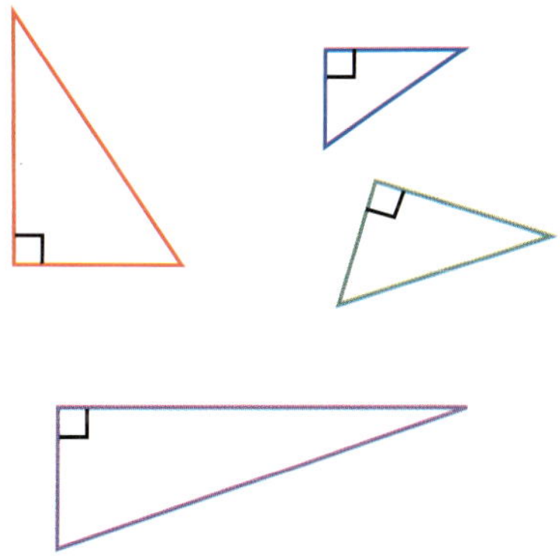	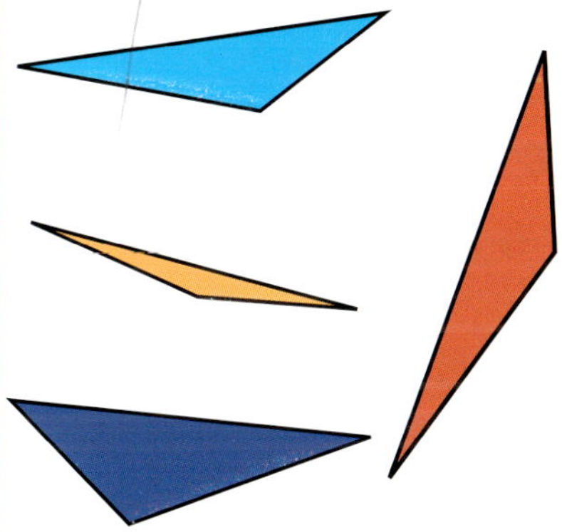

Angles are marked as congruent using tick marks on the arcs. In the same way, tick marks can show that the lengths of two line segments are equal.

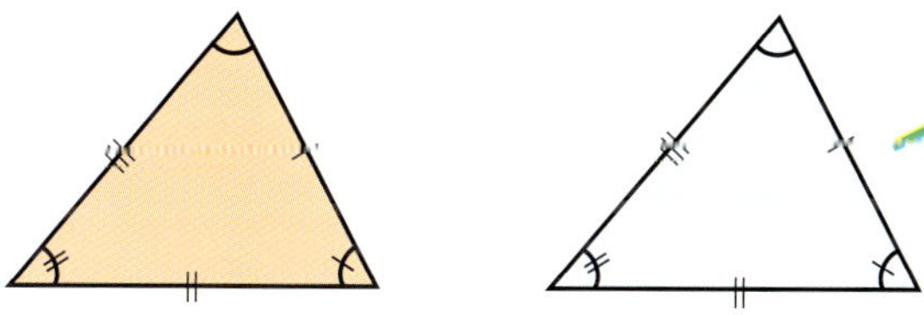

The side with one tick mark is congruent to the side in the other triangle with one tick mark.

EXPLORE!

Equilateral Triangles	Scalene Triangles	Isosceles Triangles
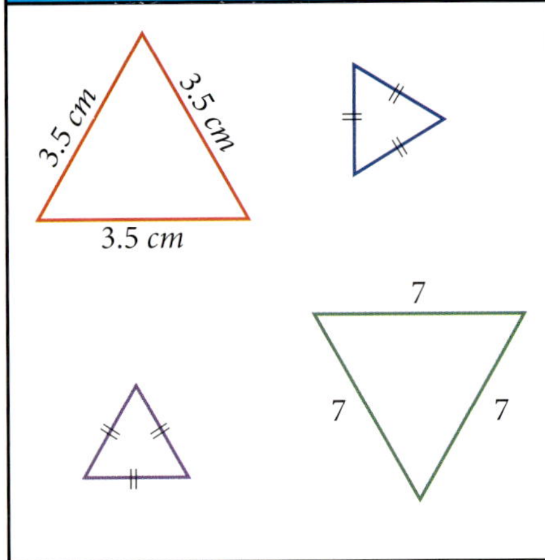	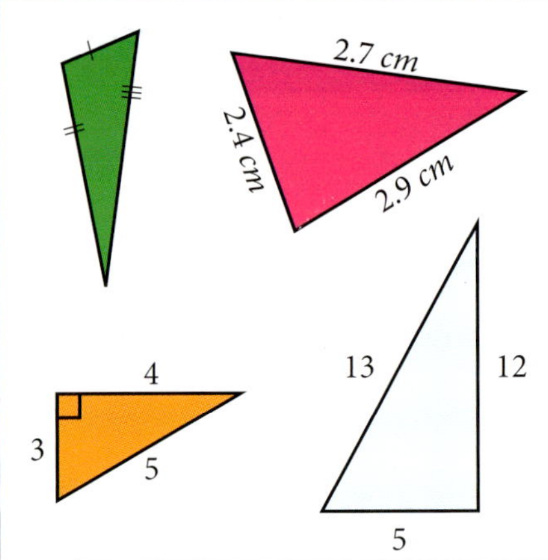	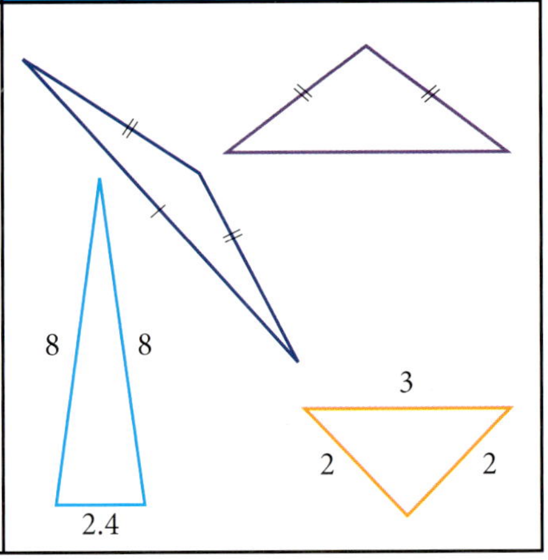

Step 1: Examine each group of triangles in the table above. List all similarities you see in each category.

Step 2: Identify how each group is different from the other groups.

Step 3: Triangles can be classified by the lengths of their sides. Triangles can be equilateral, scalene or isosceles. Based on your observations of the triangles in the table above, write a definition for each type of triangle.

Step 4: Sketch an equilateral triangle, a scalene triangle and an isosceles triangle. Label the sides in each triangle.

All sides of an **equilateral triangle** are the same length. An **isosceles triangle** has at least two sides with the same measure. A **scalene triangle** has no sides with the same measure. If a triangle is a scalene triangle, each side of the triangle will be a different length.

EXAMPLE 1

Classify each triangle using angle measures and side lengths.

a.

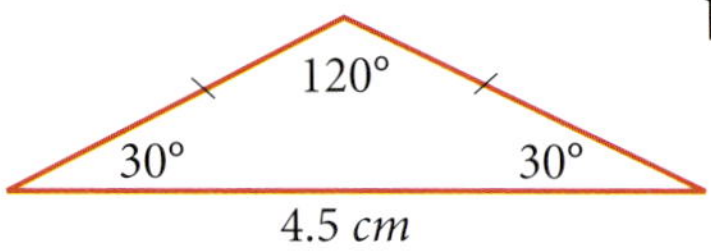

b.

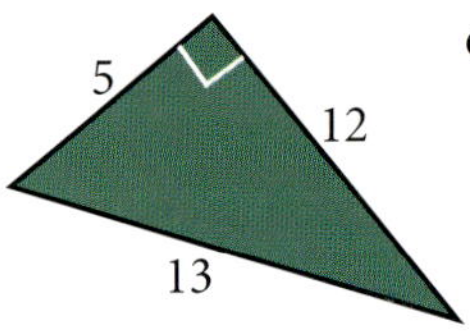

c.

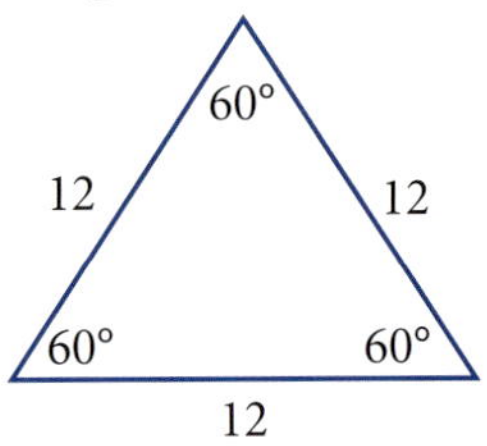

SOLUTIONS

a. Isosceles obtuse triangle. Two sides are the same length and one angle in the triangle is more than 90°.

b. Scalene right triangle. No sides are the same length and there is a right angle.

c. Acute equilateral triangle. All angles in the triangle are less than 90° and all sides are the same length.

EXAMPLE 2 **Sketch a diagram to represent an acute isosceles triangle named ΔABC.**

SOLUTION In order to be an acute triangle all angles must be less than 90°. Since the triangle is isosceles, at least two sides must be the same length. Tick marks can be used to show which sides are equal lengths in the sketch.

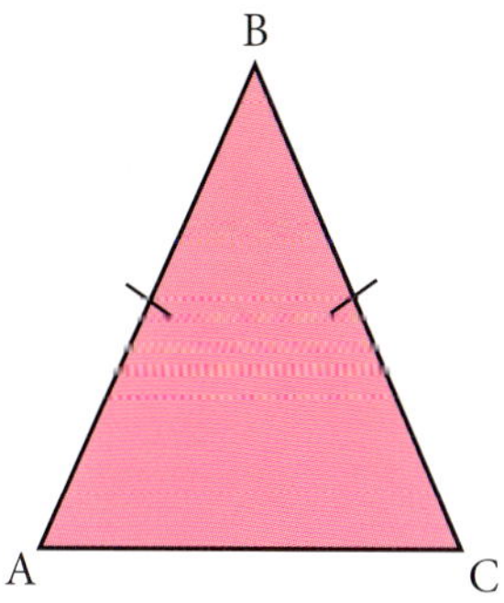

EXERCISES

1. What are the three triangle classifications based on side lengths?

2. What are the three triangle classifications based on angle measures?

Classify each triangle by its angle measures.

3.
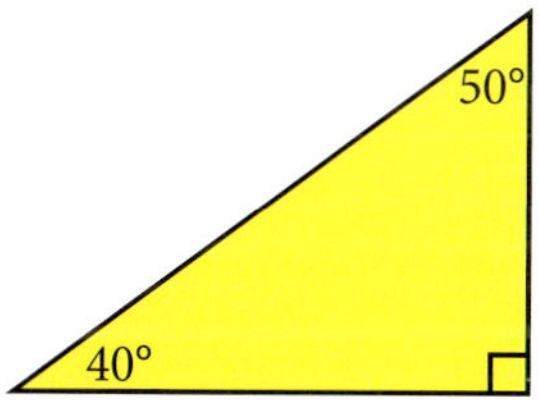

4.
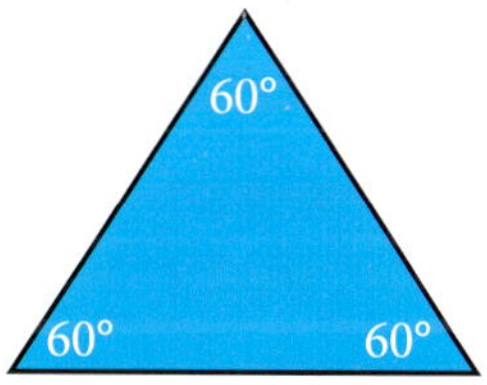

5.
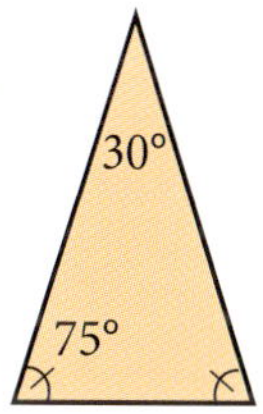

6.
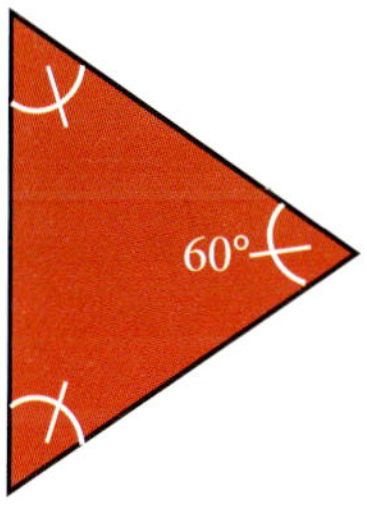

7.
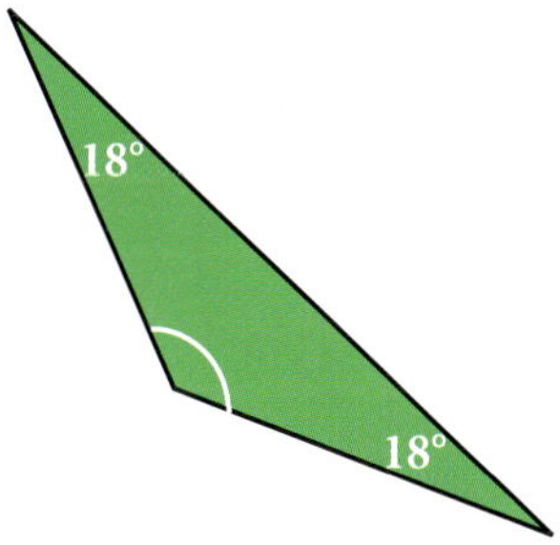

8.

Classify each triangle by its side lengths.

9.
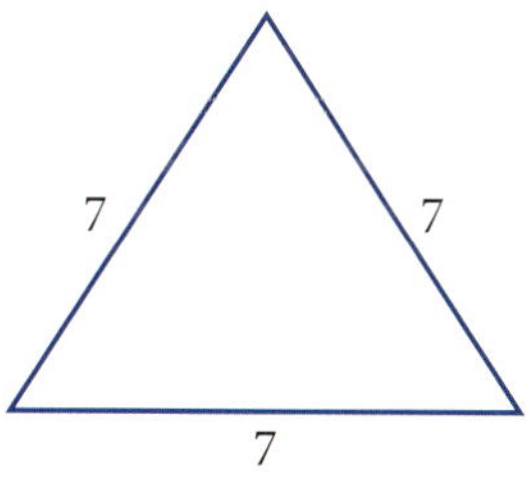

10.
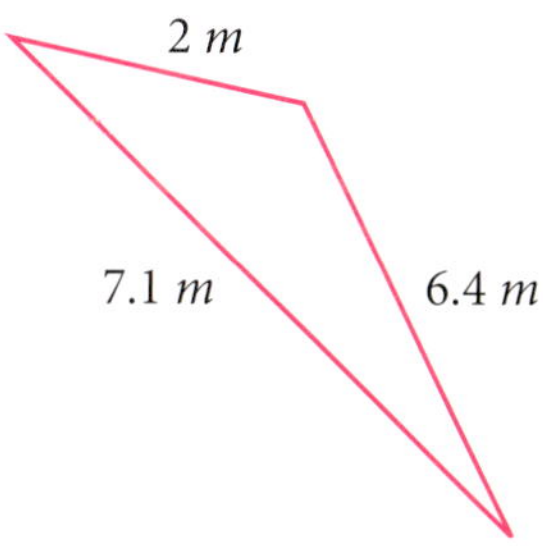

11.
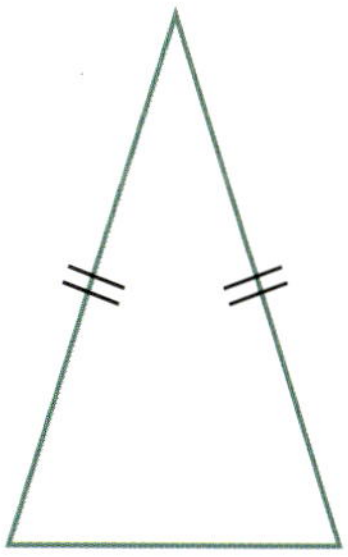

Classify each triangle by its side lengths.

12.

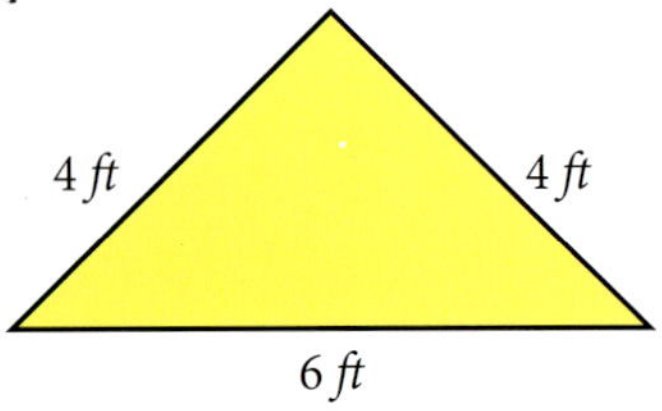

13.

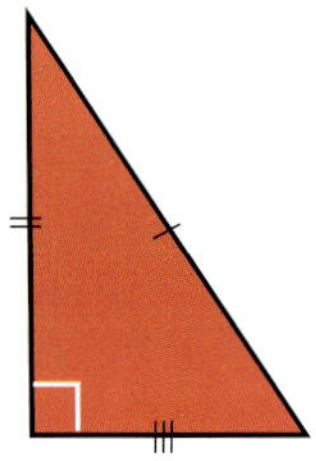

14.

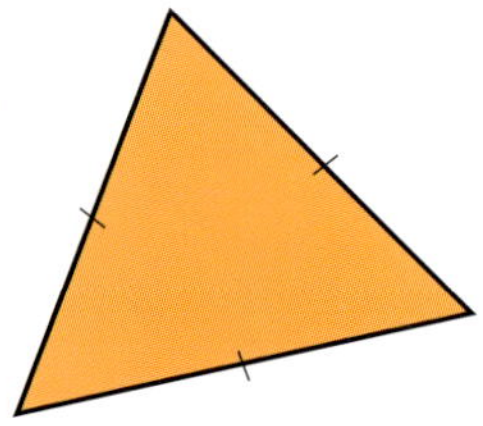

Classify each triangle using angle measures and side lengths.

15.

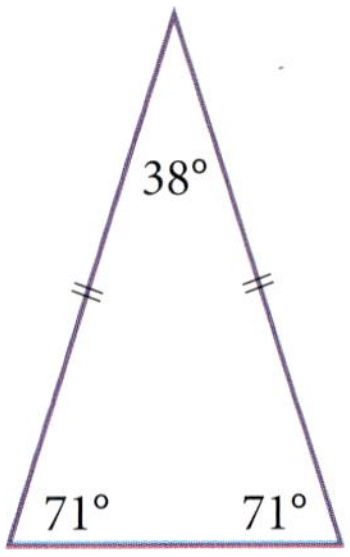

16.

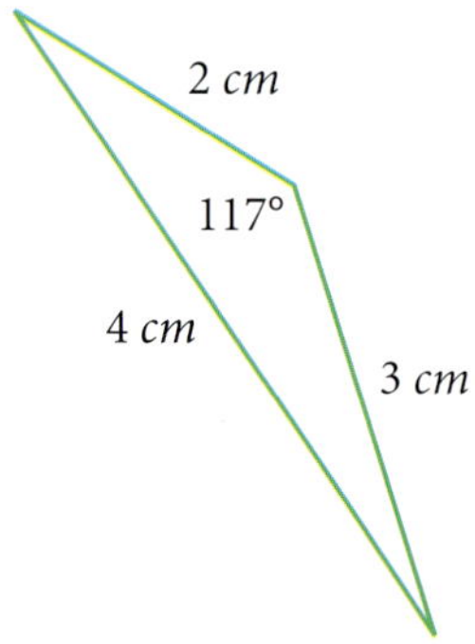

17.

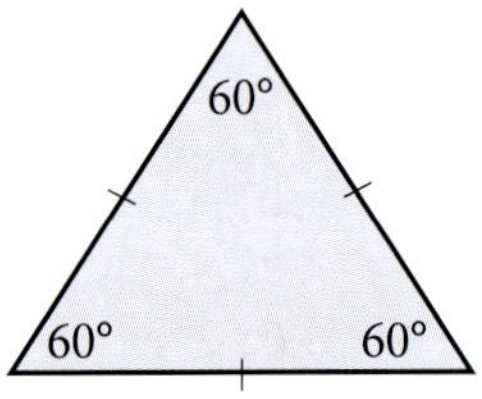

18.

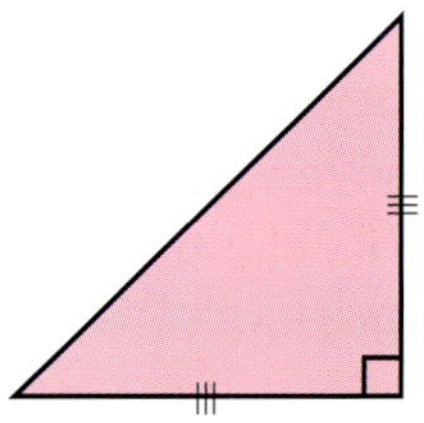

19. A sail on a sailboat has one 90° angle and the sides are three different lengths. Classify the triangle using angle measures and side lengths.

20. Another sail on the sailboat has angles that are each less than 90° and side lengths of 10 feet, 6 feet and 10 feet. Classify the sail by using angle measures and side lengths.

Draw and label a triangle to match each description.

21. Acute ΔCUT

22. Isosceles ΔSAM

23. Right ΔRGT

24. Isosceles, right ΔPOE

25. Obtuse, scalene ΔCUP

26. Acute, equilateral ΔEQU

The side lengths of a triangle are given. Classify the triangle by its sides.

27. 1, 1, 1

28. 8, 8, 10

29. 6, 8, 10

30. h, h, h

31. $5p, 3p, 5p$

32. $3j, 4j, 5j$

33. Can a right triangle have more than one right angle? Support your answer with a diagram.

34. In an obtuse triangle one angle is obtuse. What type of angle are the other two angles?

REVIEW

Name each angle relationship. Choose from: adjacent, complementary, supplementary and vertical. Use each name once. Solve for *x*.

35.

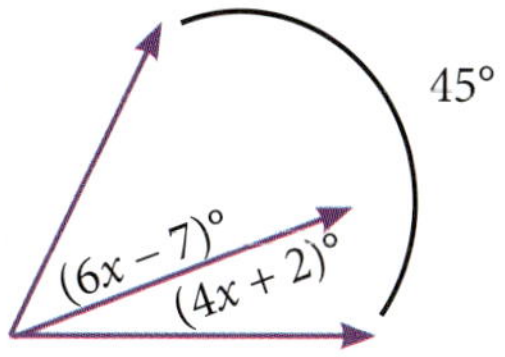

36.

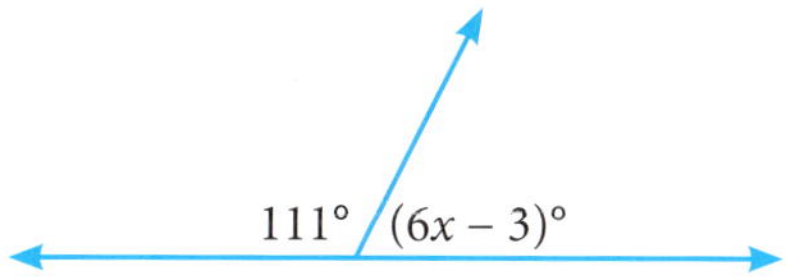

37.

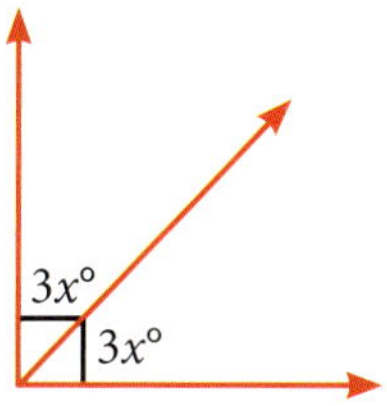

38.

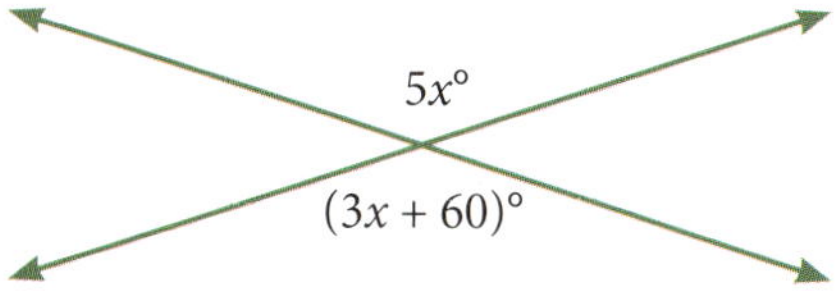

TIC-TAC-TOE ~ CLASSIFICATION GAME

Create a memory card game where players must match cards with triangles with given angle measures, side lengths and/or diagrams to other cards that classify the triangle. You can include some cards that classify only by sides or angles and others that classify by both sides and angles. Make sure each information card has a classification card to match it. Make a set of at least twelve pairs of cards. Try playing the game with a classmate. If it does not work, make the needed adjustments before turning in the game.

Example of a matching set:

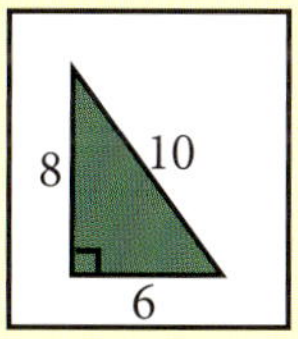

Scalene
Right
Triangle

TIC-TAC-TOE ~ HIGH SCHOOL CHOICES

Investigate the course offerings in mathematics at the high school you will be (or are) attending by looking at their course catalog or on their website. What courses are offered? How many of the courses are you required to take to graduate? Is there a sequence of courses that students must follow? What level of math must you reach to be eligible for entry into a four-year university?

Create a brochure illustrating your findings. Survey your friends to see if they have any questions about high school mathematics. If the information is not already covered in your brochure, try to locate the answer. Include the answers to their questions in a "Frequently Asked Questions" section on the back of your brochure.

ANGLE SUM OF A TRIANGLE

LESSON 1.4

Find angle measures in a triangle.

The sum of the measures of the angles inside every triangle is the same. In this lesson you will determine the angle sum of a triangle. This property of triangles will be very useful as you apply it to triangles in many situations.

EXPLORE! **ADD THEM UP**

Step 1: Draw the three triangles listed below on a blank sheet of paper. Make the triangles large enough to measure their angles with a protractor.

Right triangle
Acute triangle
Obtuse triangle

Step 2: Use a protractor to measure each angle in all three triangles. Write the measure of each angle inside the triangle.

Step 3: Find the sum of the angles in each triangle.

Step 4: Do you notice any similarities in the sums of the angles in each triangle? If possible, write a rule for the sum of the measures of the angles of any triangle.

Step 5: Compare your triangle sums and rule with a classmate. Did he/she get the same or similar results?

The sum of the angles of a triangle can also be shown using the method below. A triangle is drawn on a piece of paper and cut out. The angles are torn apart and lined up. The three angles form a straight angle. The measure of a straight angle is 180°.

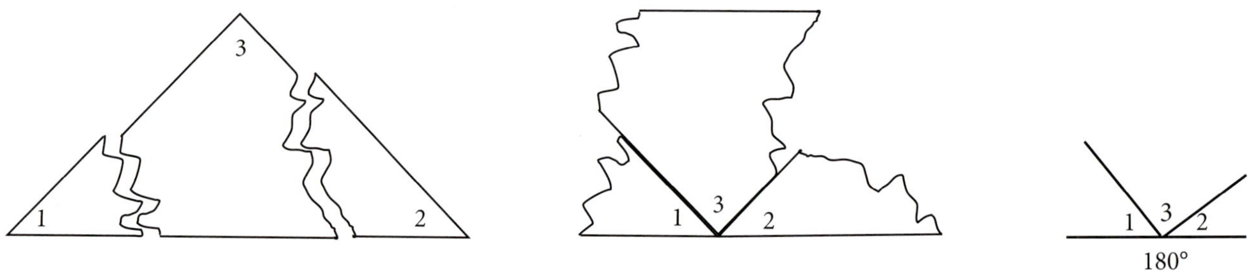

TRIANGLE ANGLE SUM

The sum of the measures of the angles inside a triangle is 180°.

EXAMPLE 1

Find the measure of the missing angle.

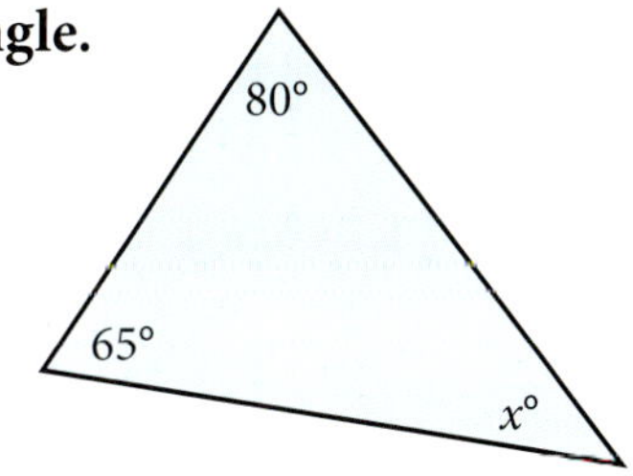

SOLUTION

The sum of the angles in a triangle is 180°. $80 + 65 + x = 180$

Combine like terms. $145 + x = 180$

Subtract 145 from each side of the equation. $-145 \quad 145$

$x = 35$

☑ $65° + 80° + 35° \stackrel{?}{=} 180°$

$180° = 180°$

The measure of the missing angle is 35°.

EXAMPLE 2

ΔYOU has the angle measures listed below.

$m\angle Y = 70°$ $\quad m\angle O = (3x - 10)°$ $\quad m\angle U = 7x°$

a. Find the value of x.

b. Find the degree measure of each angle.

SOLUTIONS

a. The angles of a triangle sum to 180°. $70 + (3x - 10) + 7x = 180$

Combine like terms. $10x + 60 = 180$

Subtract 60 from each side of the equation. $-60 \quad -60$

Divide both sides of the equation by 10. $\frac{10x}{10} = \frac{120}{10}$

The value of x is 12. $x = 12$

b. Write the given expression for each angle. $m\angle O = (3x - 10)°$ $\quad m\angle U = 7x$

Substitute 12 for x. $= 3(12) - 10$ $\quad = 7(12)$

Multiply. $= 36 - 10$ $\quad = 84$

Subtract. $= 26°$

$m\angle O = 26°$ $\quad m\angle U = 84°$

☑ $m\angle Y + m\angle O + m\angle U = 180°$

$70° + 26° + 84° \stackrel{?}{=} 180°$

$180° = 180°$

The $m\angle O = 26°$ and $m\angle U = 84°$.

EXERCISES

Find the measure of each missing angle.

1.

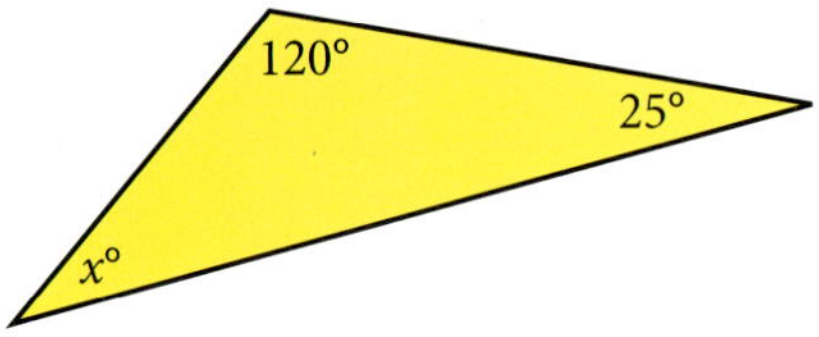

2.

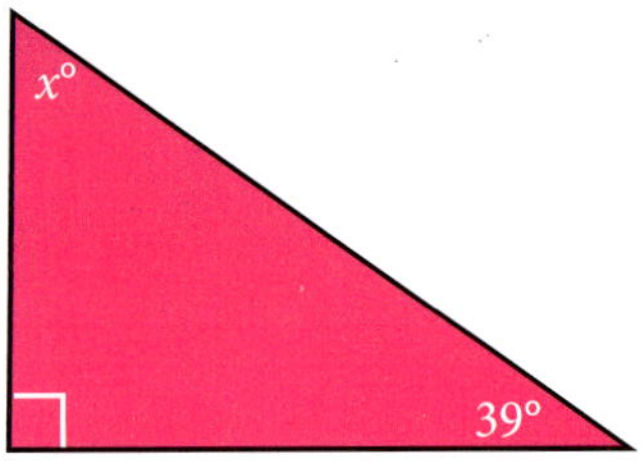

3.

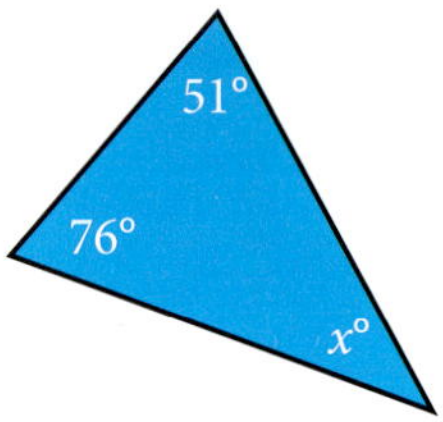

4.

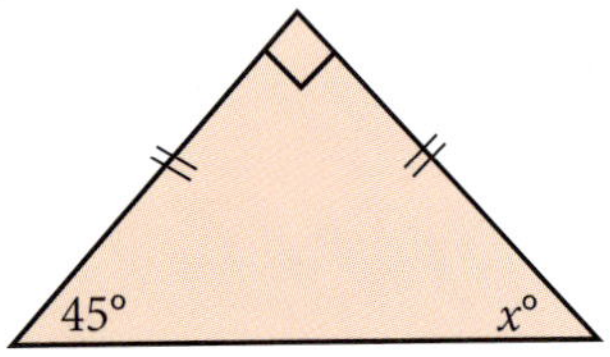

5.

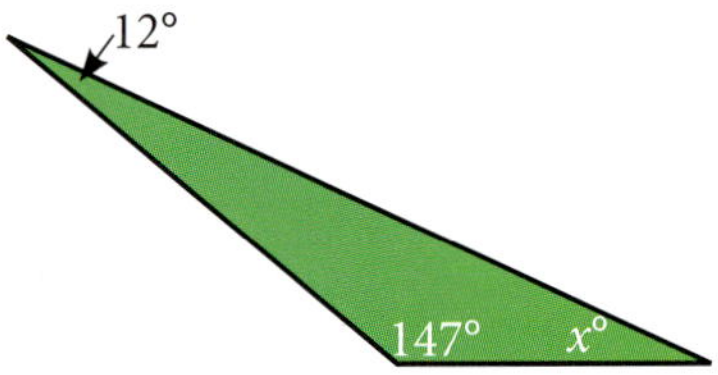

6.

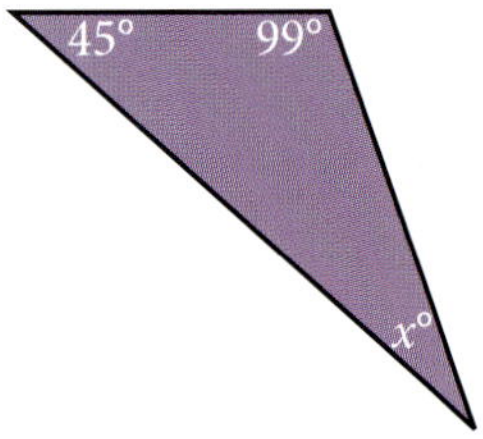

Write an equation and solve for x.

7.

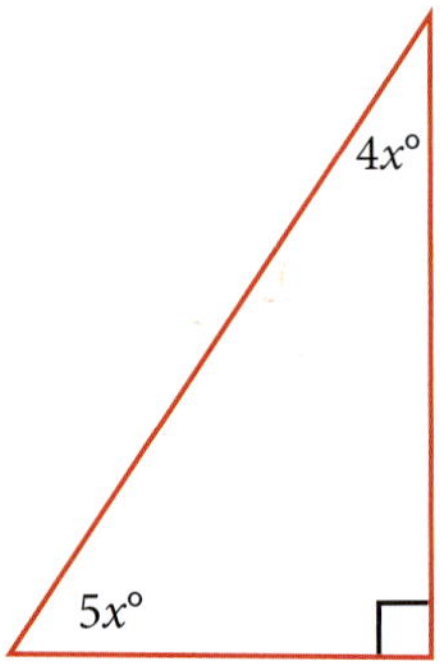

8.

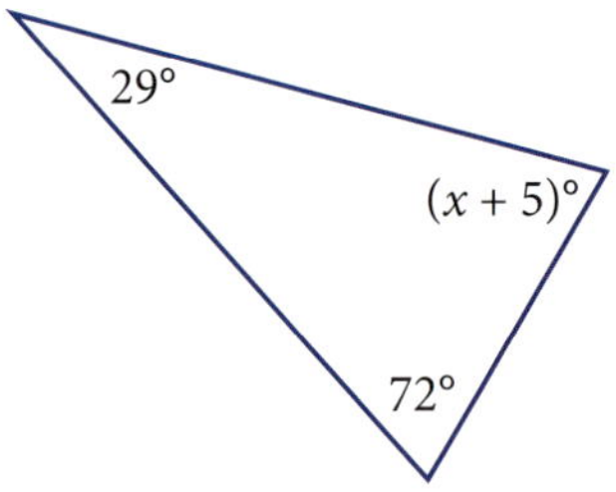

9.

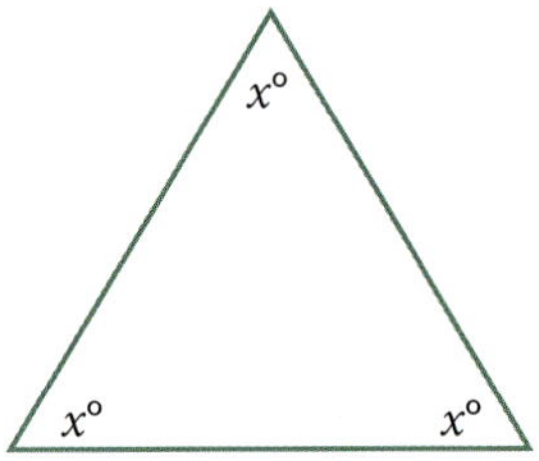

10.

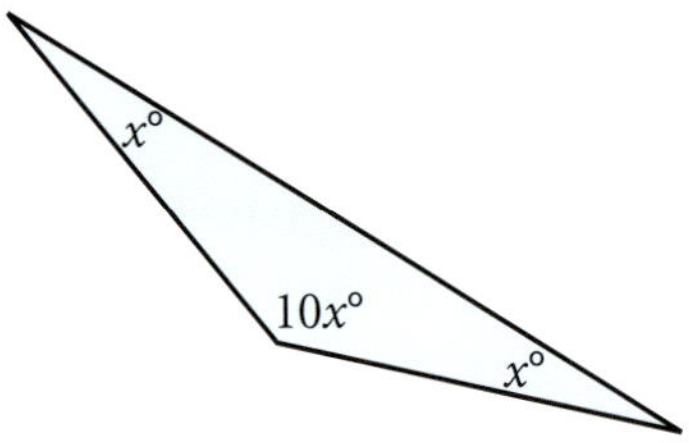

11.

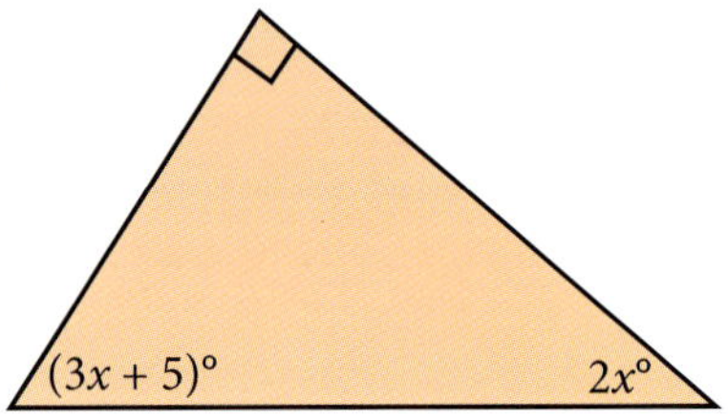

12.

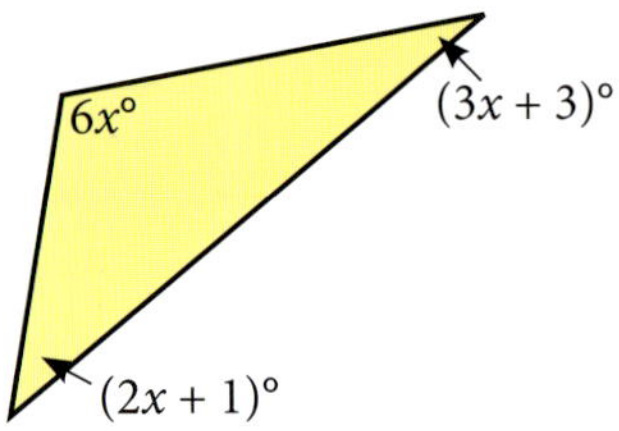

13. Use ΔAMT at the right.

a. Write an equation to find the value of x.

b. Solve for x.

c. Find the measure of each angle.

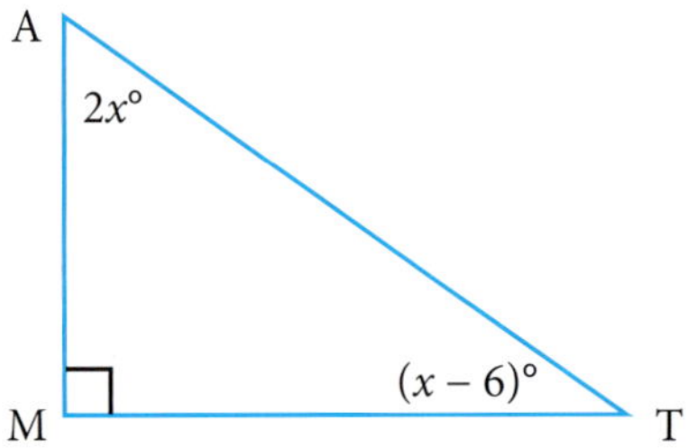

14. In ΔCUP the $m\angle C = 60°$, $m\angle U = (7 + 5x)°$ and $m\angle P = (1 + 3x)°$.

a. Write an equation and solve for x.

b. Find the $m\angle C$, $m\angle U$ and $m\angle P$.

15. ΔPRT is an isosceles triangle. The measure of $\angle P$ is $(2x + 1)°$. The other two angles each measure 42°.

a. Write an equation and solve for x.

b. Find $m\angle P$.

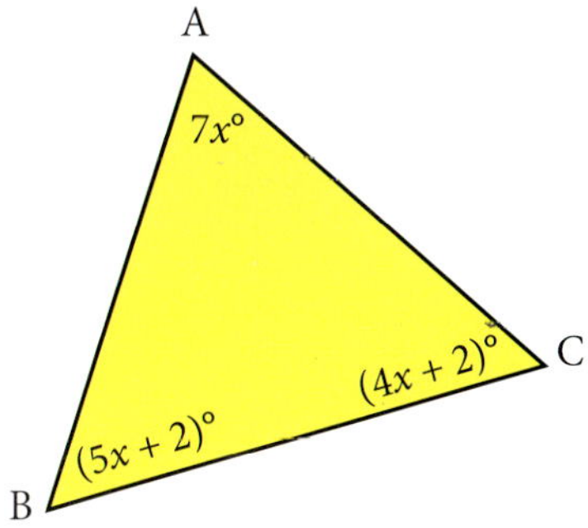

16. Jeff determined that the value of x in the triangle at the right is 11.

a. Find the value of each angle by substituting 11 for x.

b. Was Jeff's solution, $x = 11$, correct? How do you know?

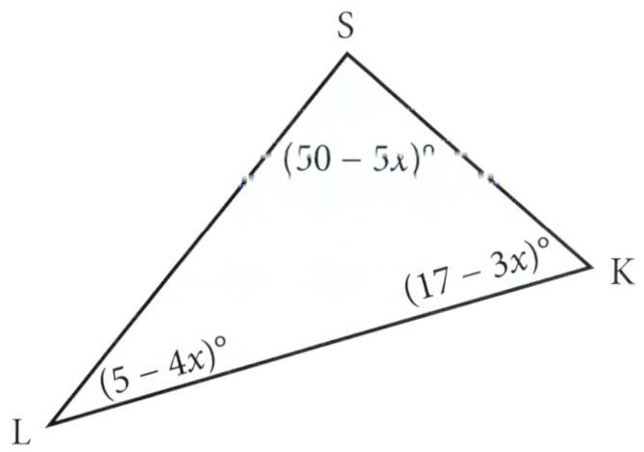

17. Siena determined that $x = -8$ in the triangle at the left.

a. Find the value of each angle by substituting -8 for x.

b. Was Siena's solution, $x = -8$, correct? How do you know?

c. Find the correct value of x.

REVIEW

Classify each triangle by its sides and angles.

18.

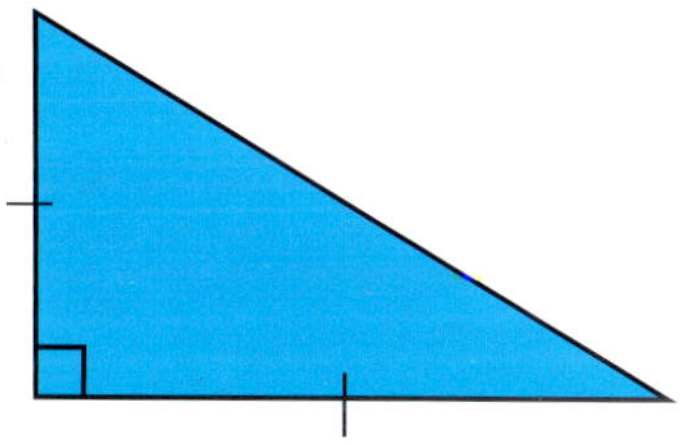

19.

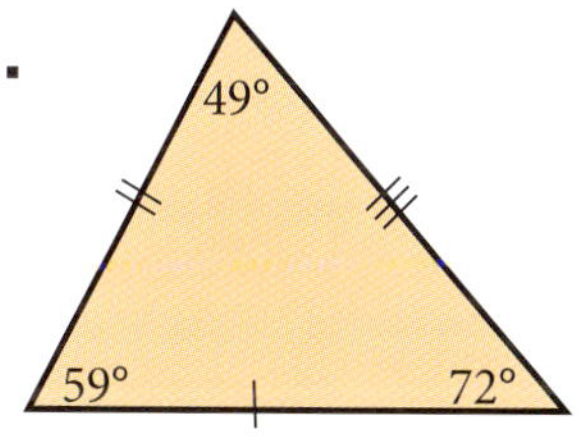

20.

Fill in each blank with the appropriate word or number.

21. Same-side interior angles add up to ______ degrees when between parallel lines.

22. Alternate interior angles are __________ to each other when between parallel lines.

23. A ____________ is the line that cuts through a set of lines.

24. Complementary angles add up to ______ degrees.

Tic-Tac-Toe ~ Four Corners

Step 1: Trace the parallelogram below.

Step 2: Connect two opposite corners in the quadrilateral with a line segment. How many triangles are formed?

Step 3: The sum of the angles in a triangle is 180°. Multiply the number of triangles formed in the parallelogram in **Step 2** by 180°. What is the sum of the angles in the parallelogram?

Step 4: Draw a square.

a. What is the degree measure of one angle in a square?

b. How many angles are in a square?

c. Find the sum of the angles in your square. Does it match your answer in **Step 3**?

Step 5: What is the degree measure of one angle in a rectangle? What is the sum of the four angles in a rectangle?

Step 6: Draw a large quadrilateral that is not a square, rectangle or parallelogram. Some examples are shown below.

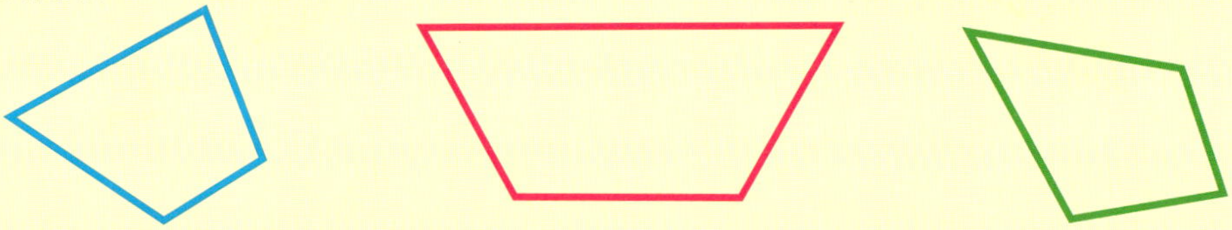

Step 7: Use a protractor to measure the angles in your quadrilateral. Add the four angles together.

Step 8: Write a rule about the angle sum of the four angles in any quadrilateral.

SPECIAL TRIANGLES

LESSON 1.5

Find measures of angles in isosceles and equilateral triangles.

You have classified triangles as equilateral, isosceles or scalene based on the lengths of their sides. When a triangle has two or more sides that are the same length, the angles in that triangle have unique properties. Complete the **Explore!** below to discover these properties.

EXPLORE! **WHAT MAKES ME SPECIAL?**

Step 1: Two equilateral triangles are drawn below. Measure the angles inside each triangle and list them on your own paper.

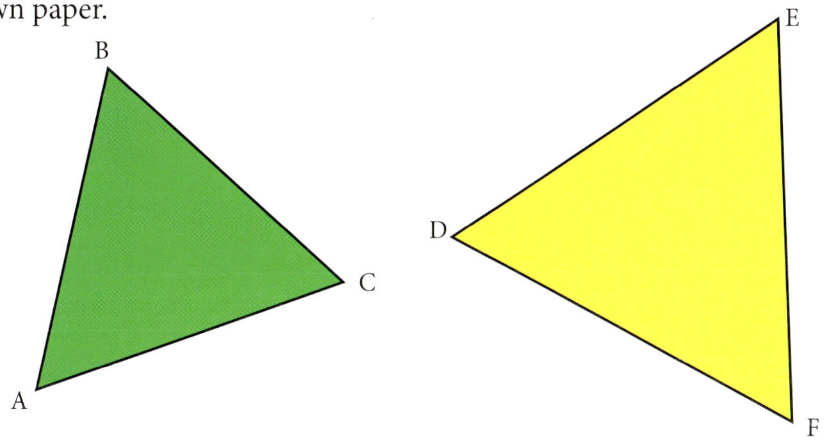

Step 2: Based on **Lesson 1.4**, what should the sum of the angle measures of each triangle equal?

Step 3: Is the sum of ∠A, ∠B and ∠C equal to 180°? If not, check your measurements. Is the sum of ∠D, ∠E and ∠F equal to 180? If not, check your measurements.

Step 4: Do you notice anything about the measure of each angle in an equilateral triangle? If so, what is your discovery?

Step 5: Use division to show how you could calculate the degree measure of an angle in an equilateral triangle.

Step 6: Use a ruler to draw two isosceles triangles. Remember that two sides must be the same length in an isosceles triangle.

Step 7: Measure the angles in your triangles. There should be two angles in each triangle that are equal to each other. Where are those angles in comparison to the two sides that are equal?

Equilateral and Isosceles Triangle Angle Properties

Equilateral Triangle: Each angle in an equilateral triangle is 60°.
Isosceles Triangle: The angles across from the congruent sides are equal in measure.

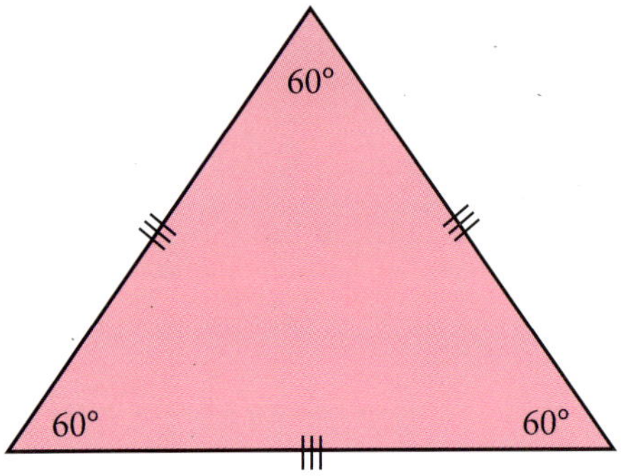

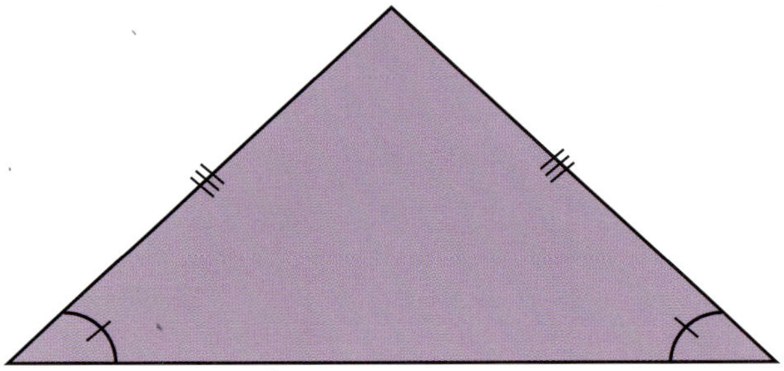

Equilateral triangles are also equiangular. **Equiangular** means that all angles have the same measure.

EXAMPLE 1

ΔMNP is an equilateral triangle. The measure of ∠M is $(2x + 6)°$. Find the value of x.

Solution

Each angle in an equilateral triangle is equal to 60°.
Set the angle equal to 60°.
Subtract 6 from each side of the equation.
Divide both sides of the equation by 2.

$$2x + 6 = 60$$
$$-6 \quad -6$$
$$\frac{2x}{2} = \frac{54}{2}$$
$$x = 27$$

The value of x is 27.

EXAMPLE 2

Find the value of x in the diagram below.

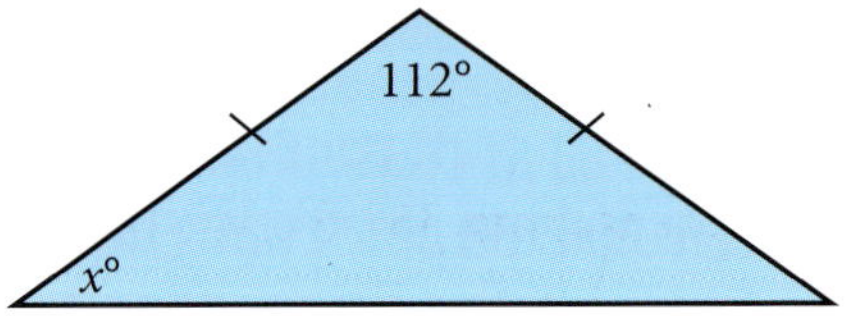

Solution

The triangle above is isosceles. The angles that are across from the congruence marks must also be equal so:

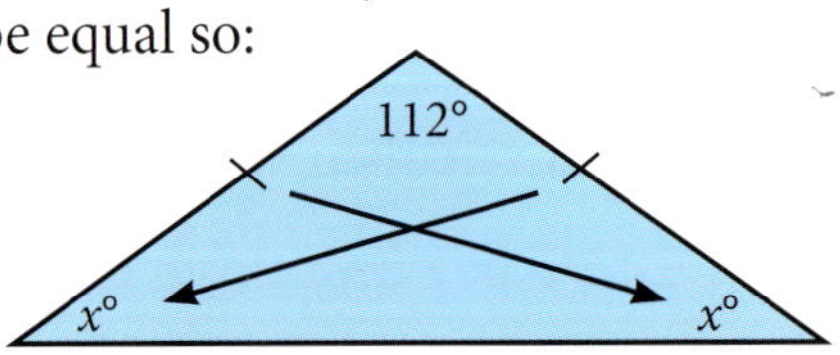

The sum of the angles in a triangle is 180°.
Combine like terms.
Subtract 112 from both sides of the equation.
Divide both sides of the equation by 2.

$$x + x + 112 = 180$$
$$2x + 112 = 180$$
$$-112 \quad -112$$
$$\frac{2x}{2} = \frac{68}{2}$$
$$x = 34$$

The value of x is 34.

EXERCISES

1. What is the measure of each angle in an equilateral triangle?

2. Which two angles in an isosceles triangle are equal? Draw a diagram to illustrate your answer.

Find the value of *x* in each diagram.

3.

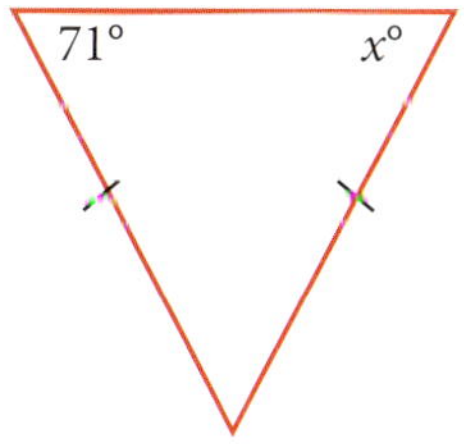

4.

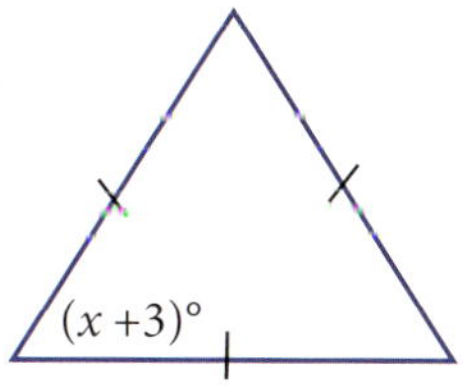

5.

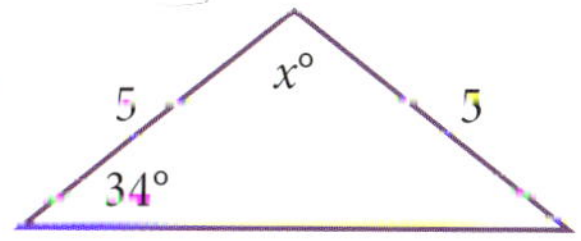

6.

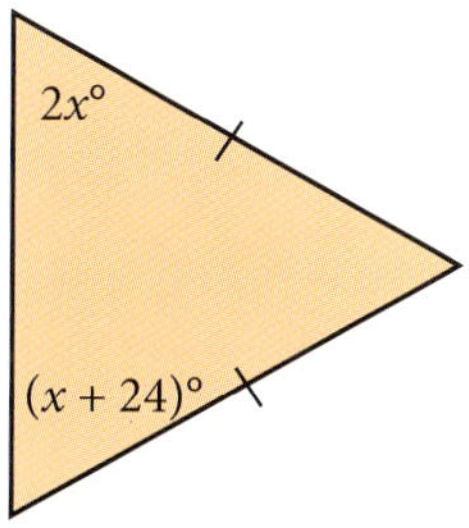

7.

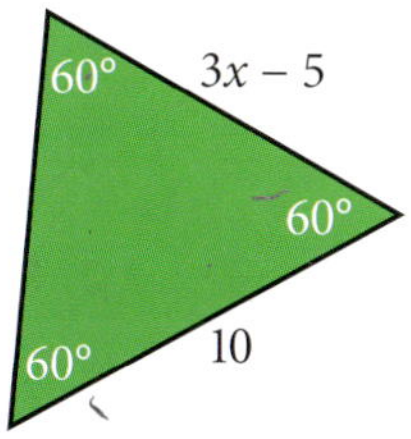

8.

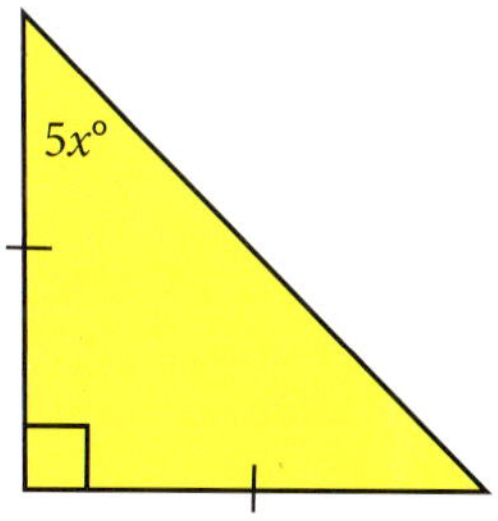

9. ΔYUM has two angles that measure 78°.
   a. Sketch a diagram of ΔYUM.
   b. Find the measure of the other angle.
   c. Classify ΔYUM by its sides and angles.

10. All three sides in ΔBET are 2 inches in length.
   a. Sketch a diagram of ΔBET.
   b. What is the measure of each angle?
   c. Classify ΔBET by its sides and angles.

11. ΔJAM is isosceles and has an angle measuring 35°. Sketch two possible diagrams of ΔJAM with the angle measures labeled.

12. One angle in an equilateral triangle is $(8x - 24)°$. Solve for x.

13. Hayley's house makes an isosceles triangle with her school and the mall as shown in the diagram below.

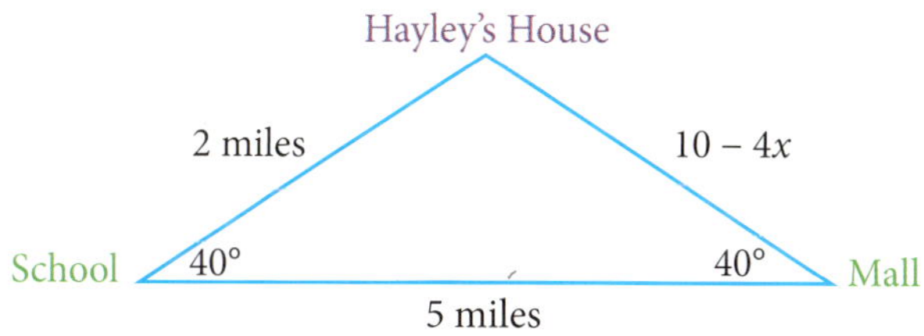

   a. How far is it from Hayley's house to the mall?
   b. Find the value of x.

14. Hans designed a photo frame shaped like an isosceles triangle. He wanted two of the sides of the frame to each be three times the length of the shortest side.

  a. Hans made the shortest side of the triangle 4 inches long. Sketch a diagram of his photo frame.

  b. The angle across from the shortest side of the frame is 20°. Determine the measures of the other two angles. Explain your reasoning.

REVIEW

Solve for *x* in each triangle.

15.

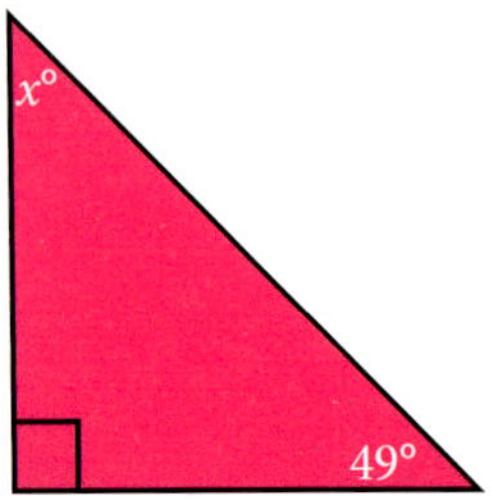

16.

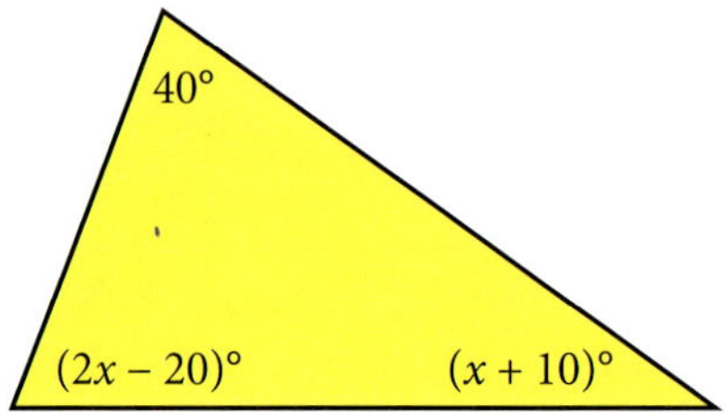

17.

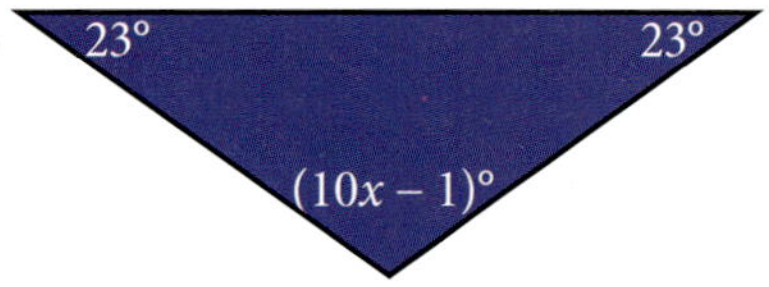

18. The complement of $\angle B$ is 37°. Find $m\angle B$.

19. The supplement of $\angle L$ is 146°. Find $m\angle L$.

20. Two angles in a triangle are 18° and 102°. Find the measure of the third angle in the triangle.

21. One angle in a linear pair is 85°. Find the measure of the other angle in the linear pair.

Tic-Tac-Toe ~ Triangle Mobile

A mobile is a type of art, often referred to as kinetic art. It hangs in space and uses balance and motion. A mobile may be more commonly recognized hanging above a baby lying in his/her crib. Create a mobile that shows all of the different things you learned about triangles in this block. Use definitions, rules and properties about triangles as information on the pieces of your mobile.

CONGRUENT AND SIMILAR TRIANGLES

LESSON 1.6

Determine if triangles are similar or congruent and find missing measures.

Two triangles that are the exact same shape and the exact same size are called **congruent figures**. Two triangles that have the exact same shape, but not necessarily the exact same size are called **similar figures**. The parts of the figures that correspond are called **corresponding parts**. Look at the two similar triangles below.

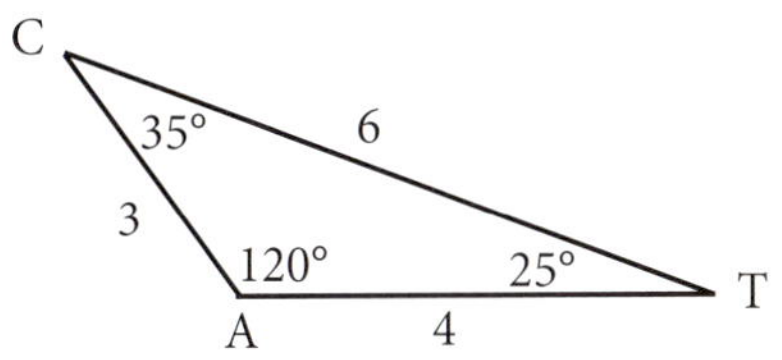

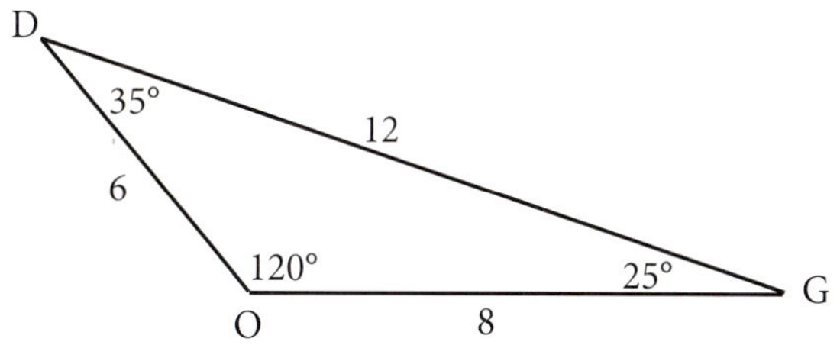

Corresponding Angles	Corresponding Sides
$\angle C$ and $\angle D$	$\overline{CT}$ and $\overline{DG}$
$\angle A$ and $\angle O$	$\overline{TA}$ and $\overline{GO}$
$\angle T$ and $\angle G$	$\overline{AC}$ and $\overline{OD}$

The corresponding angles in similar triangles are congruent. The corresponding sides are proportional. You can show two triangles are similar using the similar symbol: $\Delta CAT \sim \Delta DOG$.

EXAMPLE 1

ΔMNP is similar to ΔJKL. Find the value of *x*, *y* and *z*.

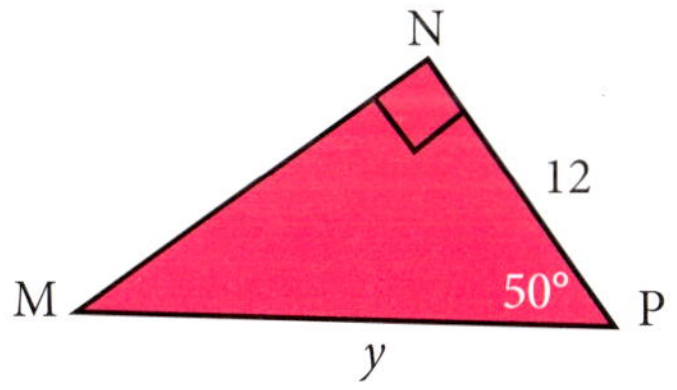

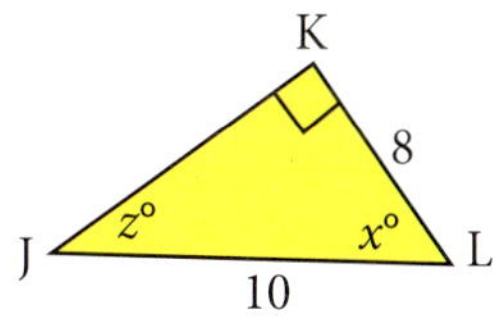

SOLUTION

Since the triangles are similar, the corresponding angles are congruent.

$$m\angle L = m\angle P$$
$$m\angle L = x° = 50°$$

The sides of congruent triangles are proportional.

$$\frac{NP}{KL} = \frac{MP}{JL}$$

Substitute the values of each known side into the proportion.

$$\frac{12}{8} = \frac{y}{10}$$

Set the cross products equal to each other.
Divide both sides of the equation by 8.

$$8y = 120$$
$$y = 15$$

To find the value of *z*, set the sum of the three angles in the triangle equal to 180°.
Subtract 140 from both sides of the equation.

$$z + 50 + 90 = 180$$
$$z + 140 = 180$$
$$-140 \quad -140$$
$$z = 40$$

Congruent and Similar Triangles

Congruent Triangles: All corresponding angles are congruent and all corresponding sides are congruent.

Similar Triangles: All corresponding angles are congruent and all corresponding sides are proportional.

If two angles in one triangle are congruent to two angles in another triangle, then the two triangles are similar. This rule is referred to as the Angle-Angle Similarity Theorem.

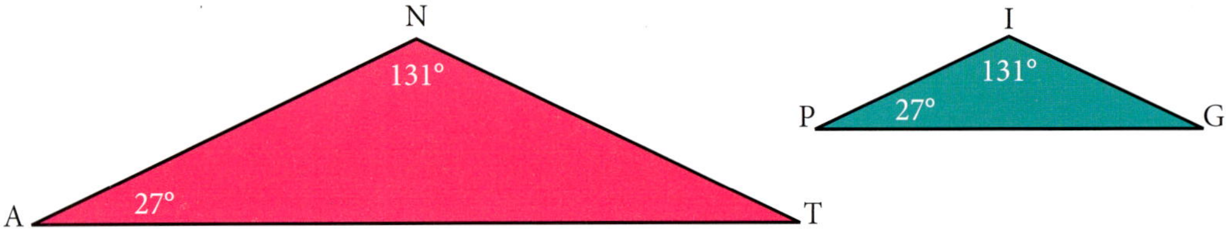

When two corresponding angles in a pair of triangles are congruent, their sides are proportional.

The triangles shown above are similar because $\angle A \cong \angle P$ and $\angle N \cong \angle I$. The Angle-Angle Similarity Theorem can be illustrated by showing that if two angles are congruent in the triangles, the third angle is also congruent.

$$\angle A + \angle N + \angle T = 180^\circ$$
$$27^\circ + 131^\circ + \angle T = 180^\circ$$
$$\angle T = 22^\circ$$

$$\angle P + \angle I + \angle G = 180^\circ$$
$$27^\circ + 131^\circ + \angle G = 180^\circ$$
$$\angle G = 22^\circ$$

Since all three angles in the triangles are congruent, the triangles are similar.

Angle-Angle Similarity Theorem

If two angles of one triangle are congruent to two angles of another triangle, the triangles are similar.

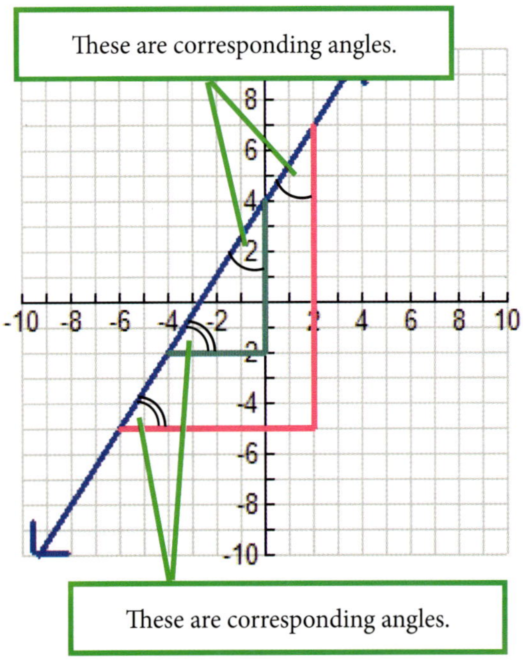

In previous books you learned about slope triangles. You were able to choose any two points on a line and form a slope triangle.

On the graph at the left, a right angle is formed in each slope triangle.

The corresponding sides of the triangles are parallel because they are either both vertical or both horizontal. When parallel lines are cut by an intersecting line called a transversal, the corresponding angles are congruent. Since two angles in the triangles are congruent, the slope triangles are similar.

EXERCISES

Find the corresponding sides and corresponding angles for each pair of figures. Complete each statement.

1.

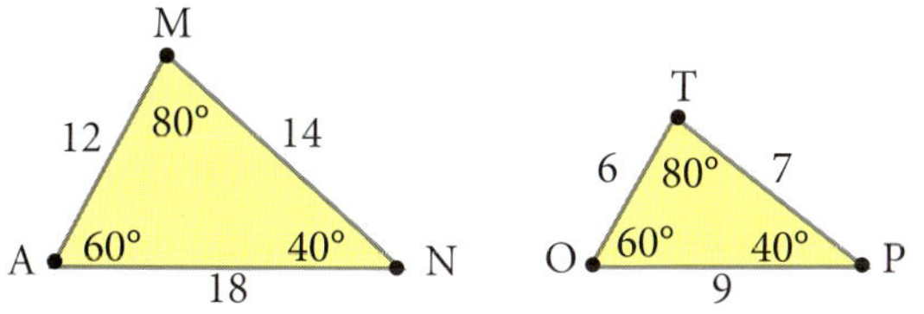

a. $\overline{MN}$ corresponds to _____ $\angle M \cong \angle$___

b. $\overline{AN}$ corresponds to _____ $\angle N \cong \angle$___

c. $\overline{MA}$ corresponds to _____ $\angle A \cong \angle$___

2.

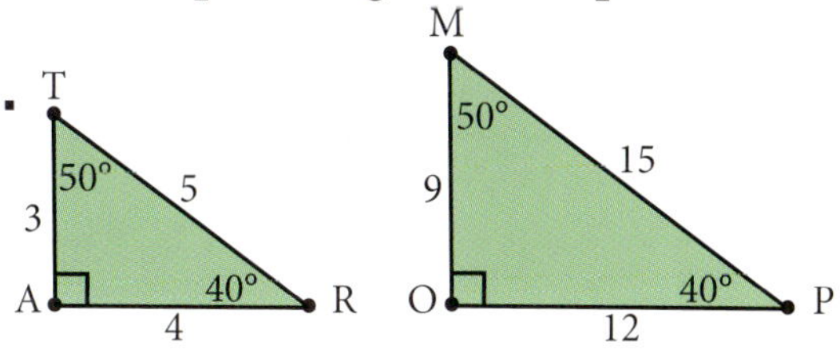

a. $\overline{TA}$ corresponds to _____ $\angle M \cong \angle$___

b. $\overline{AR}$ corresponds to _____ $\angle O \cong \angle$___

c. $\overline{TR}$ corresponds to _____ $\angle P \cong \angle$___

3. Sketch a similar triangle to the right triangle below. Include angle measures and side lengths in your drawing.

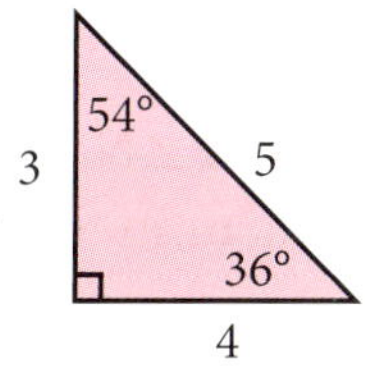

4. Explain the difference between congruent triangles and similar triangles.

Find the missing side length(s) for each set of similar figures.

5.

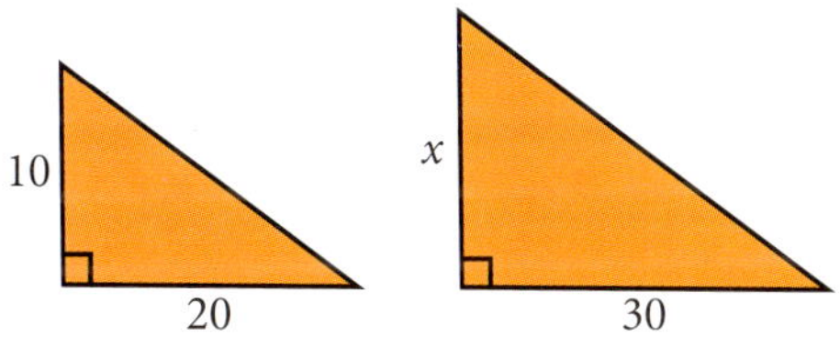

6.

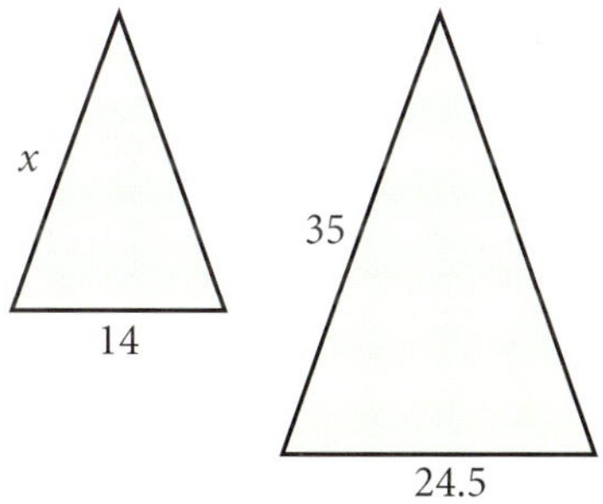

7.

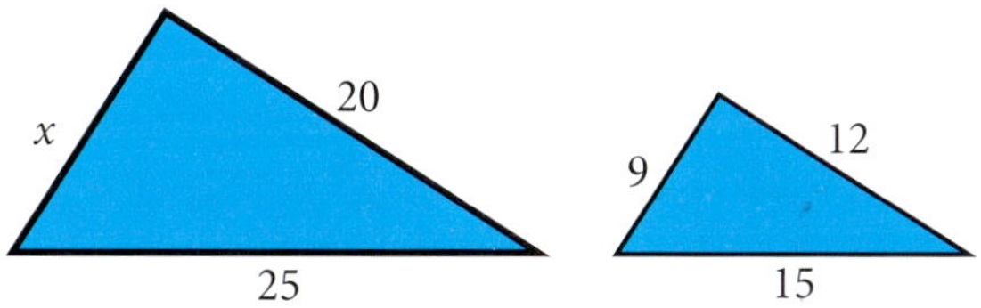

8.

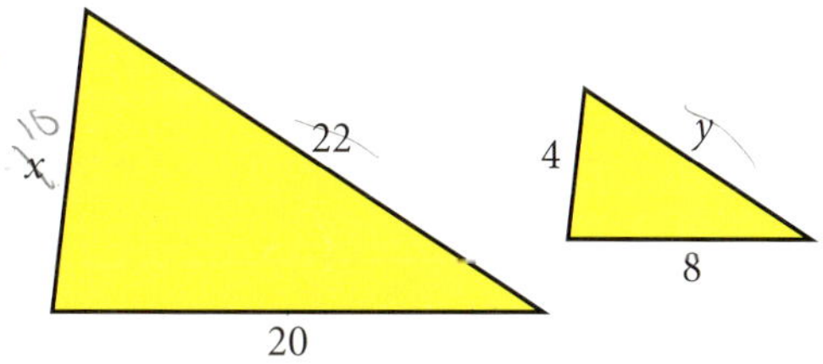

9.

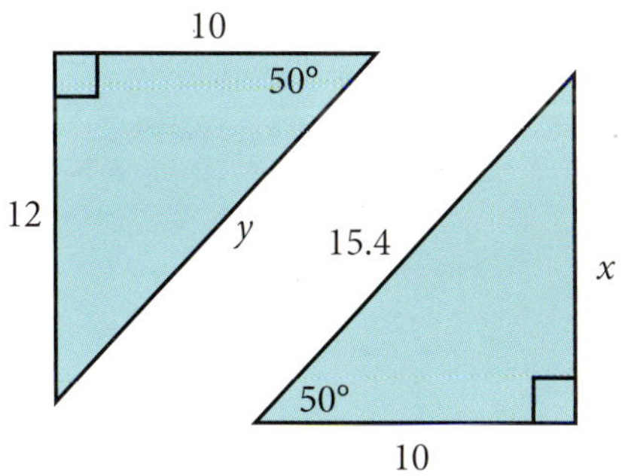

10.

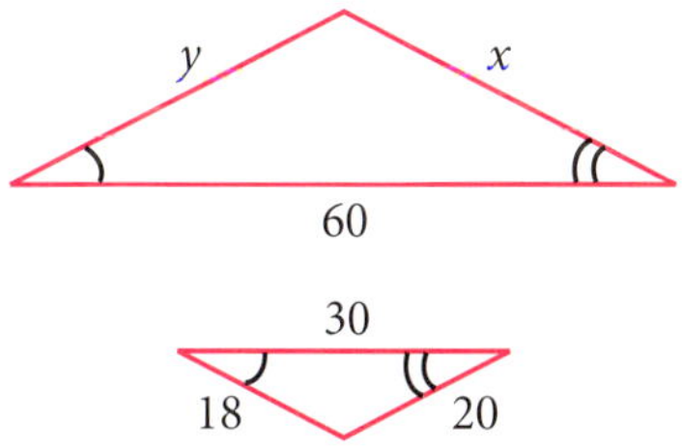

11. Which set of triangles in **Exercises 5 through 10** are congruent figures? Explain your reasoning.

12. Are the triangles below similar? Explain your answer.

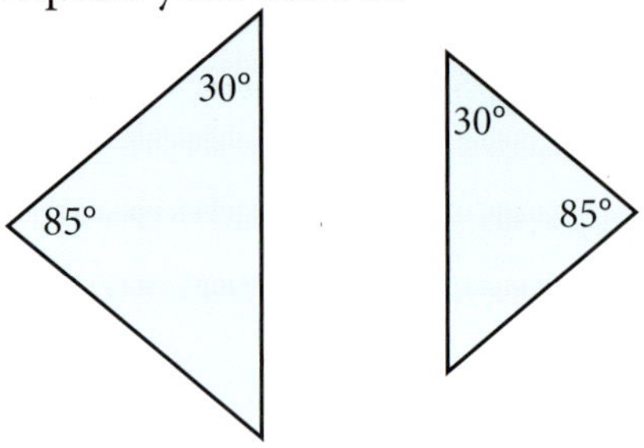

13. Are the triangles below similar? Explain your answer.

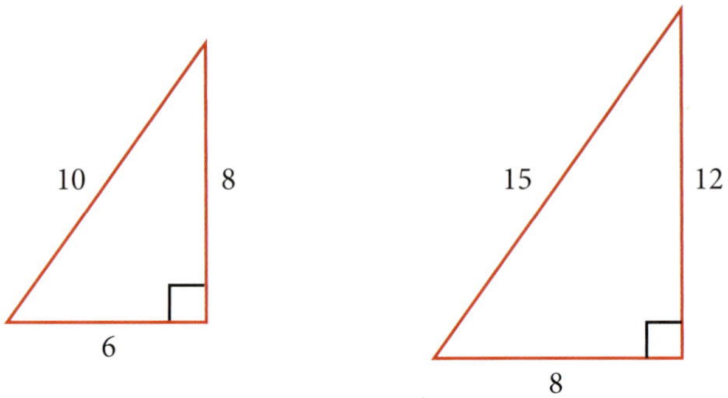

14. Which triangles below are similar to ΔRAP? Explain your reasoning for each.

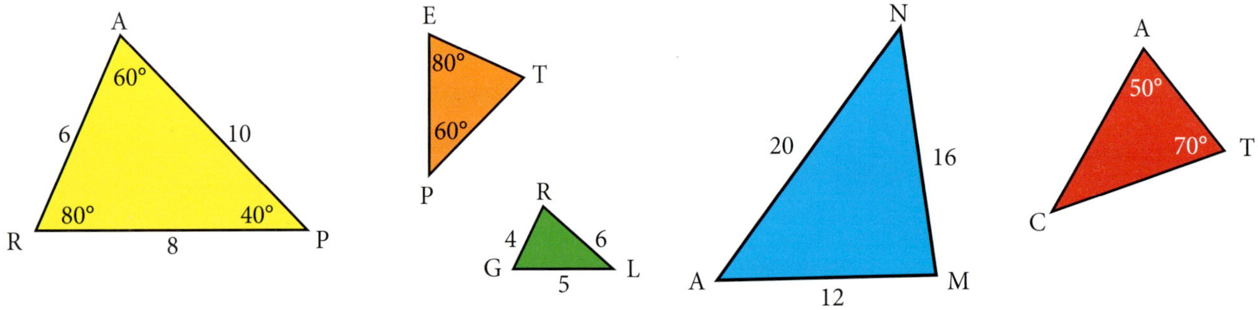

15. Explain how two slope triangles formed on the same line are similar. Draw a diagram to support your answer.

16. Sketch a pair of triangles that are congruent. Label all the angles and sides.

17. Draw two triangles of different sizes that are similar to ΔMQT. Label all angle measures and side lengths.

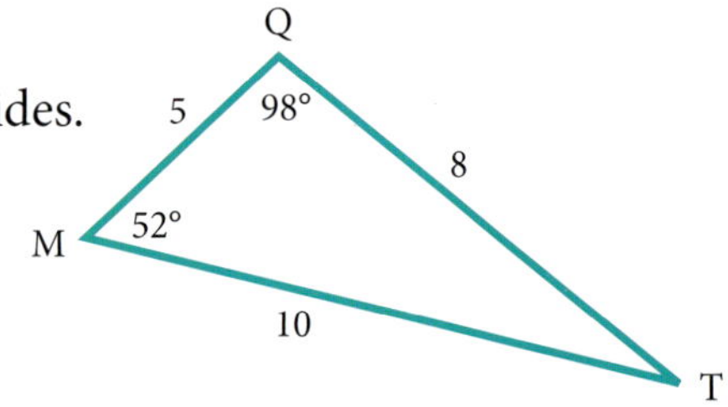

REVIEW

Name the special angle relationship between the two angles. Find x.

18.

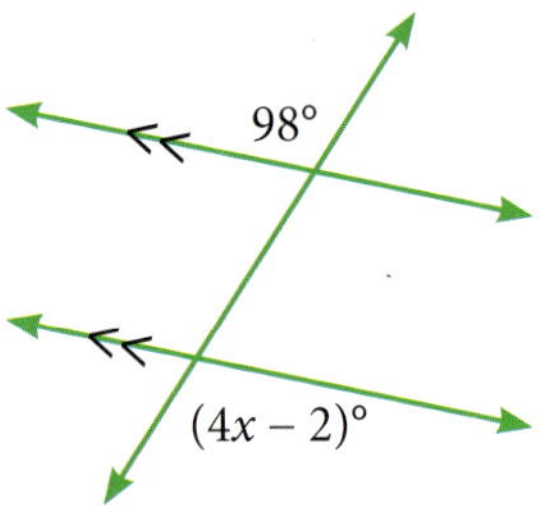

19.

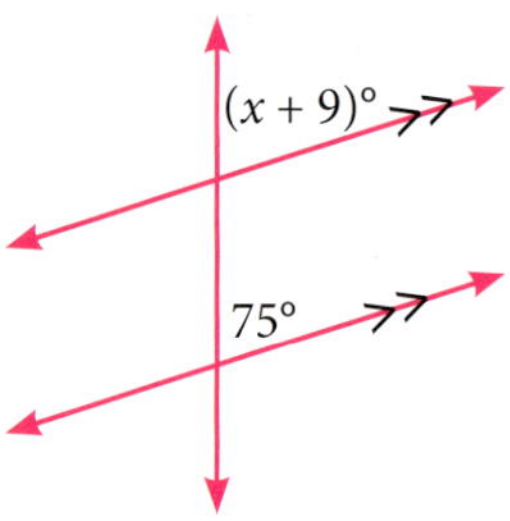

20.

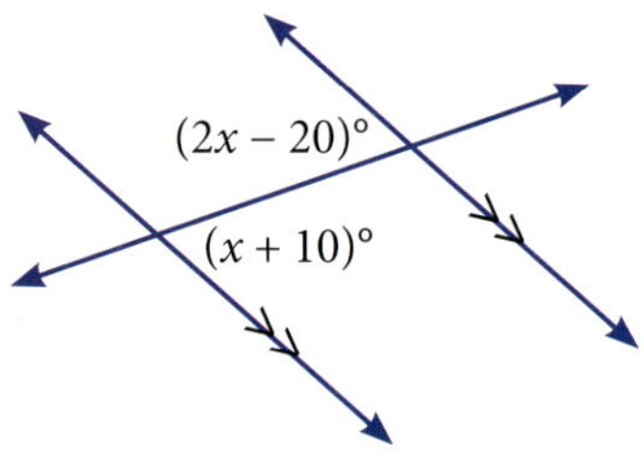

21. Two angles are supplementary. They measure $(5x - 4)°$ and $(9x + 16)°$.
  a. Write an equation and solve for x.
  b. Find the measure of each angle.

ANGLE RELATIONSHIPS

LESSON 1.7

Use special angle relationships to find angle measures.

So far in this block you have learned about parallel lines, transversals, special angle pairs and the sum of the angles in a triangle. This lesson will require you to use these different pieces of information to find angle measures, show triangles to be similar to one another and make mathematical connections.

The **exterior angle** of a triangle is an angle formed by a side and an extension of the adjacent side. Each angle of a triangle has an exterior angle that is its supplement. **Remote interior angles** are the two angles inside the triangle and not adjacent to the exterior angle.

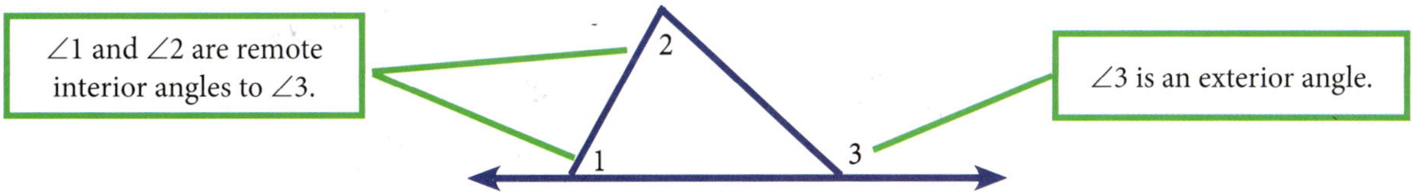

An algebraic proof of the exterior angle and remote interior angles relationship shows that the sum of the remote interior angles equals the measure of the corresponding exterior angle.

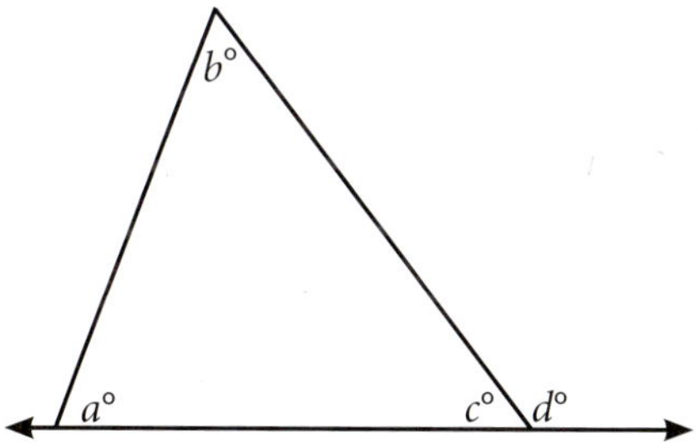

Statement	**Reason**
$a + b + c = 180°$	The sum of the angles of a triangle is 180°.
$c + d = 180°$	Angles c and d are supplementary.
$a + b + c = c + d$	Substitute $c + d$ for 180°.
$-c \quad -c$	Subtract c from both sides.
$a + b = d$	

Exterior Angles Theorem

The measure of an exterior angle of a triangle is equal to the sum of the measures of the two non-adjacent interior angles of the triangle.

$m\angle 1 + m\angle 2 = m\angle 3$

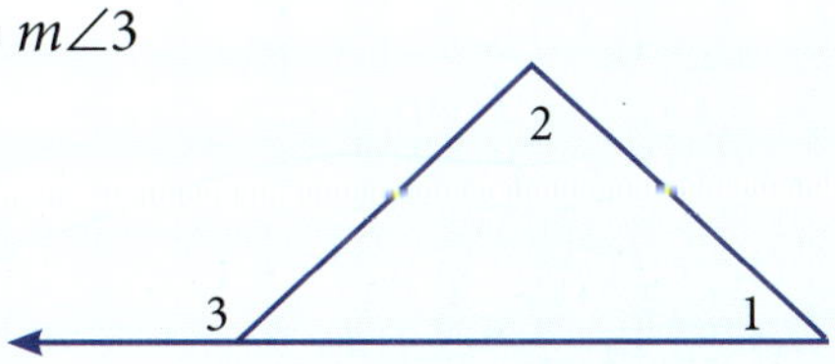

EXAMPLE 1 **Solve for *x*.**

a.

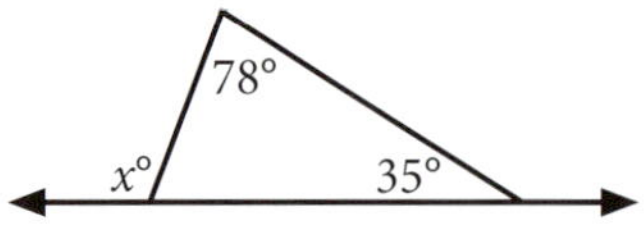

b.

SOLUTIONS The sum of the remote interior angles is congruent to the corresponding exterior angle.

a. Write an equation. $x = 78 + 35$
Add. $x = 113$

b. Write an equation. $148 = x + 26$
Subtract 26 from both sides of the equation. $\underline{-26 \qquad -26}$
$122 = x$

EXPLORE! **MAKING ANGLE CONNECTIONS**

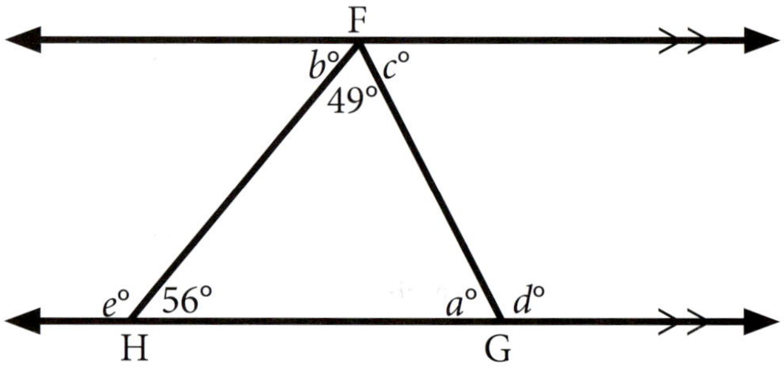

Step 1: Copy or trace the diagram at the right. Label the vertices and angles as shown.

Step 2: What is the relationship between $d°$, 49° and 56°? Find the value of d.

Step 3: Find the value of a. Support your answer with words and/or symbols.

Step 4: Name the special angle relationship between $a°$ and $c°$. What is the value of c?

Step 5: Use alternate interior angles to find the value of b.

Step 6: In **Lesson 1.4**, the angles of a triangle were separated and put together to form a straight angle. How are the angles of ΔHFG and the angles b, 49 and c connected to one another and to the "torn" triangle?

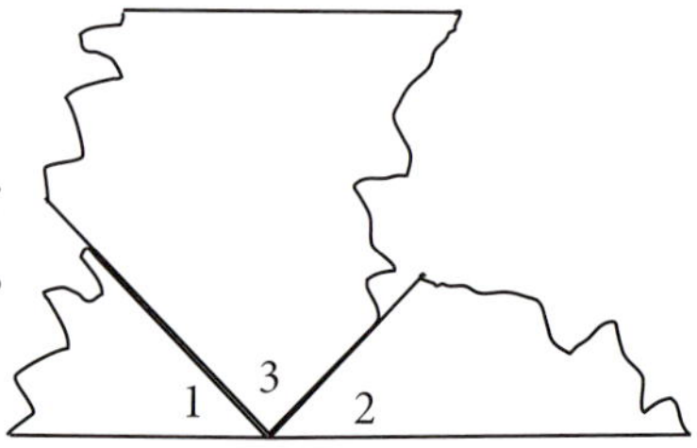

Step 7: Copy or trace the diagram below and right.

Step 8: What special angle pair describes ∠1 and ∠3?

Step 9: Name the special angle relationship between ∠2 and ∠4.

Step 10: Copy and complete each statement.
∠1 ≅ ∠___ ∠4 ≅ ∠___

Step 11: What can be concluded about ΔABC and ΔEDC? Explain.

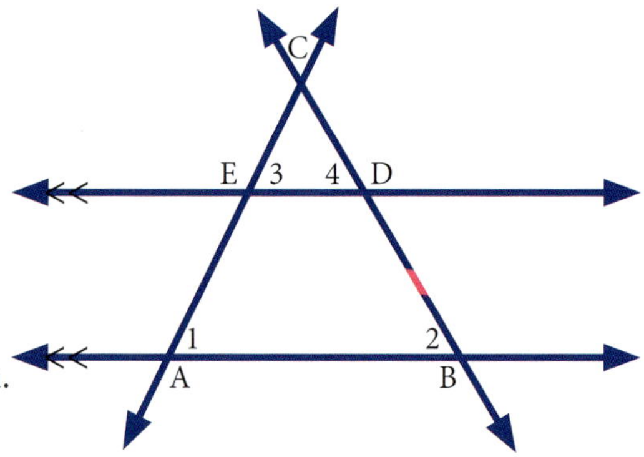

Special angle pairs can be useful when determining whether two triangles are similar. Once two angles are equal to two angles in another triangle, the two triangles are similar based on the Angle-Angle Similarity Theorem. When two triangles are similar, corresponding angles are congruent to each other and corresponding sides are proportional.

EXAMPLE 2

Show that ΔJKL ~ ΔMNL.

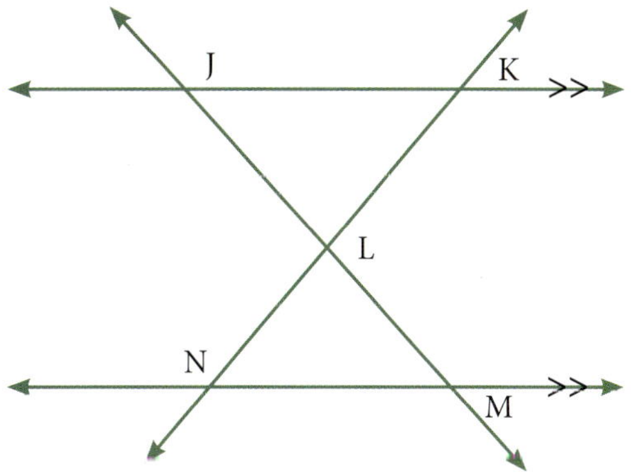

SOLUTION

∠LMN ≅ ∠LJK because they are alternate interior angles.

∠LNM ≅ ∠LKJ because they are alternate interior angles.

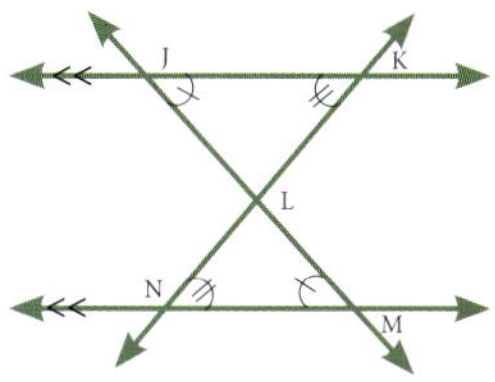

ΔJKL ~ ΔMNL because two angles in each triangle are congruent to one another. This is based on the Angle-Angle Similarity Rule.

EXAMPLE 3

Find the missing measures in the two similar triangles.

a. $m\angle HAE$

b. $m\angle Y$

c. x

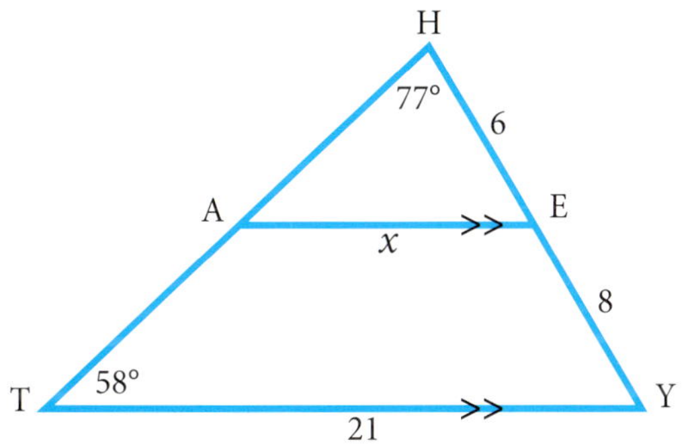

SOLUTIONS

a. Corresponding angles are congruent. $m\angle T = m\angle HAE$

Substitute 58° for ∠T. $58° = m\angle HAE$

b. The sum of three angles in a triangle is 180°. $m\angle T + m\angle H + m\angle Y = 180°$

Substitute the angle values for ∠T and ∠H. $58° + 77° + m\angle Y = 180°$

Combine like terms. $135° + m\angle Y = 180°$

Subtract 135 from both sides of the equation. $m\angle Y = 45°$

c. Write a proportion with corresponding sides. $\frac{AE}{TY} = \frac{HE}{HY}$

Fill in the known lengths. $\frac{x}{21} = \frac{6}{14}$

Add the two lengths, HE and EY, to find the total length of HY.

Set the cross products equal to each other. $126 = 14x$

Divide by 14 on both sides of the equation. $\frac{126}{14} = \frac{\cancel{14}x}{\cancel{14}}$

$9 = x$

EXERCISES

Solve for *x*.

1.

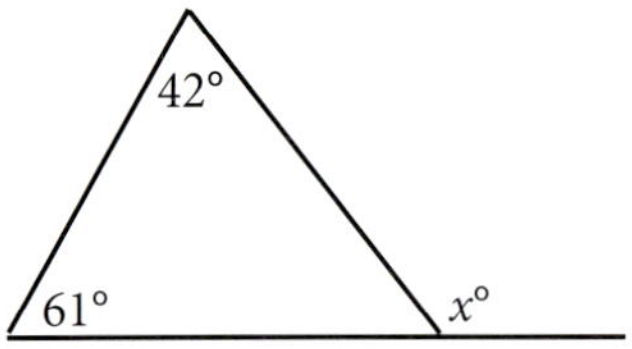

2.

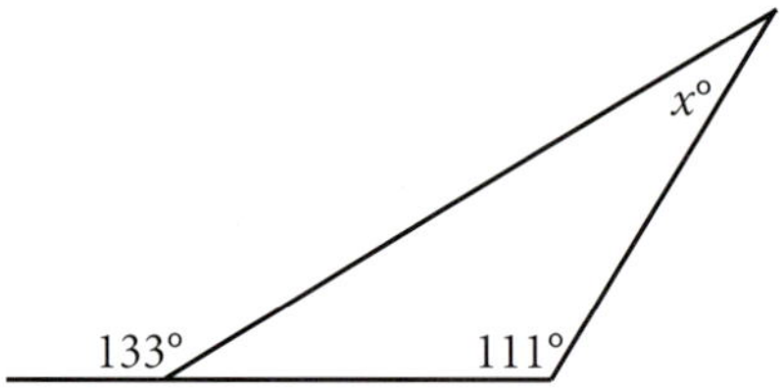

3.

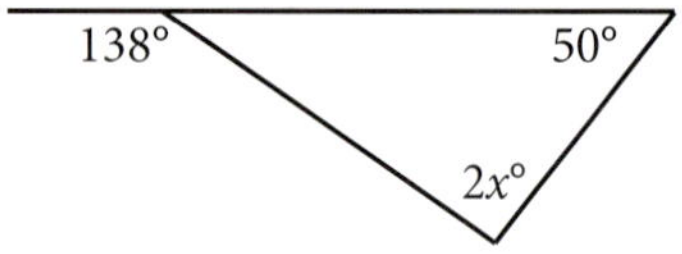

4. One of the remote interior angles of a triangle is 66°. The exterior angle is a right angle. What is the degree measure of the other remote interior angle?

5. An exterior angle of a triangle measures 78°. Give a possible pair of degree measures of the remote interior angles.

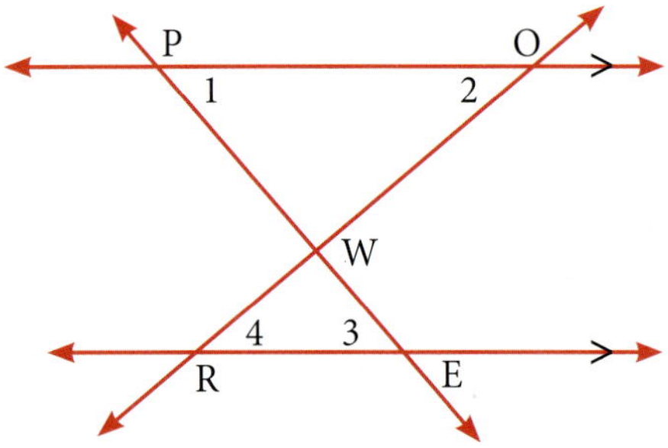

6. Use the diagram at the left.

a. Name the two pairs of angles that are congruent inside the triangles based on alternate interior angles.

b. Complete the statement: ΔPOW ~ Δ______

7. Use the diagram at the right.

a. What angle in ΔAND measures 56°?

b. What is *m*∠AND?

c. What is *m*∠AYS?

d. How do you know ΔAND ~ ΔAYS?

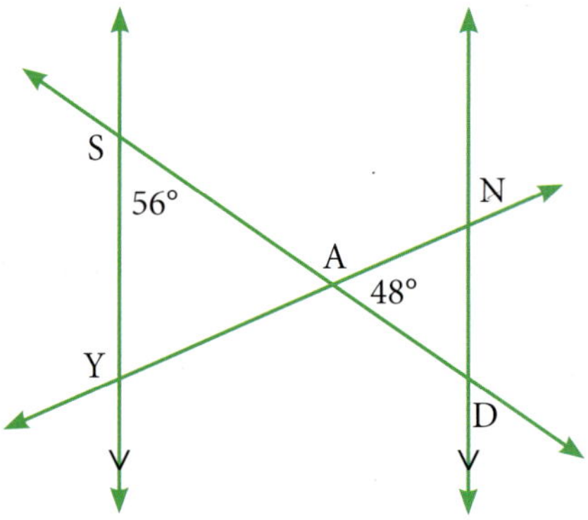

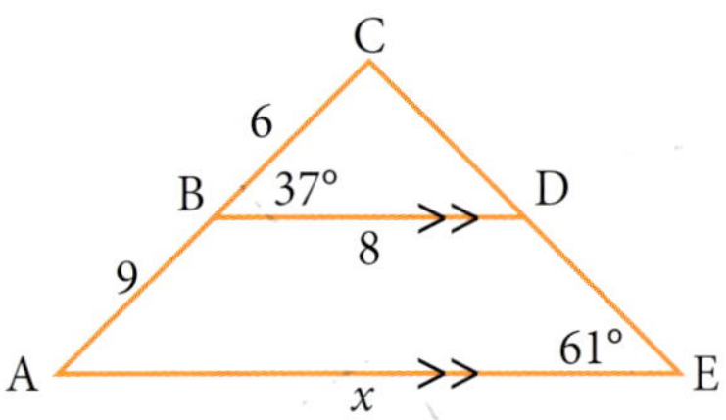

8. Use the diagram at the left.

a. What angle besides ∠CBD measures 37°?

b. What is the measure of ∠CDB?

c. What is the measure of ∠C?

d. Find the value of *x*.

Find the missing measures in each diagram.

9.

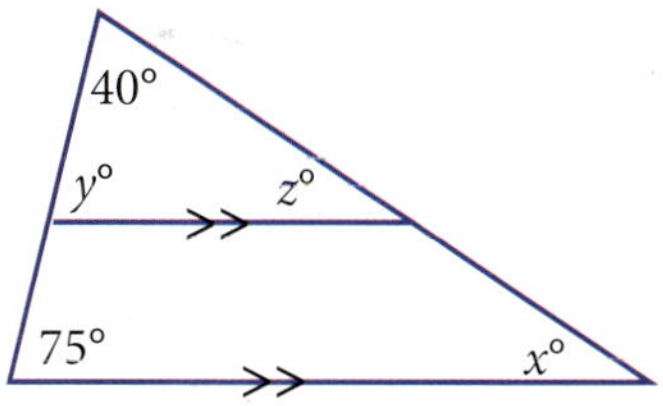

10.

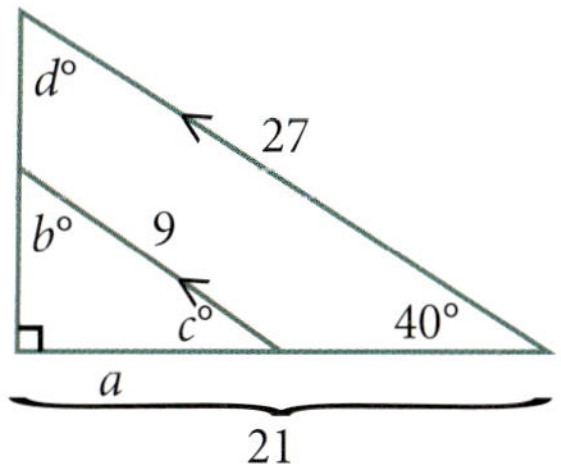

11.

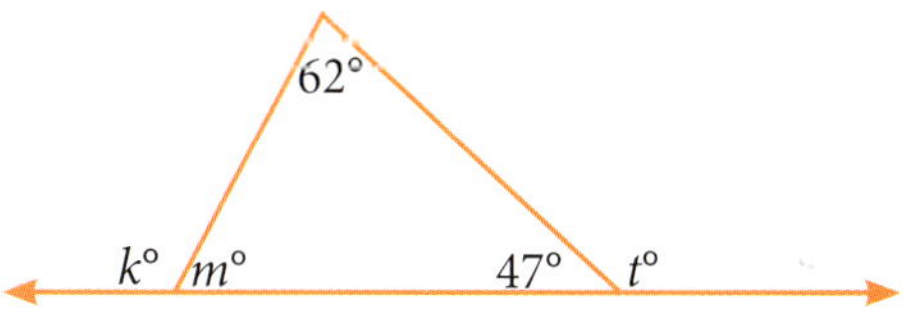

12.

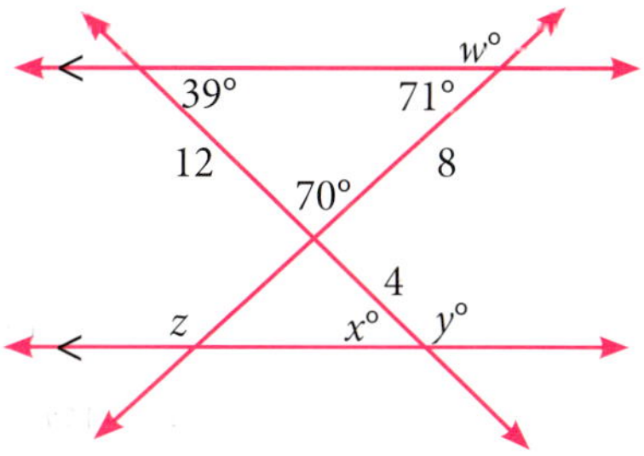

13. $\overleftrightarrow{AB} \parallel \overleftrightarrow{CD}$

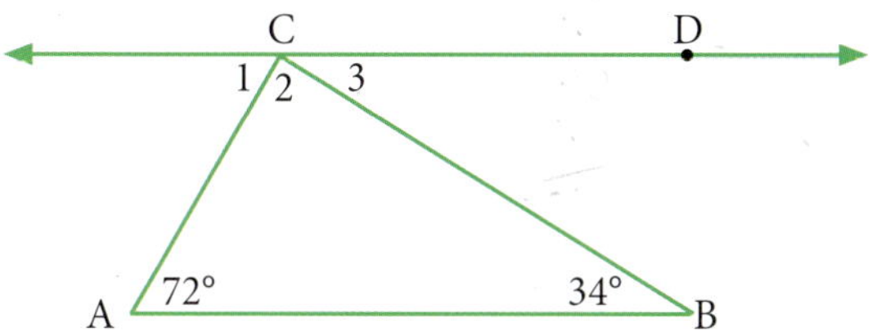

14.

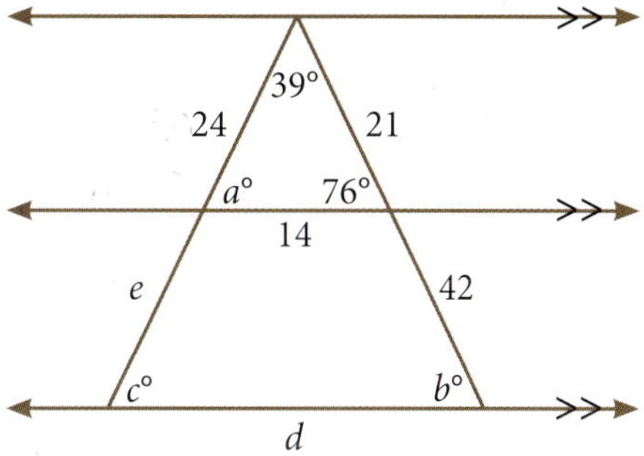

15. Use the diagram at right.

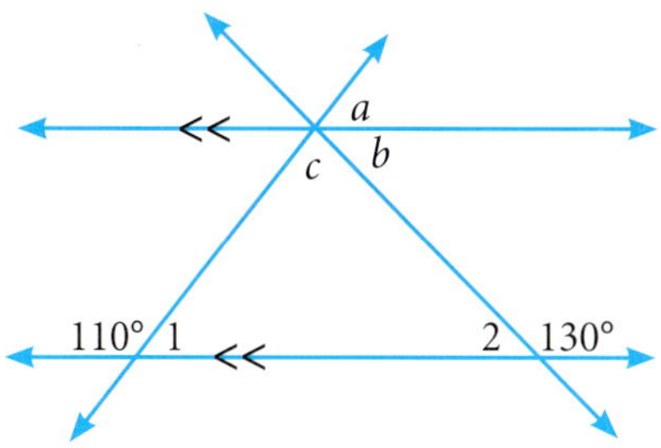

a. What is the name of the special angle relationship between $\angle 1$ and a?

b. What is the name of the special angle relationship between $\angle 2$ and b?

c. Find $m\angle 2$.

d. Find $m\angle 1$.

e. Find the measure of a.

f. Find the measure of b.

g. Find the measure of c.

h. Find the sum of $\angle 1$, $\angle 2$ and $\angle c$.

i. Find the sum of a, b and c.

j. Explain why the sums in **part h** and **part i** must be equal to one another.

16. Use the triangle at the right.

a. Write an equation and solve for x.

b. Find the measure of $\angle H$.

c. Find the measure of $\angle G$.

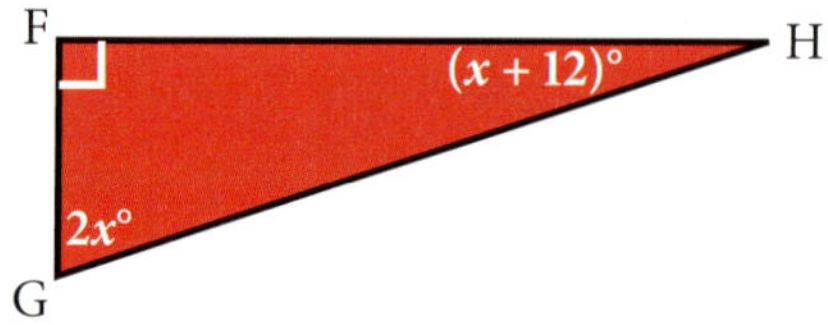

17. The beams forming the roof of a house create an isosceles triangle. Copy the diagram at right and fill in the missing angles.

18. An angle in an equilateral triangle measures $(3x - 60)°$. Find the value of x.

Tic-Tac-Toe ~ Too Tall to Know

Very tall objects are difficult to measure. This activity requires that you go outside to find the height of objects that cannot be measured using traditional means. It is important that all measurements are made within about one hour total time. It must be sunny to complete this activity.

Step 1: Begin by measuring your height in feet. Write the inches part of the measurement as a fraction over 12.

Step 2: Go outside and measure the length of your shadow in feet.

Step 3: At the same time as you measure your own shadow, measure the shadow of at least 5 objects such as a flagpole, the goal posts on a football field, a tree or a building. Measure in feet with remaining inches written as a fraction of a foot.

Step 4: Draw yourself as a stick-figure with your shadow on the ground. In the drawing, connect the end of the shadow to the top of the stick figure's head. This should make a right triangle. Label all known lengths.

Step 5: Draw triangles for each object you measured just as you did in **Step 4.** These triangles are similar to your stick figure triangle. Why is this true?

Step 6: Find the height of each object using the similar triangles.

Step 7: Explain why it was important to measure your shadow and the other object's shadow at the same time of day.

Tic-Tac-Toe ~ Constructing Segments

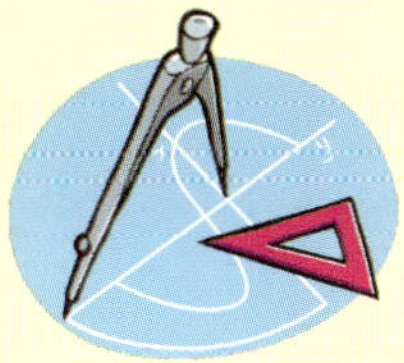

Use a compass and straightedge to duplicate segments and make combinations of segments. Use the segments below.

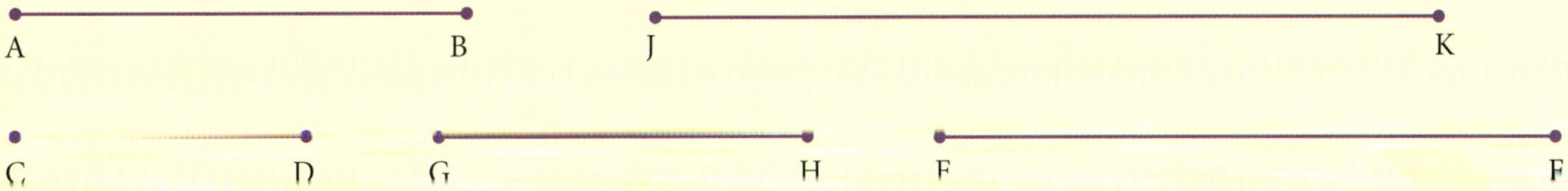

1. Duplicate $\overline{AB}$.

a. Use a straightedge to draw a segment longer than $\overline{AB}$.

b. Measure $\overline{AB}$ by placing the stylus on A and opening the compass so the pencil is on B.

c. Using this setting, place the stylus on one endpoint of your segment then make a mark that crosses your segment.

d. Label your congruent segment $\overline{XY}$.

2. Duplicate $\overline{GH}$.

3. Construct $\overline{EF} + \overline{GH}$.

a. Use a straightedge to draw a segment longer than the combined length of $\overline{EF}$ and $\overline{GH}$.

b. Measure $\overline{EF}$ by placing the stylus on E and opening the compass so the pencil is on F.

c. Using this setting, place the stylus on one endpoint of your segment, then make a mark that crosses your segment.

d. Measure $\overline{GH}$ by placing the stylus on G and opening the compass so the pencil is on H.

e. Using this setting, place the stylus on your intersection from **part c,** then make a mark that crosses your segment. This mark will be further down the segment making it longer.

f. Label the segment $\overline{EH}$ that is equal to the combined length of $\overline{EF}$ and $\overline{GH}$.

4. Construct $\overline{AB} + \overline{EF}$.

5. Construct $2 \cdot \overline{GH}$.

6. Construct $\overline{JK} - \overline{CD}$.

a. Use a straightedge to draw a segment longer than $\overline{JK}$.

b. Measure $\overline{JK}$ with a compass.

c. Using this setting, mark off the length of $\overline{JK}$ on your segment.

d. Measure $\overline{CD}$ with a compass.

e. Using this setting, place the stylus on the intersection from **part c** and turn the compass backwards to mark off the length of $\overline{CD}$. The length of $\overline{JK}$ should be reduced or taken away from.

f. Label the segment $\overline{VW}$ that represents the length of $\overline{JK}$ with $\overline{CD}$ removed from it.

7. Construct $\overline{JK} - \overline{GH}$.

8. Construct $2 \cdot \overline{EF} - \overline{JK}$.

REVIEW

BLOCK 1

Vocabulary

acute triangle
alternate exterior angles
alternate interior angles
congruent figures
corresponding angles
corresponding parts
equiangular
equilateral triangle
exterior angle
isosceles triangle
obtuse triangle
remote interior angle
right triangle
same-side interior angles
scalene triangle
similar figures
transversal

Recognize and apply properties of alternate exterior and alternate interior angles.
Recognize and apply properties of corresponding and same-side interior angles.
Classify triangles based on sides and angles.
Find angle measures in a triangle.
Find measures of angles in isosceles and equilateral triangles.
Determine if triangles are similar or congruent and find missing measures.
Use special angle relationships to find angle measures.

Lesson 1.1 ~ Alternate Exterior and Interior Angles

Name the special angle pair relationship between ∠1 and ∠2. Identify whether or not the angles are congruent.

1.

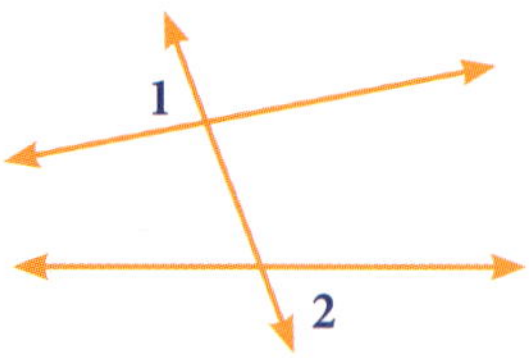

2.

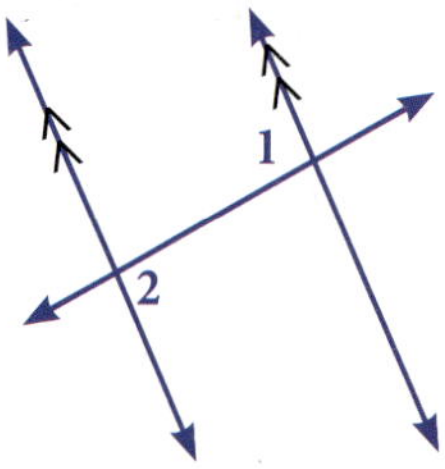

3.

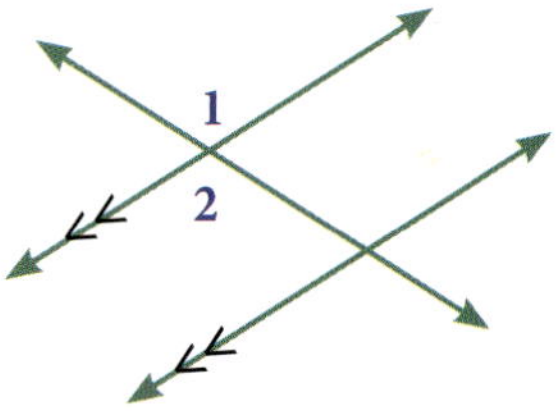

Solve for x. Then find the measure of each angle. Check your solution.

4.

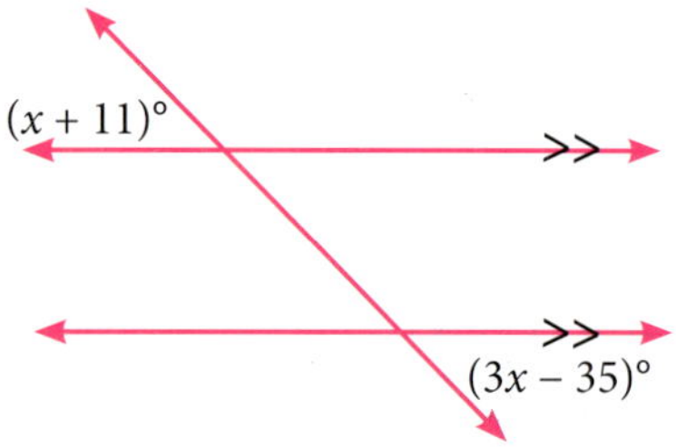

5.

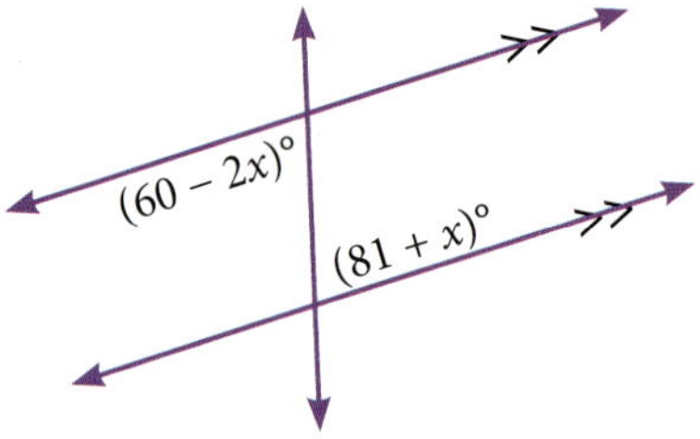

6.

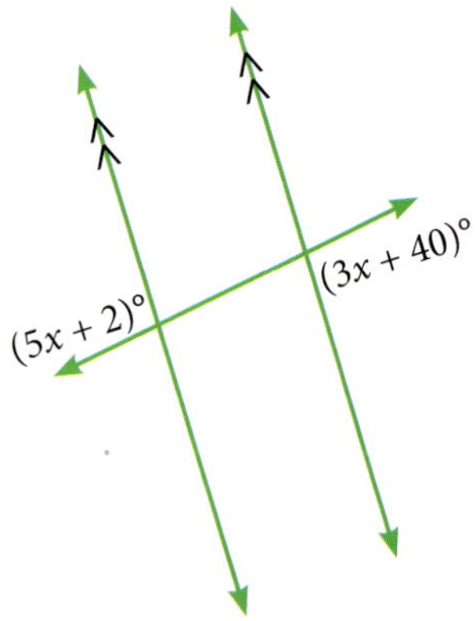

Lesson 1.2 ~ Corresponding and Same-Side Interior Angles

Name the special angle pair relationship between $\angle 1$ and $\angle 2$.

7.

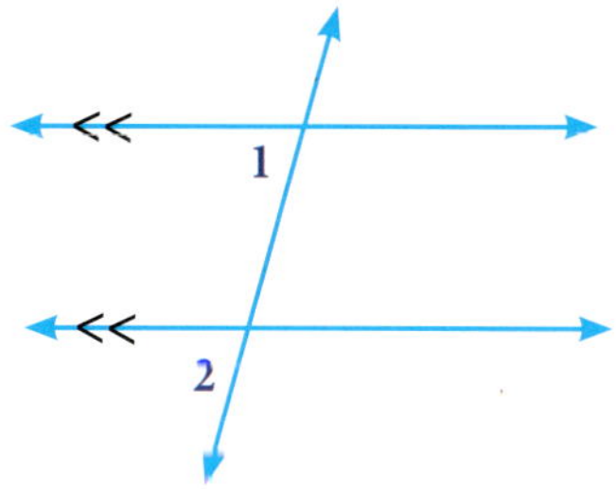

8.

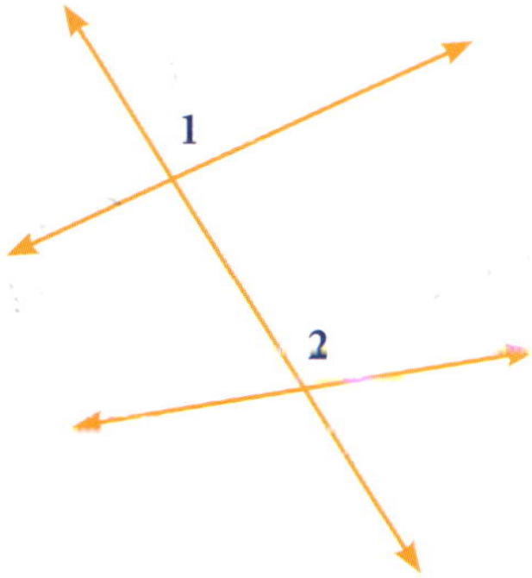

9.

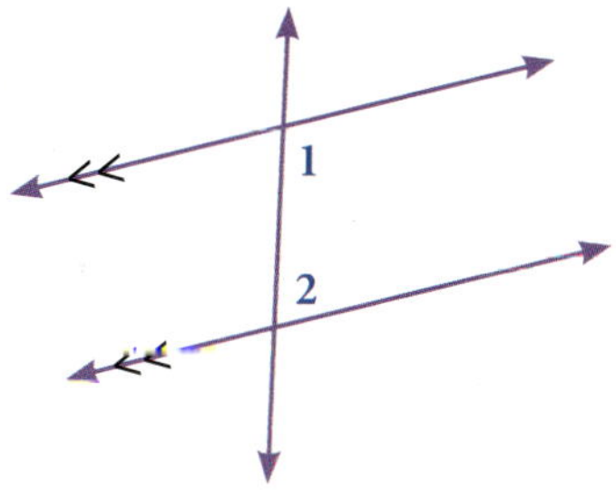

10. Copy the table below. List each special angle pair from the blue box in one category. Assume lines are parallel for *angle pairs.

Linear Pairs	Vertical Angles
*Alternate Exterior Angles	Complementary Angles
*Alternate Interior Angles	*Same-Side Interior Angles
*Corresponding Angles	Supplementary Angles

Have a sum of 90°	Have a sum of 180°	Are equal in measure

Solve for *x*. Then find the measure of each angle. Check your solution.

11.

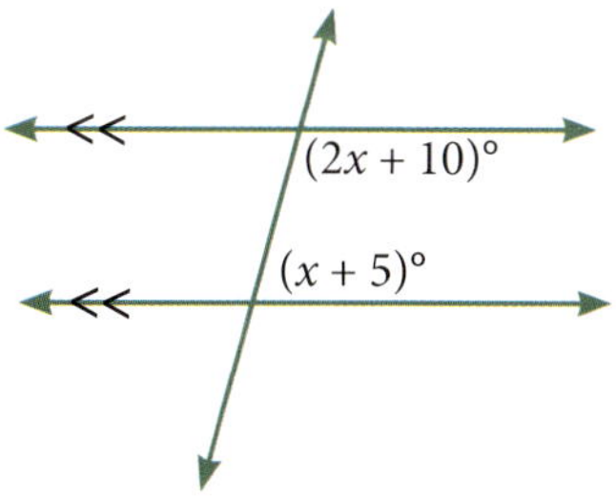

12.

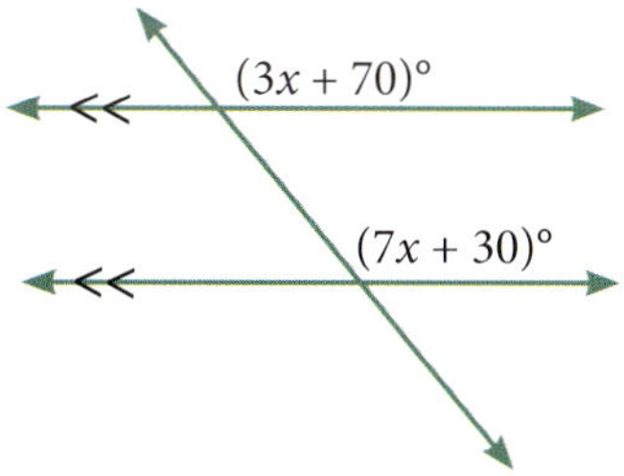

13.

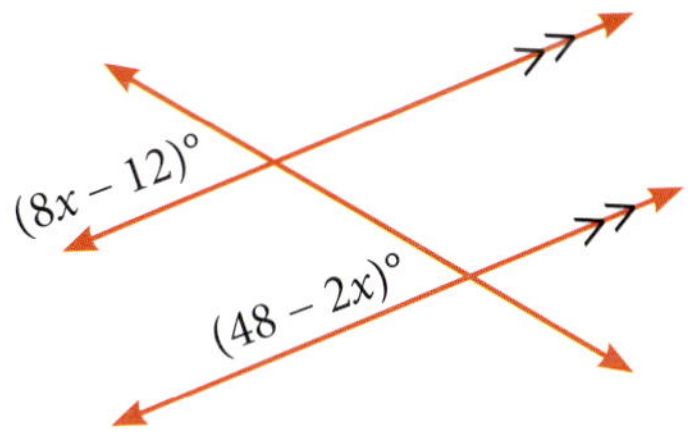

Lesson 1.3 ~ Classifying Triangles

Classify each triangle using angle measures and side lengths.

14.

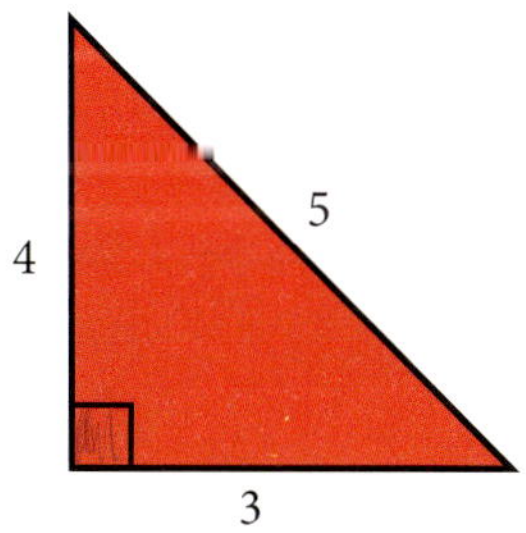

15.

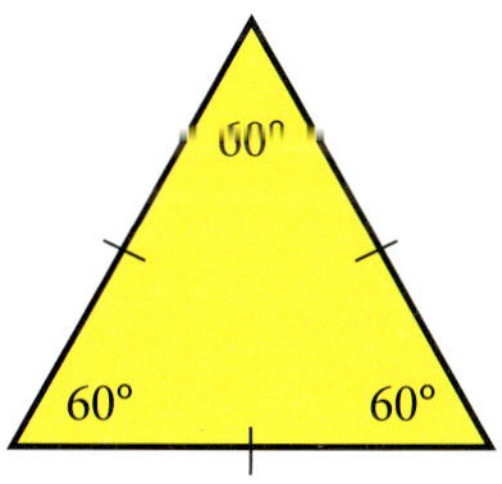

16.

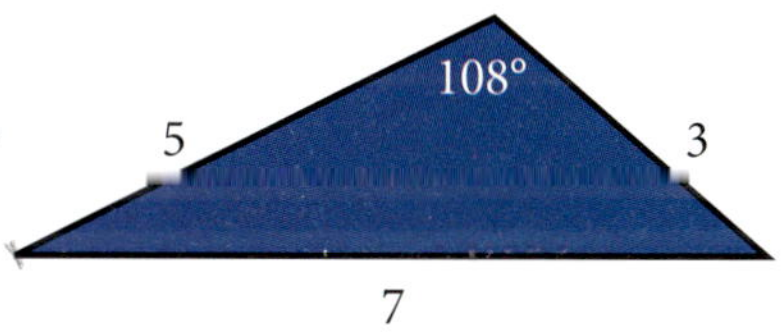

Classify each triangle using angle measures and side lengths.

17.

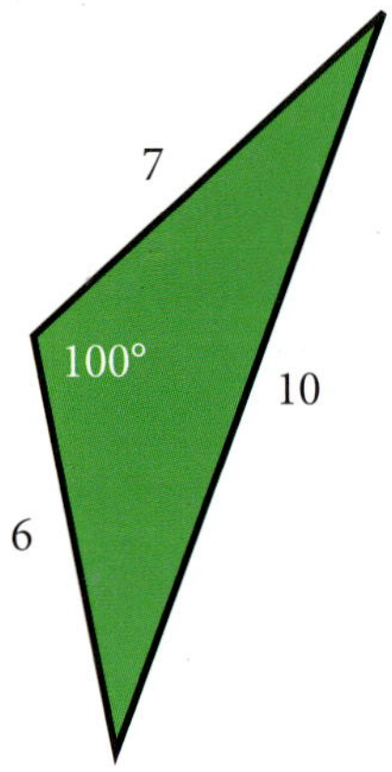

18.

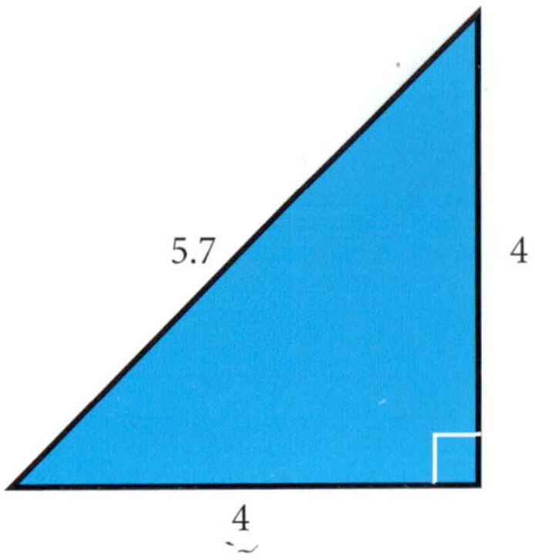

19.

Sketch and label a triangle to match each description.

20. Scalene ΔGED

21. Obtuse ΔPTA

22. Acute isosceles ΔRAM

Lesson 1.4 ~ Angle Sum of a Triangle

Solve for *x*.

23.

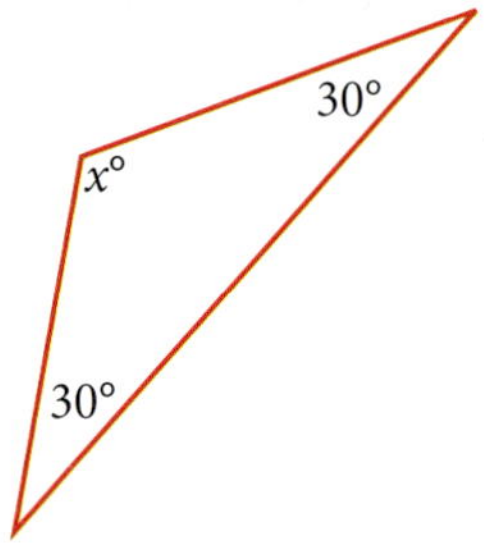

24.

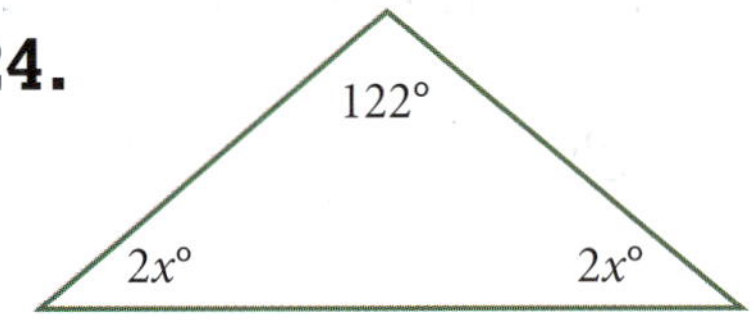

25.

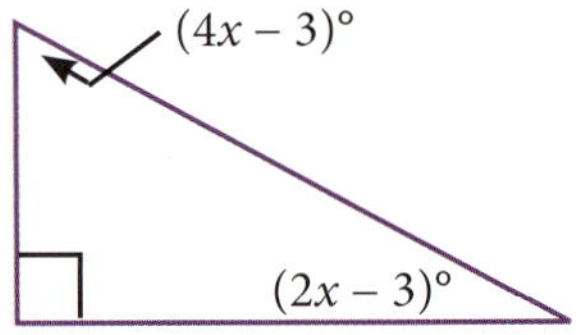

26. Use ΔAPE at the right.

a. Write an equation to find the value of x.

b. Solve for x.

c. Find $m\angle A$ and $m\angle P$.

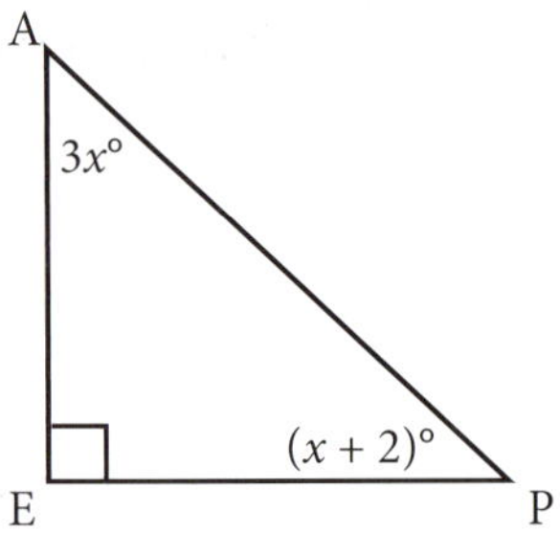

27. In ΔCAR, $m\angle A = (15 - 3x)°$ and $m\angle R = (5 - 2x)°$. The measure of ∠C is 80°.

a. Sketch and label a diagram of ΔCAR.

b. Write an equation and solve for x.

c. Find $m\angle A$ and $m\angle R$.

28. Nancy determined that the value of x in the triangle at the right is 20.

a. Find the measure of each angle by substituting 20 for x.

b. Was Nancy's solution of $x = 20$ correct? How do you know?

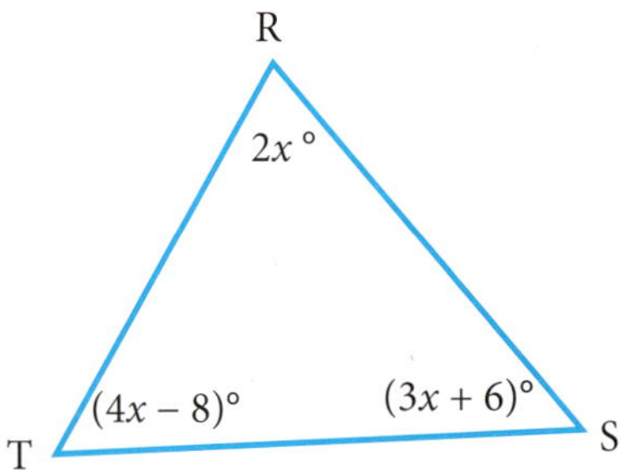

Lesson 1.5 ~ Special Triangles

Solve for *x*.

29.

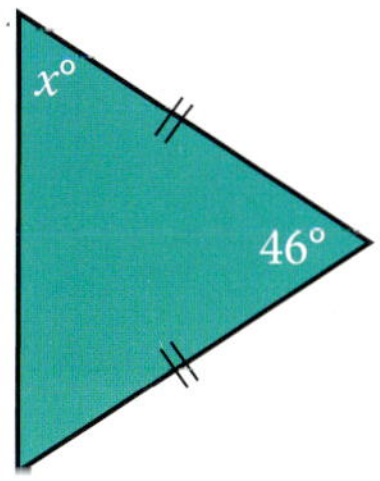

30.

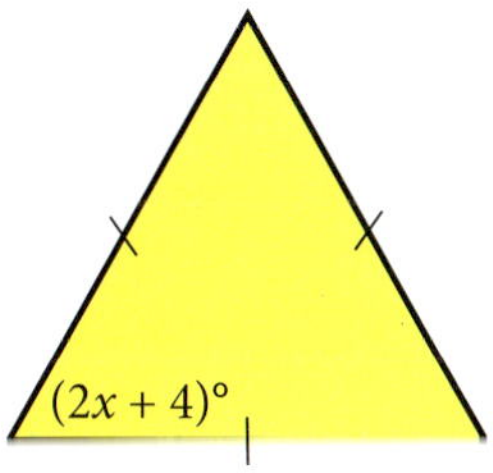

31.

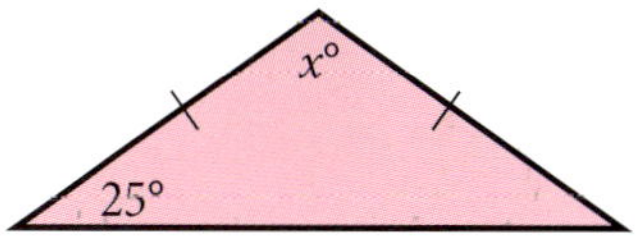

32. ΔTWL has two angles that measure 63°.

a. Sketch a diagram of ΔTWL.

b. Find the measure of the third angle.

c. What type of triangle is ΔTWL based on its side lengths and angle measures?

33. One angle in an equilateral triangle is $(6x + 12)°$. Solve for x.

34. Francisco has a triangular window in his bedroom. It is the shape of an equilateral triangle.

a. What are the measures of each angle in the window?

b. He has equal-sized pieces of framing he wants to put around the window. It takes 6 pieces of framing to cover one side of the window. How much will it take to go around the entire window?

35. Shannon's home makes an isosceles triangle with her school and the park as shown in the diagram below.

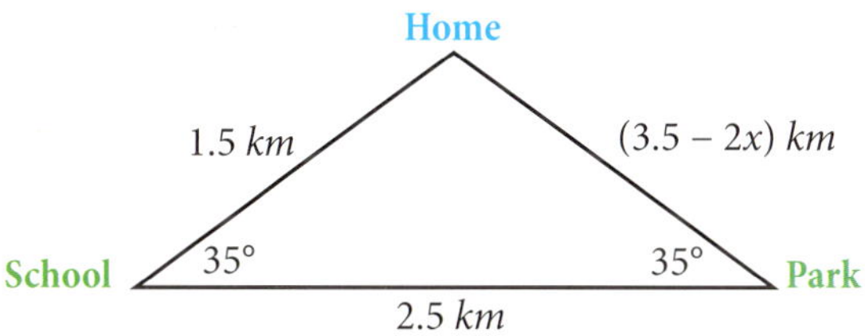

a. How far is it from Shannon's home to the park?

b. Find the value of x.

Lesson 1.6 ~ Congruent and Similar Triangles

36. $\Delta PLA \sim \Delta NET$. List the corresponding angles and sides.

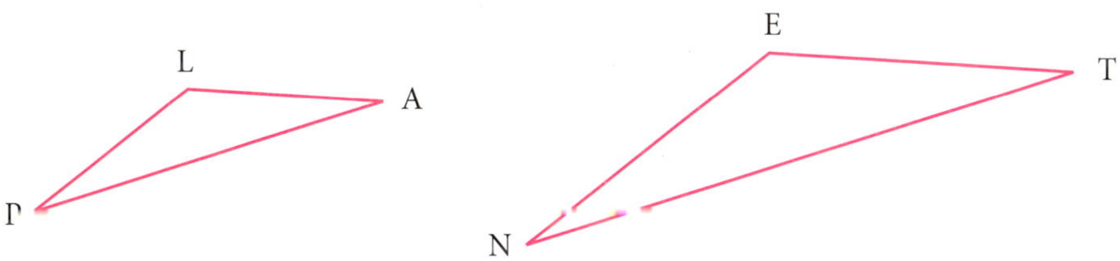

Solve for x in each set of similar triangles.

37.

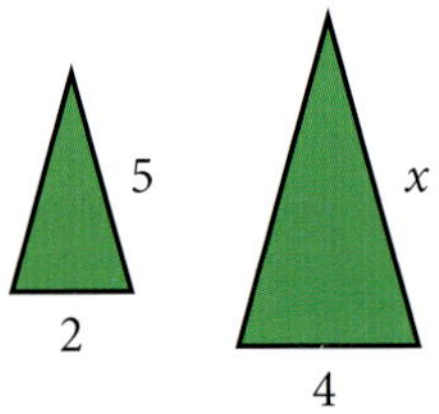

38.

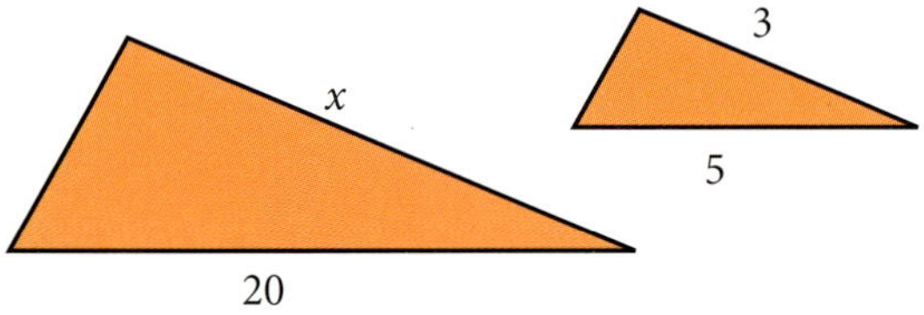

39.

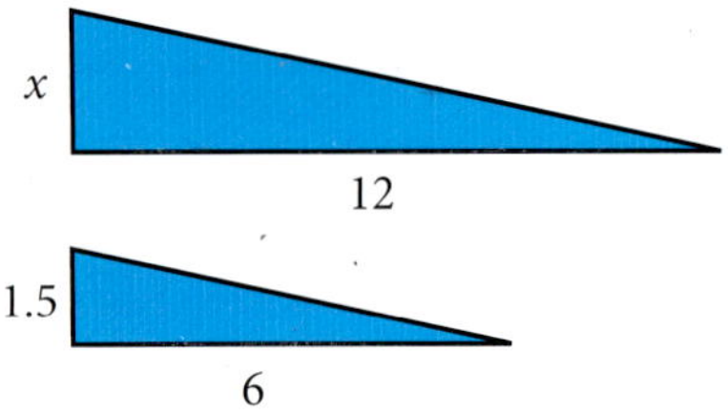

40.

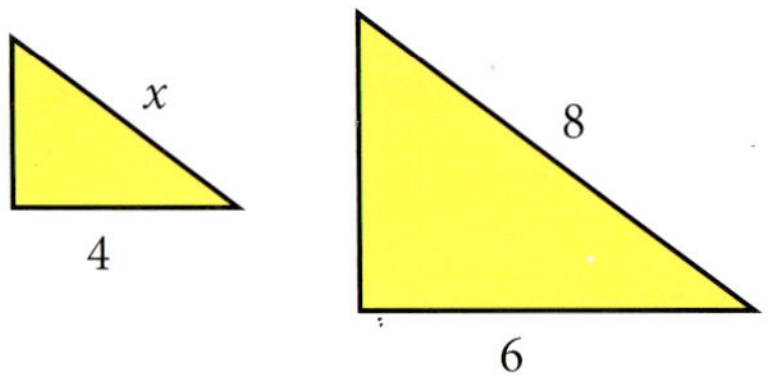

41. Are the triangles below similar? Explain your answer.

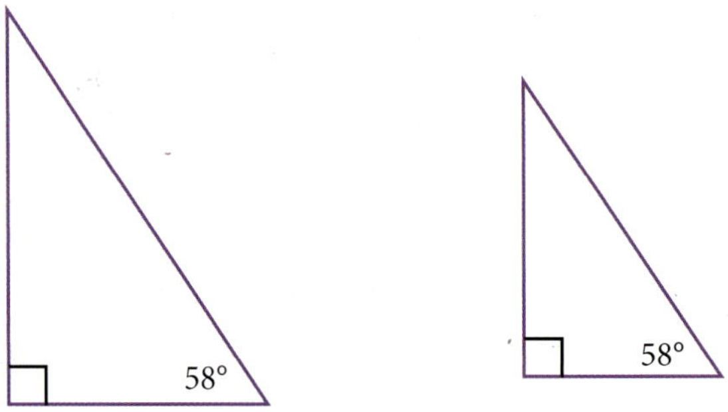

Lesson 1.7 ~ Angle Relationships

42. Use the diagram at the right.

a. Which angle in ΔWSB measures 49°?

b. What is $m\angle WAM$?

c. What is $m\angle BWS$?

d. Which property can you use to say $\Delta WAM \sim \Delta WBS$?

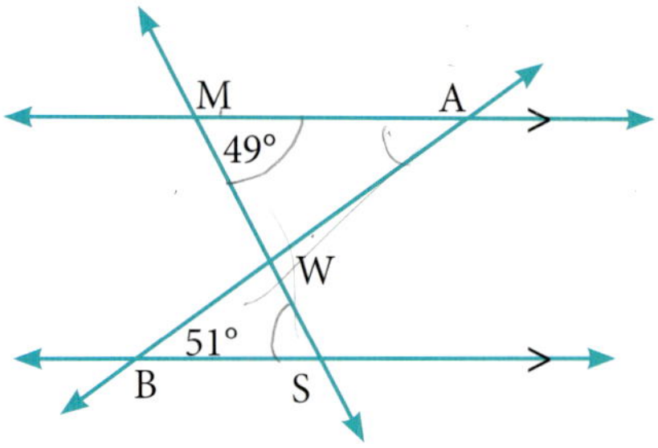

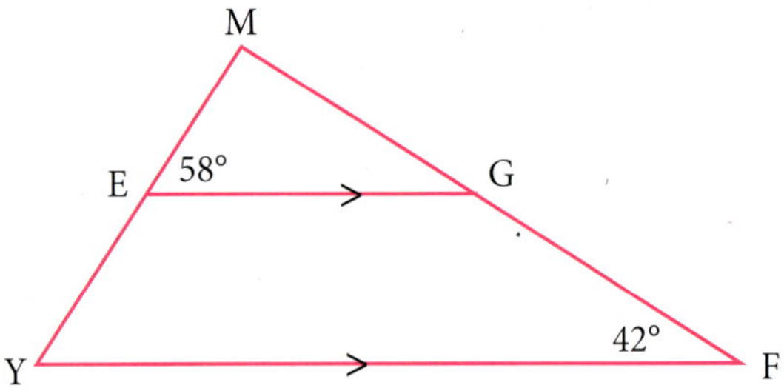

43. Use the diagram at the left.

a. Which angle besides $\angle YFG$ measures 42°?

b. What is the measure of $\angle FYE$?

c. What is the measure of $\angle M$?

d. Fill in the blank: $\Delta MYF \sim \Delta$_____.

44. Find the value of *a*, *b*, *c*, *d* and *e*.

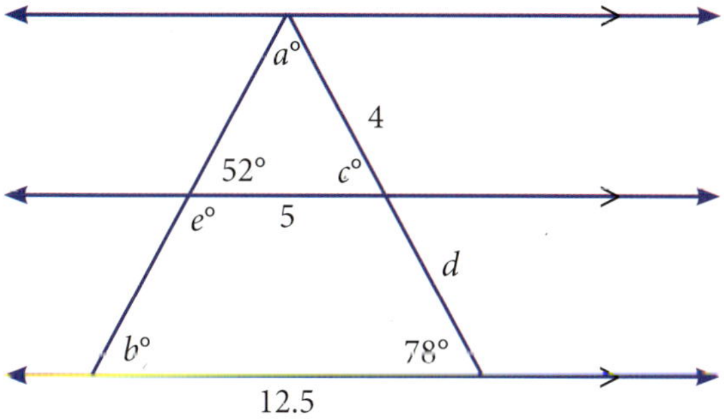

45. Find $m\angle 1$, $m\angle 2$ and $m\angle 3$.

$\overleftrightarrow{WX} \parallel \overleftrightarrow{ZY}$

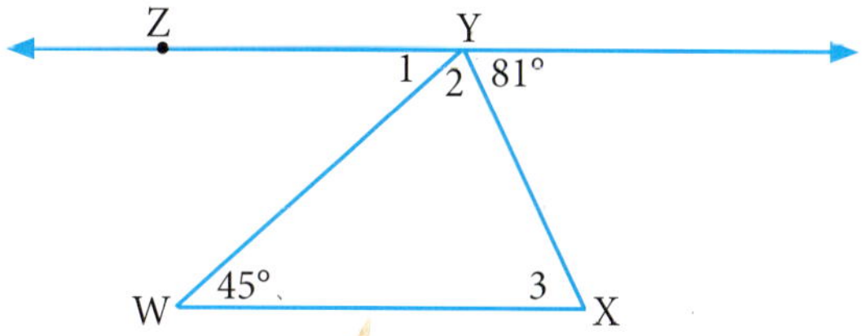

46. Use the diagram at right.

a. Name three angles that have a sum of 180°.

b. Name another set of three angles that have a sum of 180°.

c. What is the relationship between the sum of angles *a* and *b* and the sum of angles 1 and 2?

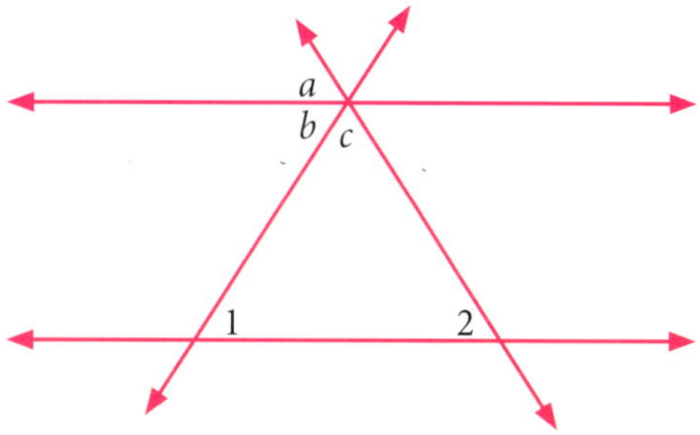

Find the values of *x* and *y* in each pair of similar triangles.

47.

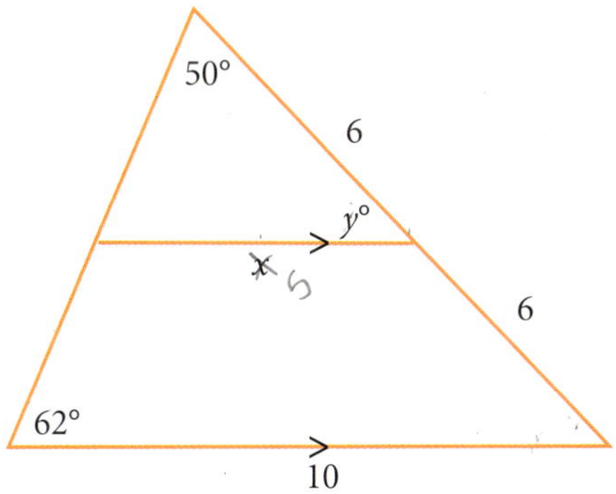

48.

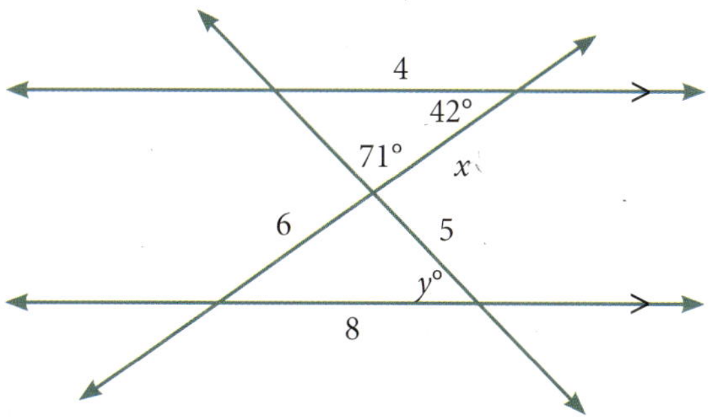

Tic-Tac-Toe ~ Transversal Collage

Step 1: Find at least 10 photographs or pictures of transversals. You can take photographs, locate and print pictures from the internet or cut pictures from magazines or newspapers.

Step 2: Identify the transversal and special angle pairs in each picture.

Step 3: Make a collage to display the pictures.

Tic-Tac-Toe ~ Sum of Interior Angles

The sum of the three angles in a triangle is 180°. Discover how to find the sum of the angles of any polygon based on its number of sides.

1. Copy and complete the chart.

Polygon name and number of sides	Diagram	Number of triangles	Sum of degree measures in the triangles	Conclusion: Angle sum
Quadrilateral 4 sides	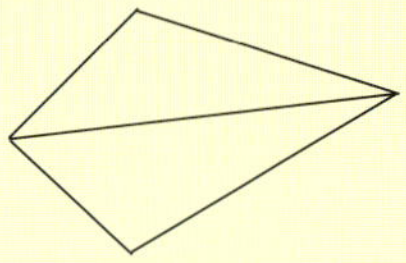	2	180° + 180° or 2(180)°	360°
Pentagon 5 sides	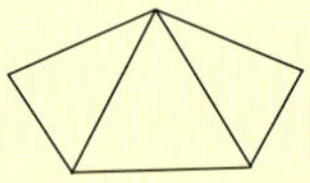	3		
Hexagon 6 sides	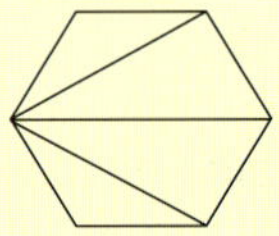			
Heptagon 7 sides				
Octagon 8 sides				
Nonagon 9 sides				
Decagon 10 sides				

2. Does a pattern form in the angle sums? How does the pattern relate to the number of sides of the polygon?

3. Write a formula to calculate the degree measure of any polygon based on its number of sides, *n*.

4. Find the degree measure of these polygons:

a. Dodecagon (12 sides) **b.** 15-gon **c.** 24-gon **d.** 41-gon

CAREER FOCUS

Alden
Architect

I am an architect. Architects are designers. Designs can be as small as a piece of furniture or as large as a city. The primary focus for most architects is buildings. Architects use a process to complete projects which includes determining what a client wants, designing it and drawing up plans. Architects also help oversee the bidding process and actual construction of a project.

Math is used everyday in architecture. Proportion, scale and dimension are important components architects use to create a plan. Architects also have to estimate how much a project will cost. If a client cannot afford to build what the architect has designed, then the design is of no use. Architects prepare budgets and cost estimates throughout the entire architectural process to avoid this problem.

Architects have to do math calculations to account for every square inch of space. Calculations have to be very precise to make sure the buildings designed will not fall over.

People need a Bachelor's degree in Architecture from a college or university to become an architect. This usually takes about five years. After getting a degree, people do on-the-job training. They must pass tests to become a licensed architect. Architects often acquire a Master's degree in Architecture.

The average starting salary for a college graduate architect is $35,000 to $45,000 per year depending on where you live in the country. A mid-level architect makes between $55,000 to $75,000 per year. A senior architect or a firm principal can make $80,000 to $100,000 or more.

Architecture offers something new every day. Whether it is putting together a design proposal, preparing a set of construction drawings, overseeing construction in the field or picking materials for a new project, architects are constantly being challenged with new tasks. Also, as a building goes up, an architect gets to see their hard work paying off. It is truly a rewarding profession.

CORE FOCUS ON GEOMETRY

BLOCK 2 ~ THE PYTHAGOREAN THEOREM

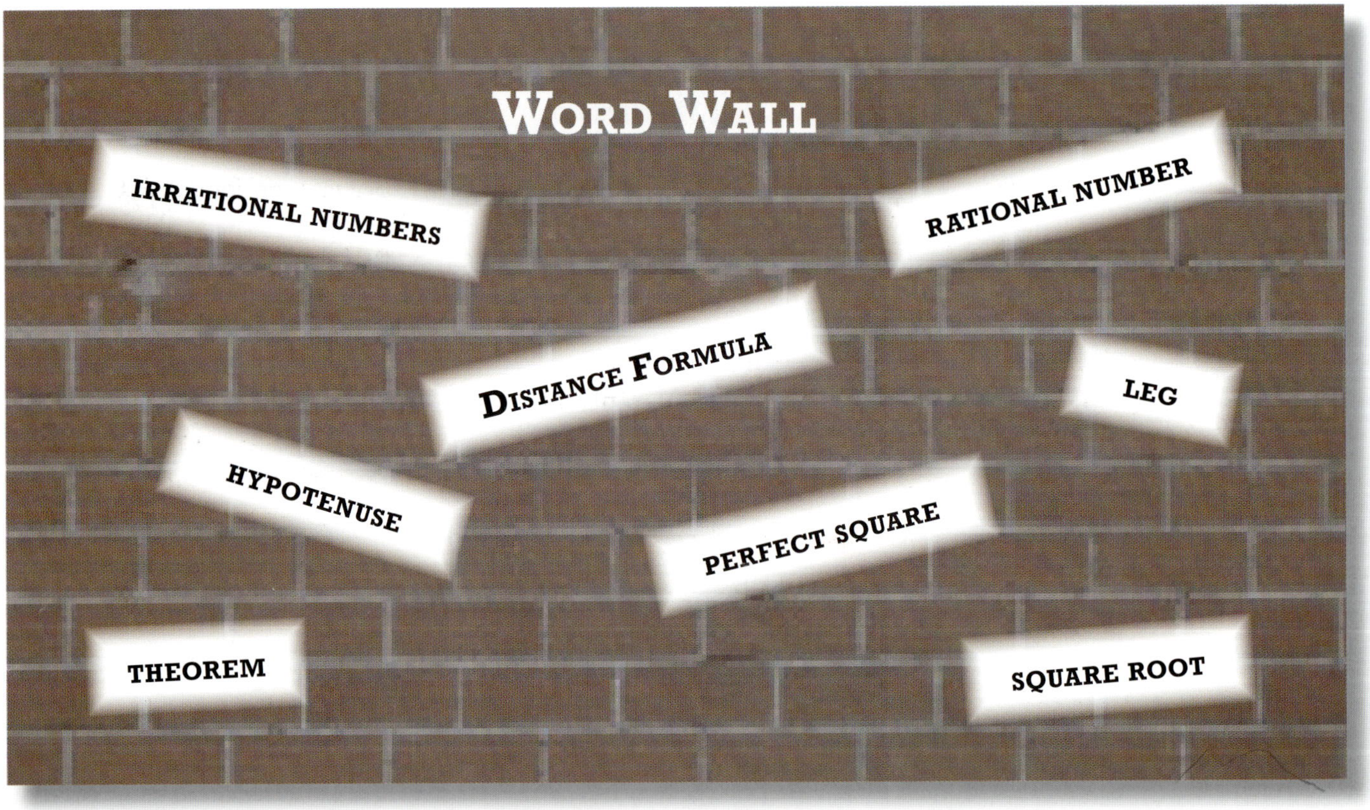

BLOCK 2 ~ THE PYTHAGOREAN THEOREM

TIC-TAC-TOE

Simplifying Roots

Simplify square roots using prime factorization.

See page 56 for details.

Vocab Booklet

Create a booklet with important vocabulary and formulas from the textbook.

See page 80 for details.

Percent Error

Estimate the values of square roots and determine your percent error.

See page 75 for details.

Perfect Square Patterns

Examine the first 25 perfect squares. Make a prediction about the possible last digits of perfect squares.

See page 51 for details.

Prove It!

Examine different ways of proving the Pythagorean Theorem. Write a short paper about your findings.

See page 62 for details.

Creating Triples

Create primitive Pythagorean triples and then find multiples.

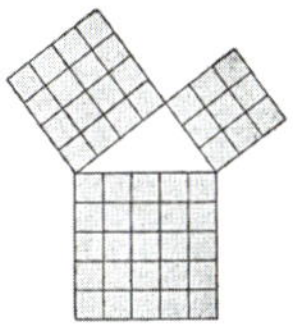

See page 66 for details.

Applications

Make a worksheet with real-world applications of the Pythagorean Theorem.

See page 71 for details.

3D Coordinate Plane

Find the distance between two points on a three-dimensional coordinate plane.

See page 81 for details.

An Irrational Skit

Write a skit where the characters (rational and irrational numbers) reveal information about who they are.

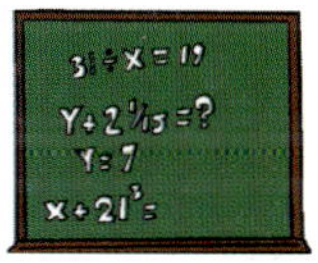

See page 62 for details.

PERFECT SQUARES

LESSON 2.1

Recognize and find the values of perfect squares.

The area of a square is the product of its base and height. Since the base and the height are equal in a square, the formula can also be written using *s* as the length of a side.

$$\text{Area of a square} = s \cdot s = s^2$$

s^2 is read as "s squared."

When the side of a square has an integer length, its area is a perfect square. A **perfect square** is the value of an integer to the second power. Some examples of perfect squares are shown below.

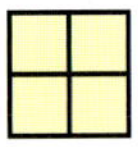

Area = s^2
Area = 2^2
Area = $2 \cdot 2 = 4$

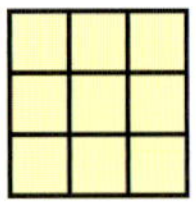

Area = s^2
Area = 3^2
Area = $3 \cdot 3 = 9$

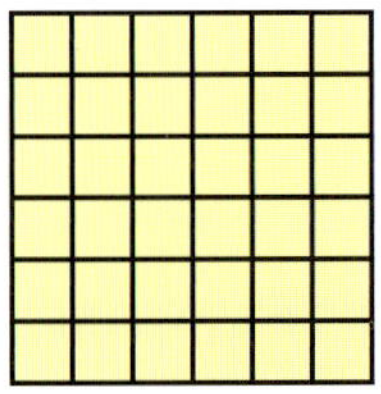

Area = s^2
Area = 6^2
Area = $6 \cdot 6 = 36$

EXPLORE! SQUARES THAT ARE PERFECT

Step 1: On grid paper, draw a square with an area of 16 units^2. What is the length of one side of this square?

Step 2: Draw a square with an area of 49 units^2. What is the length of one side of this square?

Step 3: Draw a square with an area of 12 units^2.
- **a.** What is the approximate length of one side of this square? How did you determine this?
- **b.** Is 12 a perfect square? Explain your answer.

Step 4: Copy and complete the table through a side length of 13.

Side Length of Square	Area in Exponential Form	Perfect Square
1	1^2	1
2	2^2	4
3		
4		

Step 5: Use the table to determine the side length of a square with the following areas:
- **a.** 100 square units
- **b.** 64 square units
- **c.** 25 square units

In the **Explore!** you found the square root of a number by working backwards from the area to find the side length of each square. The **square root** of a number is one of the two equal factors of the number. The product of the two equal factors is the number. A square root is written using a symbol called a radical sign, $\sqrt{}$. The radical sign is used to indicate the positive square root.

$$\sqrt{25} = 5 \text{ since } 5 \cdot 5 = 25$$

Read "square root of 25."

EXAMPLE 1

Find the value of each square root.

a. $\sqrt{9}$ **b.** $\sqrt{121}$

SOLUTIONS

a. $\sqrt{9} = 3$ because $3 \cdot 3 = 9$.

$3 \cdot 3 = 3^2$
Three squared is nine, so the square root of nine is three.
$\sqrt{9} = \sqrt{3^2} = 3$

b. $\sqrt{121} = 11$ because $11 \cdot 11 = 121$.

Square roots are useful when solving equations with a variable that is squared. Just as addition is the inverse operation of subtraction, finding the square root is the inverse operation of squaring a number. As always, check answers by substituting the solution into the original equation.

When solving an equation that involves a variable to the second power, the value of the variable could be positive or negative. For example, if $x^2 = 4$, then $x = 2$ or -2 because both values squared equal 4. This is written $x = \pm 2$ and is read "x equals plus or minus 2".

EXAMPLE 2

Solve each equation.

a. $x^2 = 9$

b. $x^2 - 21 = 28$

c. $3y^2 = 300$

SOLUTIONS

a. Write the equation.
Square root both sides of the equation.

$$x^2 = 9$$
$$\sqrt{x^2} = \pm\sqrt{9}$$
$$x = \pm 3$$

$\sqrt{x^2} = x$

When you take the square root of both sides, you must put the symbol "±" in front of the expression on the other side of the equals sign.

$3 \cdot 3 = 9$ and $(-3)(-3) = 9$

$x = \pm 3$

☑ Check the answer.

$(3)^2 \stackrel{?}{=} 9$ $\quad$ $(-3)^2 \stackrel{?}{=} 9$

$9 = 9$ $\quad$ $9 = 9$

b. Write the equation.
Add 21 to both sides of the equation.

Square root both sides of the equation.

$$x^2 - 21 = 28$$
$$\underline{+21 \quad +21}$$
$$x^2 = 49$$
$$\sqrt{x^2} = \pm\sqrt{49}$$
$$x = \pm 7$$

$7 \cdot 7 = 49$ and $(-7)(-7) = 49$

$x = \pm 7$

☑ Check the answer.

$7^2 - 21 \stackrel{?}{=} 28$ $\quad$ $(-7)^2 - 21 \stackrel{?}{=} 28$

$49 - 21 \stackrel{?}{=} 28$ $\quad$ $49 - 21 \stackrel{?}{=} 28$

$28 = 28$ $\quad$ $28 = 28$

EXAMPLE 2 SOLUTIONS (CONTINUED)

c. Divide both sides of the equation by 3.

$$\frac{3y^2}{3} = \frac{300}{3}$$

$$y^2 = 100$$

Square root both sides of the equation.

$$\sqrt{y^2} = \pm\sqrt{100}$$

$y = \pm 10$.

$$y = \pm 10$$

$10 \cdot 10 = 100$ and $(-10)(-10) = 100$

☑ Check the answer.

$$3(10)^2 \stackrel{?}{=} 300 \qquad 3(-10)^2 \stackrel{?}{=} 300$$
$$3(100) \stackrel{?}{=} 300 \qquad 3(100) \stackrel{?}{=} 300$$
$$300 = 300 \qquad 300 = 300$$

EXERCISES

Find each value.

1. 4^2 **2.** 9^2 **3.** 1^2

4. $\sqrt{49}$ **5.** $\sqrt{9}$ **6.** $\sqrt{4}$

7. $\sqrt{36}$ **8.** $\sqrt{81}$ **9.** $\sqrt{25}$

10. $\sqrt{1}$ **11.** $\sqrt{64}$ **12.** $\sqrt{144}$

13. Monique designed a square patio. She made the sides of the patio 15 feet. What is the area of the patio?

14. Jack is preparing a garden space. He has enough fertilizer to cover 144 square feet. He wants to make a square garden. What should be the length of each side?

15. A square has an area of 64 square meters.
 a. Find the length of one side of the square.
 b. What is the perimeter of the square?

16. A square pizza has an area of 121 square inches. Find the length of one side of the pizza.

17. The area of a circle is 49π square units. The formula for the area of a circle is πr^2. What is the radius of the circle?

Solve each equation. Check each answer.

18. $x^2 = 49$

19. $8 + x^2 = 89$

20. $-2h^2 = -72$

21. $35 = m^2 - 29$

22. $3y^2 - 2 = 430$

23. $75 = 2d^2 + 57$

24. $25 - x^2 = -24$

25. $9x^2 + 1 = 10$

26. $x^2 + x^2 + x^2 = 75$

REVIEW

List all the possible names for each angle.

27.

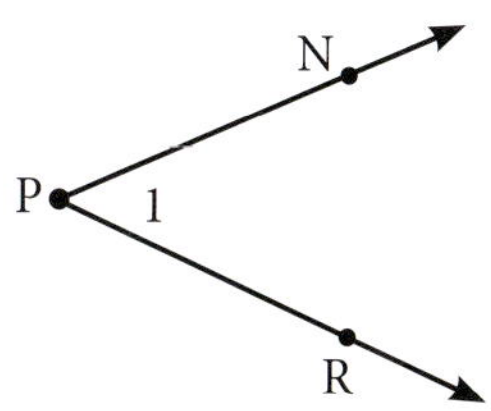

28.

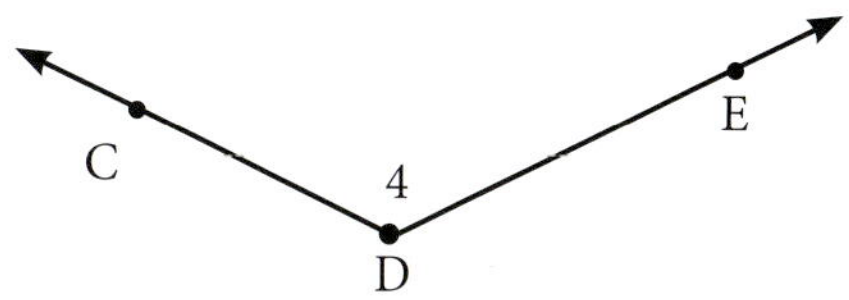

Classify each triangle by its sides and angles.

29.

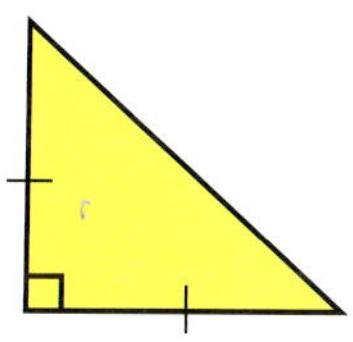

30.

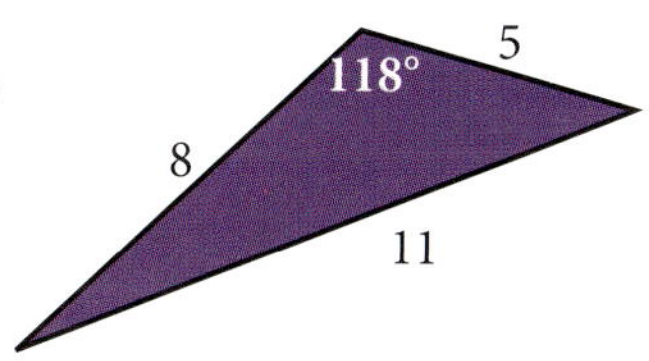

31.

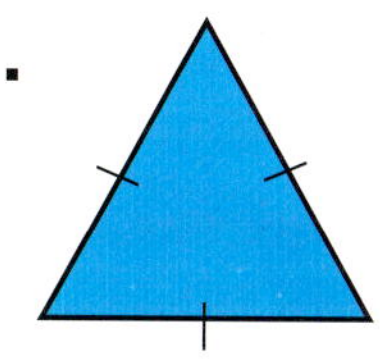

Tic-Tac-Toe ~ Perfect Square Patterns

1. Make a poster showing the first 25 perfect squares.
2. There are only six possibilities for the last digit of a perfect square. Based on what you listed, what are the six possible digits?
3. Make as many predictions as you can about the relationship of the last digit of the number being squared to the last digit of the perfect square. Verify each prediction with an additional perfect square that was not on your original list.
4. Organize all your predictions on your poster.

ESTIMATING SQUARE ROOTS

LESSON 2.2

Estimate the values of square roots.

The square root of a perfect square is always a **rational number**, or a number that can be written as a fraction with integers in the numerator and denominator. Other examples of rational numbers are 3, 0.25, $0.\overline{6}$ and $4\frac{1}{2}$ because they can be written $\frac{3}{1}$, $\frac{1}{4}$, $\frac{2}{3}$ and $\frac{9}{2}$, which are fractions using integers in the numerator and denominator.

When you square root a number that is not a perfect square, the answer will be an irrational number. **Irrational numbers** are numbers that cannot be expressed as a fraction with integers in the numerator and denominator. Irrational numbers can also be recognized in decimal form when the decimal does not repeat or terminate.

Estimating the value of square roots is an important skill. In order to estimate the value of a square root, you must be able to compare the number under the radical symbol to the perfect squares.

In the **Explore!** in the last lesson, you were asked to find the side length of a square with an area of 12 square units. This side length is not a whole number but is between the whole numbers 3 and 4 because $\sqrt{12}$ is between $\sqrt{9} = 3$ and $\sqrt{16} = 4$.

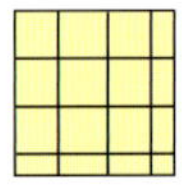

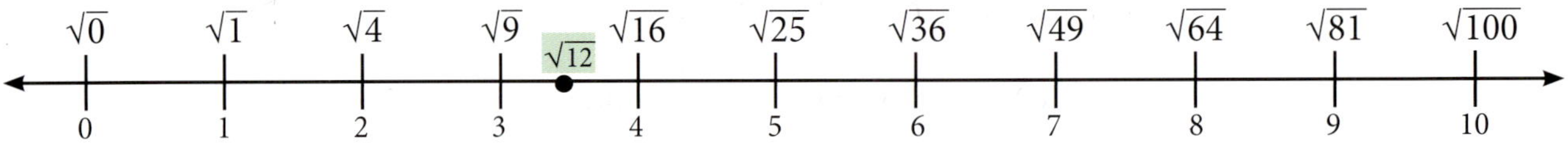

EXAMPLE 1

Determine the two positive integers that each value is between.

a. $\sqrt{35}$

b. $\sqrt{83}$

SOLUTIONS

a. 35 is between the perfect squares 25 and 36.

$\sqrt{25} = 5$ and $\sqrt{36} = 6$

$\sqrt{35}$ is <u>between 5 and 6</u>.

b. 83 is between the perfect squares 81 and 100.

$\sqrt{81} = 9$ and $\sqrt{100} = 10$

$\sqrt{83}$ is <u>between 9 and 10</u>.

When estimating square roots, you can provide an even more accurate estimate than the closest two integers by determining which integer the perfect square is closer to. Estimate $\sqrt{12}$ again. The number 12 is approximately halfway between the perfect squares 9 and 16. This means $\sqrt{12}$ will be about halfway between 3 and 4. A good estimate would be $\sqrt{12} \approx 3.5$.

Estimating Square Roots

1. Determine which two perfect squares the number under the radical sign is between.
2. Determine the integers that represent the square root of those perfect squares.
3. Write a statement using the integer values: "$\sqrt{}$ is between ___ and ___."
4. Determine if the value of the square root is closer to the first or second integer. Estimate its value.

A calculator is a helpful tool when working with square roots of non-perfect squares. Estimating gives you a good idea of the approximate value, but a calculator will give you a better approximation. The **Explore!** below shows how to find the value of square roots using a calculator.

EXPLORE! **CALCULATE IT!**

Step 1: Locate the $\boxed{x^2}$ button on a calculator. Use your calculator to find the value of 29^2 using the $\boxed{x^2}$ button.

Step 2: What is another way to find 29^2 using a calculator? Check your answer to **Step 1** using this method.

Step 3: Locate the $\boxed{\sqrt{}}$ button on a calculator. Describe where it is located in relation to the $\boxed{x^2}$ button on the calculator.

Step 4: Find the value of $\sqrt{841}$. Some calculators require you to press the $\boxed{\sqrt{}}$ button before entering the number. Other calculators require you to enter the number first and then press the $\boxed{\sqrt{}}$ button. Experiment with a calculator and find which key strokes are necessary.

Step 5: How does your answer for $\sqrt{841}$ compare to your work done in **Steps 1 and 2**? Is 841 a perfect square?

Step 6: Use a calculator to find the value of $\sqrt{12}$. Write the entire number that shows on the calculator screen.

Step 7: $\sqrt{12}$ is an irrational number. Irrational numbers are decimals that do not repeat or terminate. Why does the number on the calculator screen end?

Step 8: Instructions will often state, "Round to the nearest __________." The solution you get after rounding is an approximation. Round $\sqrt{12}$ to the nearest tenth.

Step 9: For each of the following, estimate the value of the square root to the nearest tenth using the method in the blue box above. Then use a calculator to find the value of each square root to the nearest tenth.

a. $\sqrt{8}$ **b.** $\sqrt{27}$ **c.** $\sqrt{42}$

EXAMPLE 2 **Use a number line to approximate the value of $\sqrt{24}$ to the nearest tenth. Use a calculator to find its value to the nearest tenth.**

SOLUTION 24 is between the perfect squares 16 and 25. $\sqrt{24}$ is between 4 and 5.

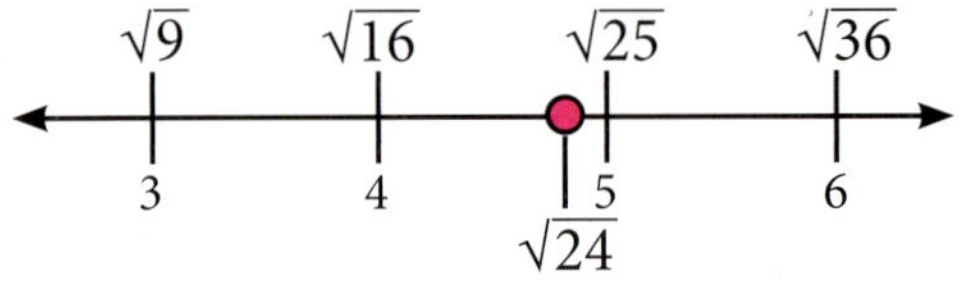

$\sqrt{24}$ is very close to 5, so a good estimation would be $\sqrt{24} \approx 4.9$.

Calculator method.	$\sqrt{24} \approx 4.898979486$
Round to the nearest tenth.	$\sqrt{24} \approx 4.898979486 \approx 4.9$

EXERCISES

Determine the two positive integers each value is between.

1. $\sqrt{61}$ **2.** $\sqrt{8}$ **3.** $\sqrt{29}$

4. $\sqrt{3}$ **5.** $\sqrt{138}$ **6.** $\sqrt{83}$

7. Your teacher asks you to estimate the value of $\sqrt{50}$.
 a. What two integers is $\sqrt{50}$ between?
 b. Which integer is $\sqrt{50}$ closer to?
 c. Estimate the value of $\sqrt{50}$ to the nearest tenth.
 d. Use a calculator to determine $\sqrt{50}$ to the nearest tenth. Was your estimate close?

Estimate each square root to the nearest tenth.

8. $\sqrt{34}$ **9.** $\sqrt{5}$ **10.** $\sqrt{71}$

11. $\sqrt{103}$ **12.** $\sqrt{20}$ **13.** $\sqrt{89}$

Use a calculator to approximate each square root. Round to the nearest tenth.

14. $\sqrt{19}$ **15.** $\sqrt{60}$ **16.** $\sqrt{199}$

17. $\sqrt{367}$ **18.** $\sqrt{2}$ **19.** $\sqrt{15}$

20. A tarp is used to keep the infield of a baseball field from getting wet. The infield is a square. The corners of the square tarp are anchored at each base. The tarp is 8,100 square feet.

   a. What is the distance between first and second base?

   b. Is 8,100 a perfect square? Why or why not?

21. A square parking lot has an area of 12,996 square yards.

   a. What is the length of one side of the parking lot?

   b. How many yards of fence will it take to enclose the entire parking lot?

22. An acre of land is 43,560 square feet. Tariq bought a square lot that is exactly one acre. What is the length of one side of his lot to the nearest foot?

23. Josie wants to tile one square section of her kitchen counter. She bought 850 square tiles with side lengths of 1 inch.

   a. What is the longest length one side of the square section of counter can be? Round to the nearest inch.

   b. How many tiles will she have left over?

Solve for x. Round answers to the nearest tenth.

24. $x^2 + 11 = 37$

25. $4x^2 = 60$

26. $96 = 2x^2 - 6$

27. $20 = 10 + 5x^2$

28. $x^2 + x^2 = 40$

29. $2 - 3x^2 = -103$

REVIEW

Solve for x.

30.

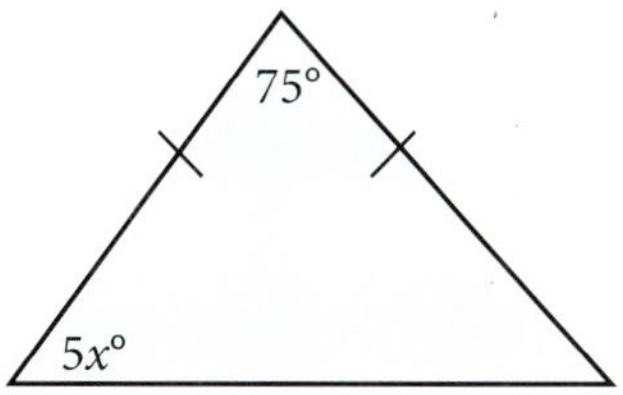

31.

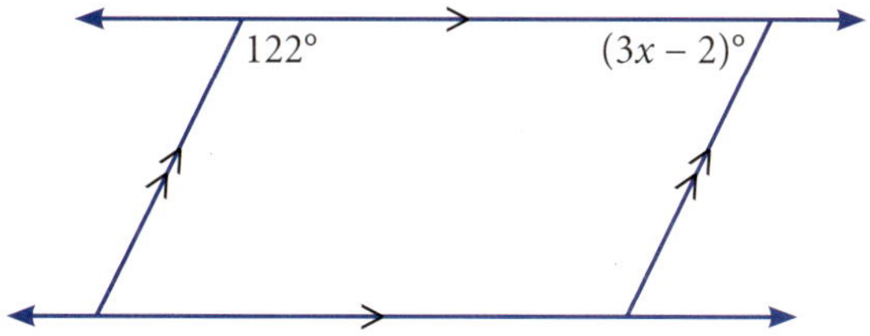

32.

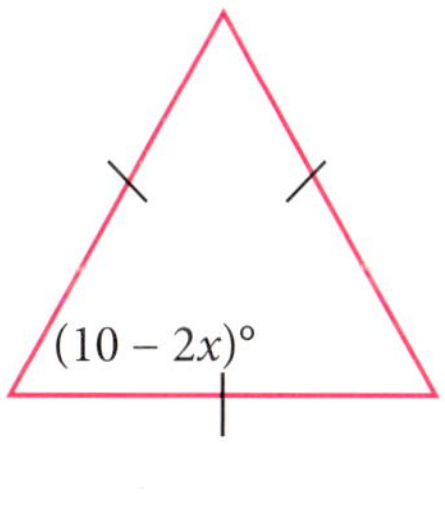

33.

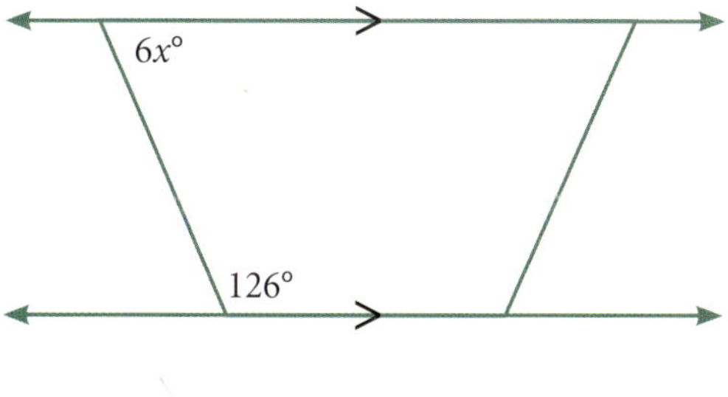

Tic-Tac-Toe ~ Simplifying Roots

Engineers and scientists often work with exact numbers rather than decimal approximations. Other professionals also use exact numbers in their work so they don't have rounding errors. In higher level math classes, you will be asked to give exact answers with square roots. This is done by finding the square root of any perfect squares under the radical sign and moving those values outside the radical sign. Look at the examples below to see how square roots can be simplified.

Example 1: Simplify $\sqrt{24}$.

Write the prime factorization of 24.	$24 = 2 \cdot 2 \cdot 2 \cdot 3$
Circle pairs of numbers that are equal.	$24 = 2 \cdot (2 \cdot 2) \cdot 3$
Bring each pair of equal numbers outside the radical sign as a single number since $\sqrt{2 \cdot 2} = \sqrt{4} = 2$.	$\sqrt{24} = 2\sqrt{2 \cdot 3}$
Multiply the numbers left under the radical sign.	$\sqrt{24} = 2\sqrt{6}$

Example 2: Simplify $\sqrt{180}$.

Write the prime factorization of 180.	$180 = 2 \cdot 2 \cdot 3 \cdot 3 \cdot 5$
Circle pairs of numbers that are equal.	$180 = (2 \cdot 2) \cdot (3 \cdot 3) \cdot 5$
Bring each pair of equal numbers outside the radical sign as a single number.	$\sqrt{180} = 2 \cdot 3\sqrt{5}$
Multiply the numbers outside the radical sign.	$\sqrt{180} = 6\sqrt{5}$

Simplify each square root.

1. $\sqrt{8}$ **2.** $\sqrt{18}$ **3.** $\sqrt{75}$ **4.** $\sqrt{450}$

5. $\sqrt{147}$ **6.** $\sqrt{60}$ **7.** $\sqrt{300}$ **8.** $\sqrt{540}$

THE PYTHAGOREAN THEOREM

LESSON 2.3

Use the Pythagorean Theorem to find missing side lengths in right triangles.

In a right triangle, the side opposite the right angle is called the **hypotenuse**. The hypotenuse is always the longest side. In the figure below, c represents the hypotenuse of the right triangle.

The two sides of the right triangle that form the right angle are called **legs**. In the figure below, a and b represent the legs of the right triangle.

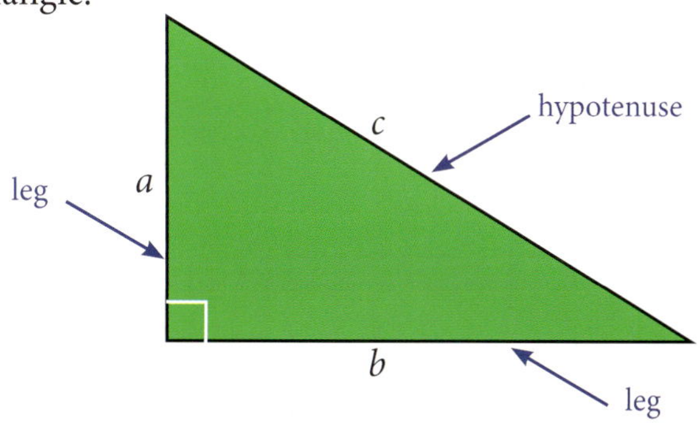

EXPLORE! **A RULE FOR RIGHT TRIANGLES**

The lengths of the legs and the hypotenuse of a right triangle have a special relationship.

Step 1: Draw a right triangle on a piece of paper. Make one of the legs 3 centimeters long and the other leg 4 centimeters long.

Step 2: Copy the chart below. Each measurement is written in centimeters.

Short Leg, Long Leg, Hypotenuse	Square of Short Leg	Square of Long Leg	Square of Hypotenuse
3, 4, ___	$3^2 = 9$	$4^2 = 16$	
5, 12, ___			
6, 8, ___			
8, 15, ___			

Step 3: Measure the length of the hypotenuse of the triangle you drew in **Step 1**. Round the measurement to the nearest centimeter. Fill in the blank in the first column with this measurement.

Step 4: Find the square of the hypotenuse. Fill in the last column as shown in the middle columns.

Step 5: Draw another right triangle with the two legs given in the next row of the table. Measure the hypotenuse to the nearest centimeter. Complete the row.

Step 6: Repeat the process for the remaining two right triangles.

Step 7: Look at your chart. Do you see any patterns or relationships between the squares of the two legs and the hypotenuse? If so, explain.

Step 8: Sum the squares of the legs of each triangle. What do you notice about the sum and the square of the hypotenuse?

Step 9: Create a rule for the lengths of the legs of a right triangle in relation to the hypotenuse length. Use a and b to represent the two legs and c to represent the hypotenuse.

In the **Explore!** you discovered the special relationship between the side lengths of a right triangle. This relationship was discovered and used in Egypt and Babylonia as far back as 1500 BC. Pythagoras (ca. 569 – 475 BC) was a famous Greek philosopher known as "the father of numbers". Pythagoras proved the relationship between the legs and hypotenuse of a right triangle. Once a relationship in mathematics has been proved, it is called a **theorem**. Pythagoras is credited with the discovery of the Pythagorean Theorem. The Pythagorean Theorem is used today in many professions including engineering, construction, architecture, navigation, art and graphic design.

The Pythagorean Theorem

In a right triangle, the sum of the squares of the lengths of the legs is equal to the square of the length of the hypotenuse.

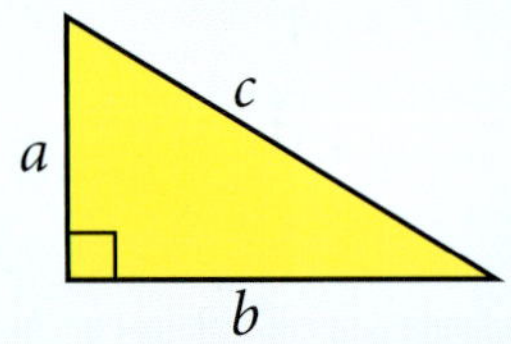

$$a^2 + b^2 = c^2$$

EXAMPLE 1

Find the length of the hypotenuse in a right triangle with leg lengths of 9 and 12 units.

Solution

Write the Pythagorean Theorem.	$a^2 + b^2 = c^2$
Substitute the lengths of the legs for a and b.	$9^2 + 12^2 = c^2$
Simplify by squaring.	$81 + 144 = c^2$
Add.	$225 = c^2$
Square root both sides of the equation.	$\pm\sqrt{225} = \sqrt{c^2}$
	$\pm 15 = c$
	$15 = c$

Since the Pythagorean Theorem is used to find a length, only the positive root is needed.

The length of the hypotenuse is 15 units.

EXAMPLE 2 **A baseball diamond is a square. Baseball regulations require the bases to have 90 feet between them. What is the shortest distance between home plate and second base to the nearest foot?**

SOLUTION Draw a diagram.

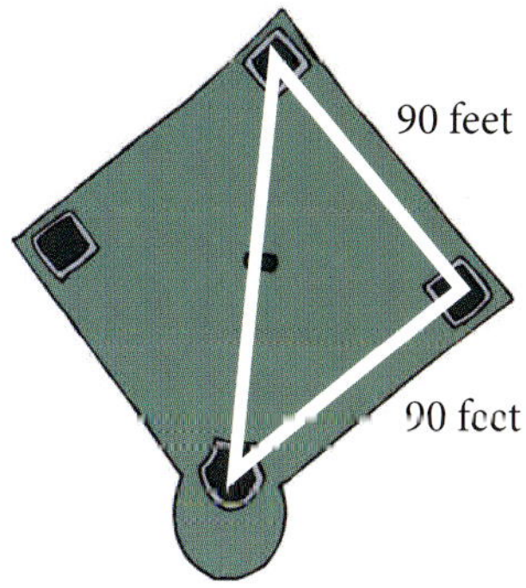

Write the Pythagorean Theorem.	$a^2 + b^2 = c^2$
Substitute the lengths of the legs for a and b.	$90^2 + 90^2 = c^2$
Simplify by squaring.	$8100 + 8100 = c^2$
Add.	$16200 = c$
Square root both sides of the equation.	$\pm\sqrt{16200} = \sqrt{c^2}$
Round to the nearest integer.	$127 \approx c$

The answer must be positive because it is a length.

The shortest distance from second base to home plate is about 127 feet.

EXAMPLE 3 **Find the missing side length. Round to the nearest tenth.**

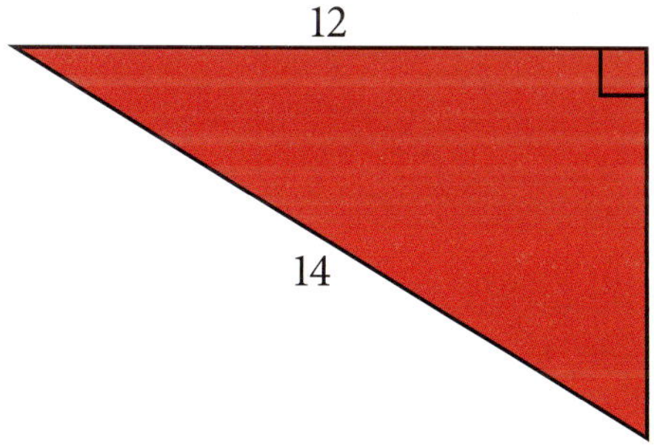

It does not matter if you substitute the given leg length for a or b.

SOLUTION

The unknown side is a leg of the right triangle.	
Write the Pythagorean Theorem.	$a^2 + b^2 = c^2$
Substitute the lengths for a leg and the hypotenuse.	$12^2 + b^2 = 14^2$
Simplify by squaring.	$144 + b^2 = 196$
Subtract 144 from both sides of the equation.	$-144 \qquad -144$
	$b^2 = 52$
Square root both sides of the equation.	$\sqrt{b^2} = \sqrt{52}$
Round to the nearest tenth.	$b \approx 7.21$
	$b \approx 7.2$

The missing side length is about 7.2 units.

EXERCISES

Find the length of the hypotenuse in each right triangle. If necessary, round to the nearest tenth.

1.

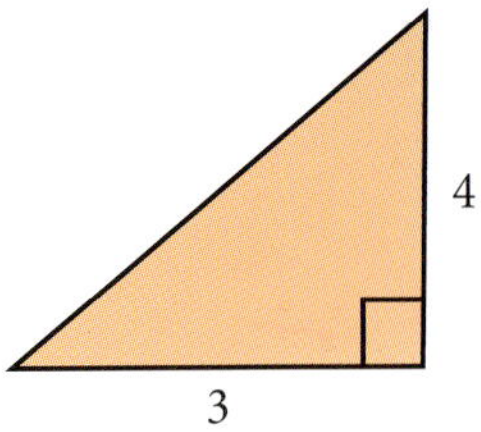

2.

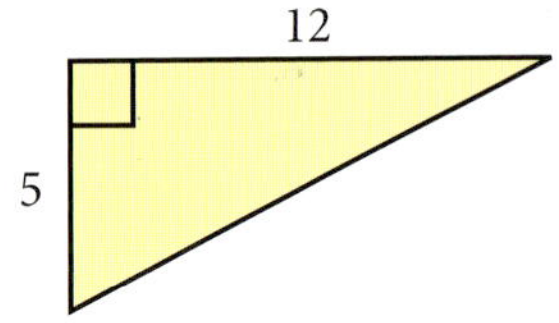

3.

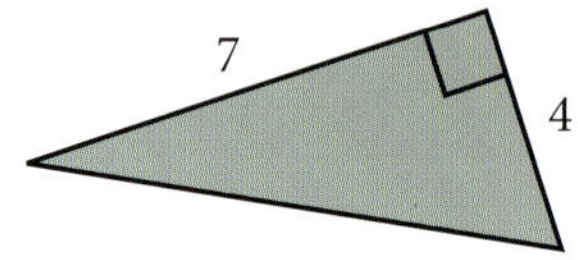

4.

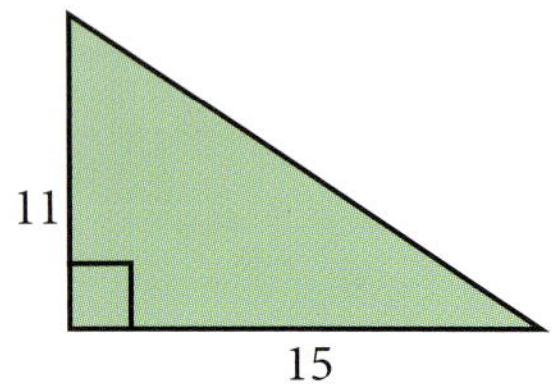

5.

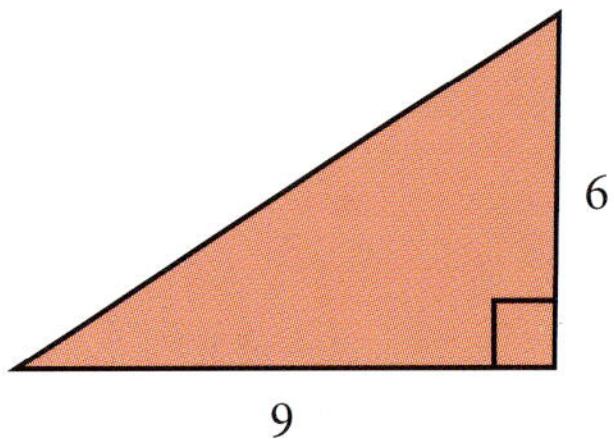

6.

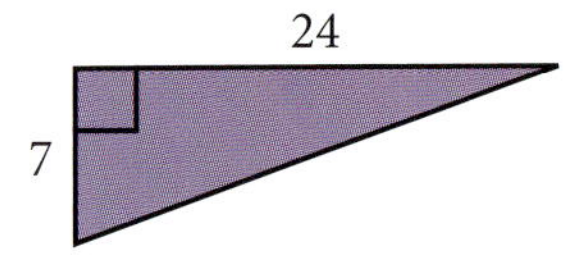

7. Shawn needs to reach a windowsill that is 10 feet above the ground. He placed his ladder 4 feet from the base of the wall. It reached the base of the window.

- **a.** Draw a diagram of the right triangle formed by Shawn's ladder, the ground and the wall.
- **b.** Find the length of Shawn's ladder.

8. According to federal guidelines, wheelchair ramps must be built with a 1-foot vertical rise for every 12 feet of horizontal distance.

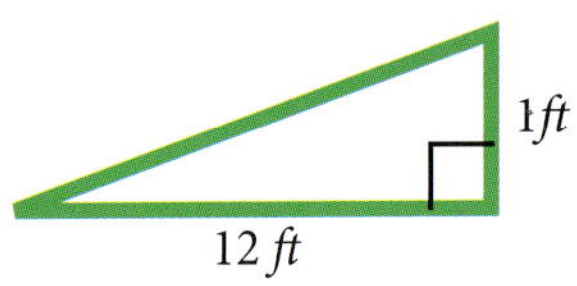

- **a.** What is the length of a ramp that rises 1 foot? Round to the nearest hundredth.
- **b.** If Max needs to build a ramp to reach a 2-foot porch, what horizontal distance will it cover?
- **c.** How long will the hypotenuse of Max's ramp be? Round to the nearest hundredth.

9. Find the length of the hypotenuse given the two leg lengths.

- **a.** 3, 4, ____
- **b.** 7, 24, ____
- **c.** 15, 36, ____

Find the value of x. If necessary, round to the nearest tenth.

10.

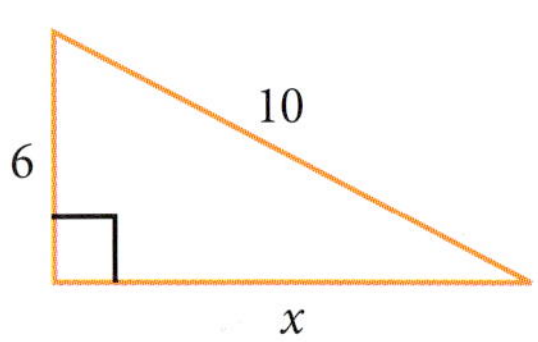

11.

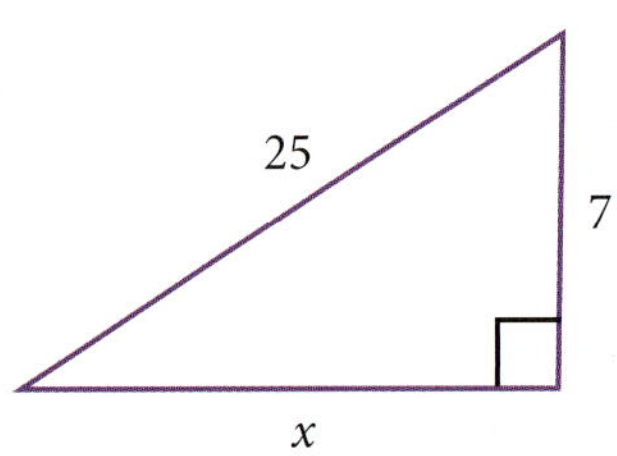

12.

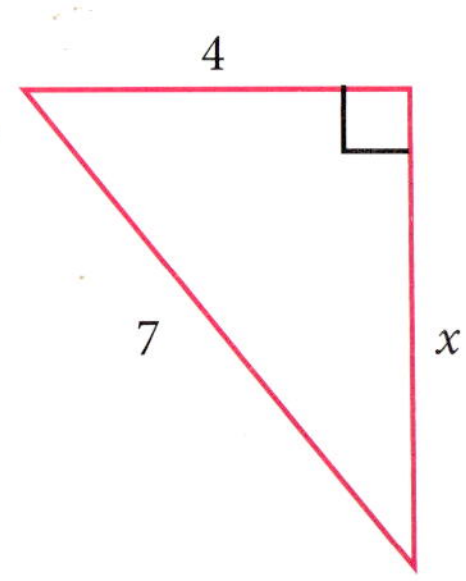

Find the value of x. If necessary, round to the nearest tenth.

13.

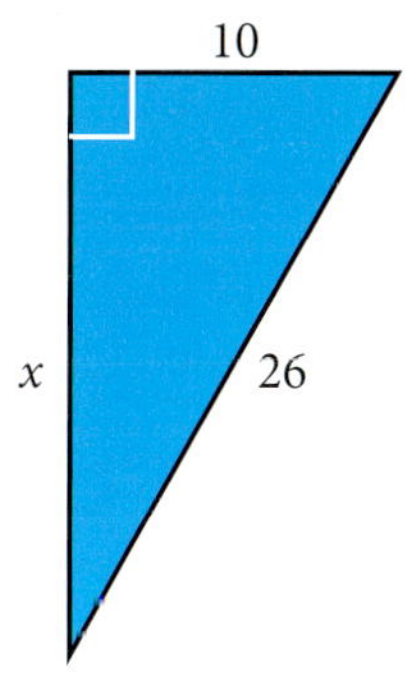

14.

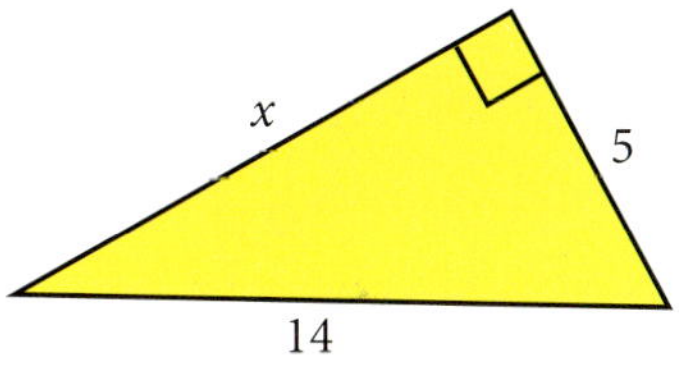

15.

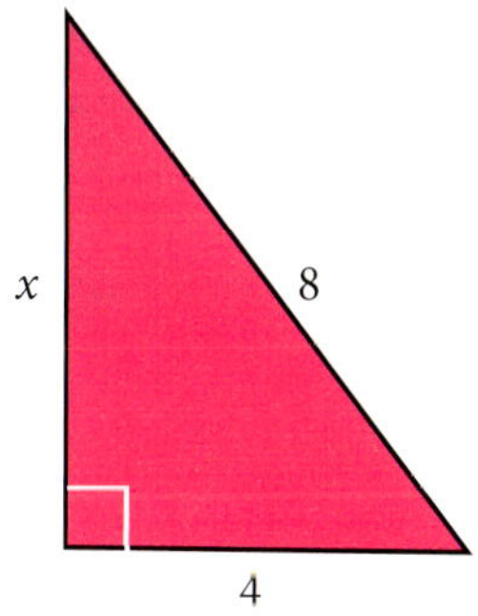

For Exercises 16–19, draw a diagram for each situation. Solve for the unknown length. Round to the nearest tenth, if necessary.

16. One of the largest swimming pools in the world is in Garden City, Kansas. It is rectangular and measures 67 meters by 100 meters. If you swim diagonally across the pool, how many meters would you swim?

17. A piece of paper measures 8.5 by 11 inches. What is the length of its diagonal?

18. The area of a square is 49 square feet.

a. What is the length of each side of the square?

b. What is the length of the square's diagonal?

19. A square has a diagonal that measures 18 inches. What is the length of one side of the square?

20. Juan wants to see the Grand Canyon, so he is taking a vaction in Arizona. He drove south from his house for 280 miles. Then, he drove east 64 miles.

a. Draw a diagram illustrating Juan's trip.

b. How many total miles did Juan travel?

c. If a road was built directly from Juan's home to the Grand Canyon, how long would it be? Round to the nearest mile.

d. Approximately how much shorter would Juan's trip be if he was able to take the direct route?

21. An airplane took off from a runway and traveled 4 miles through the air. The airplane reached an altitude, or height off the ground, of 13,200 feet.

a. Convert 13,200 feet to miles. (1 mile = 5,280 feet)

b. Copy the triangle below and label it using lengths in miles. Put an x on the unknown side.

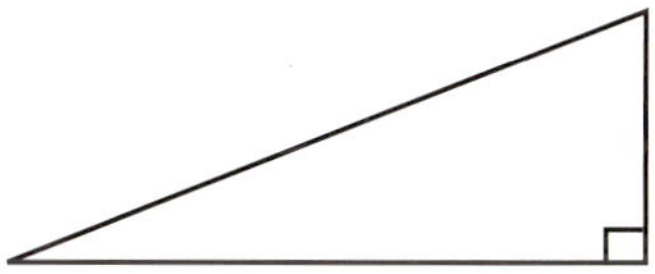

c. A car traveled on the ground from the point where the airplane took off and was directly below the airplane. How far did the car travel? Round to the nearest tenth of a mile.

REVIEW

22. Which of the following numbers are perfect squares?

20	49	9	15	100	88	1	60	144

Determine the two positive integers that each value is between.

23. $\sqrt{18}$

24. $\sqrt{7}$

25. $\sqrt{45}$

Solve for x. If necessary, round answers to the nearest tenth.

26. $5x^2 = 90$

27. $120 = 2x^2 + 10$

28. $12 - 3x^2 = -63$

Tic-Tac-Toe ~ Prove It!

The Pythagorean Theorem can be shown in many different ways. In the **Explore!** in **Lesson 2.3**, you drew right triangles to discover that $a^2 + b^2 = c^2$. Research other ways that mathematicians "prove" the Pythagorean Theorem works. Write a one to two page paper explaining the method(s) you discovered. Also include in your paper more about the background of the Pythagorean Theorem.

Tic-Tac-Toe ~ An Irrational Skit

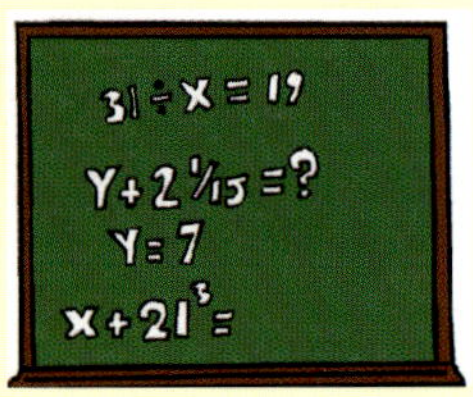

Write a skit where the characters are rational and irrational numbers. Include a variety of rational numbers as characters (fractions, terminating decimals, repeating decimals, integers, etc). The characters should reveal qualities about themselves through the skit dialogues. The skit should also have a well-developed plot. Include a title page for your skit.

CONVERSE OF THE PYTHAGOREAN THEOREM

LESSON 2.4

Determine if three side lengths create a right triangle.

If a triangle is a right triangle, then the Pythagorean Theorem can be used to find the measure of a missing side. The converse of the Pythagorean Theorem makes it possible to determine if a triangle is a right triangle when the lengths of the sides are known.

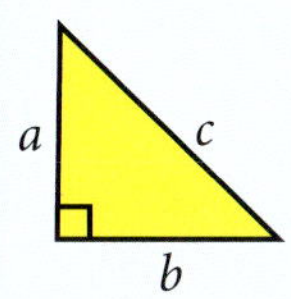

THE CONVERSE OF THE PYTHAGOREAN THEOREM

If $a^2 + b^2 = c^2$, then the triangle is a right triangle.

EXPLORE! **MAKE IT RIGHT**

Cory and Paco each chose three side lengths they believe will form a right triangle. Cory chose the lengths 8, 10 and 18. Paco chose 5, 12 and 13.

Step 1: Begin with Cory's set of numbers: 8, 10 and 18.

a. Which measure is the hypotenuse? How do you know?

b. Which measures represent the legs of the triangle?

c. Substitute the values into the Pythagorean Theorem for a, b and c. Simplify the equation. Does one side of the equation equal the other side of the equation? If so, it is a right triangle.

$$a^2 + b^2 \stackrel{?}{=} c^2$$

Step 2: Repeat the process in **Step 1** using Paco's set of numbers: 5, 12 and 13. Do Paco's side lengths form a right triangle?

Step 3: Find at least one other set of measures that will form a right triangle.

Three positive integers that work in the Pythagorean Theorem are called Pythagorean triples. Recognizing the common Pythagorean triples will save you time when you find them in problems or real-world situations. Some of the common sets of Pythagorean triples are:

3, 4, 5 **5, 12, 13** **8, 15, 17** **7, 24, 25**

9, 12, 15 **10, 24, 26** **6, 8, 10**

Notice that the Pythagorean triples in the second row are multiples of a Pythagorean triple in the top row. You can create an infinite number of Pythagorean triples by multiplying all numbers in a Pythagorean triple by a constant.

PYTHAGOREAN TRIPLES

A Pythagorean triple is a set of three positive integers such that $a^2 + b^2 = c^2$.

EXAMPLE 1 **A triangle has side lengths of 4, 10 and 9 inches. Determine if this triangle is a right triangle.**

SOLUTION

The largest measure is the hypotenuse. The other two measures are the legs. (It does not matter which leg is a and which leg is b.)	$c = 10$ $a = 4$ $b = 9$
Write the Pythagorean Theorem.	$a^2 + b^2 = c^2$
Substitute the given values for the hypotenuse and legs.	$4^2 + 9^2 \stackrel{?}{=} 10^2$
Simplify by squaring.	$16 + 81 \stackrel{?}{=} 100$
Check to see if one side of the equation equals the other side.	$97 \neq 100$

A triangle with side lengths of 4, 10 and 9 is not a right triangle.

EXERCISES

Determine if the given lengths form a right triangle. Support your answer with words and/or symbols.

1. 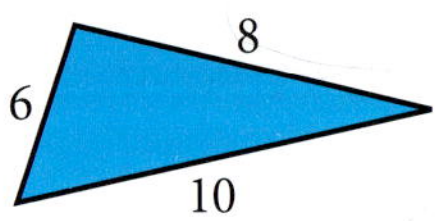

2. 21, 42, 15

3. 35, 37, 12

4. 2.0, 2.1, 2.9

5. 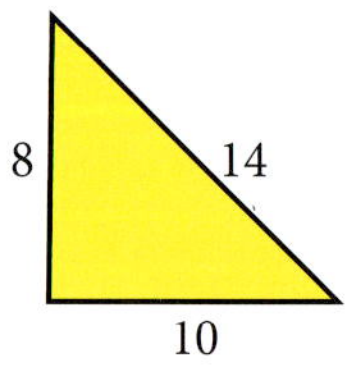

6. 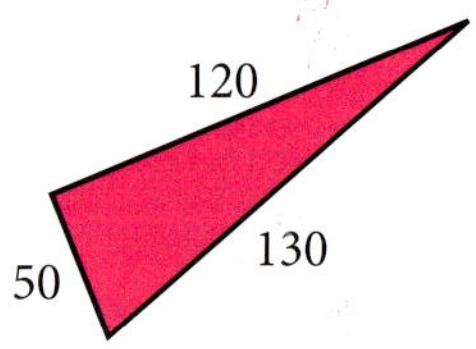

7. Determine if each statement is true or false.

a. Pythagorean triples are sets of positive whole numbers that work in the Pythagorean Theorem.

b. Decimals can be in Pythagorean triples as long as they make a true statement in the Pythagorean Theorem.

c. A Pythagorean triple can include simplified fractions.

Determine if each set of numbers is a Pythagorean triple.

8. 8, 15, 17

9. 3, 5, 7

10. 4.1, 4.0, 9.0

11. 30, 40, 50

12. 21, 72, 75

13. $\frac{3}{4}$, 1, $\frac{5}{4}$

14. Use the Pythagorean triple containing the integers 5, 12, 13.
- **a.** Multiply each number in the set by 2.
- **b.** Verify that the new set of numbers from **part a** form a right triangle.
- **c.** Create two more multiples of the Pythagorean triple 5, 12, 13.

15. Use the Pythagorean triple 3, 4, 5.
- **a.** Multiply each number in the set by $\frac{1}{2}$.
- **b.** Is the new set of numbers a Pythagorean triple? Why or why not?
- **c.** Use the converse of the Pythagorean Theorem to determine if the new set of measures will form a right triangle.

16. Michael wants to determine if an old window frame is still rectangular. The height of the frame is 24 inches, the base is 20 inches and the diagonal is 27 inches long. Is the old window frame rectangular?

17. A boy leaned a ladder against his clubhouse. The ladder was 8.5 feet long. It hit the top of the clubhouse 7.5 feet off the ground. He placed the base of the ladder 4 feet from the building. Does his clubhouse make a right angle with the ground? Show work to verify the answer.

18. Use two Pythagorean triples that are multiples to compare areas.
- **a.** Draw a 3, 4, 5 triangle.
- **b.** Draw a 6, 8, 10 triangle.
- **c.** Calculate the area of each triangle. (Area of a Triangle = $\frac{1}{2}bh$)
- **d.** How many times larger is the area of the 6, 8, 10 triangle compared to the 3, 4, 5 triangle?

REVIEW

Solve for x and y. If necessary, round to the nearest tenth.

19.

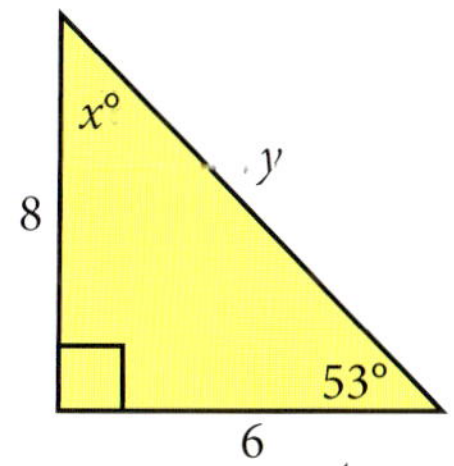

20.

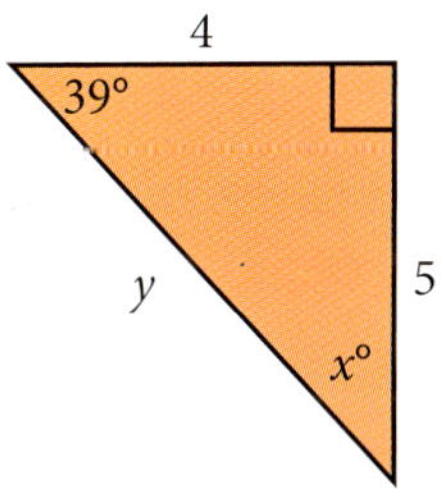

21.

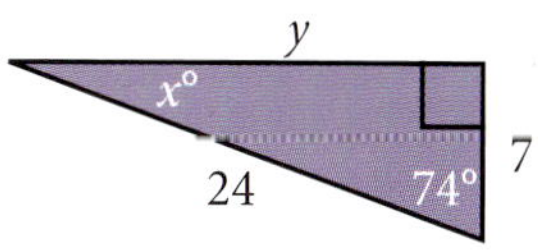

22.

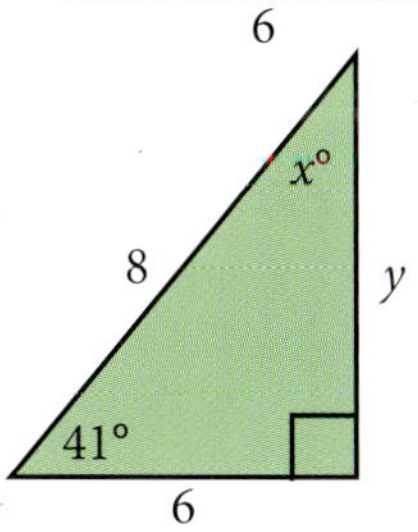

23.

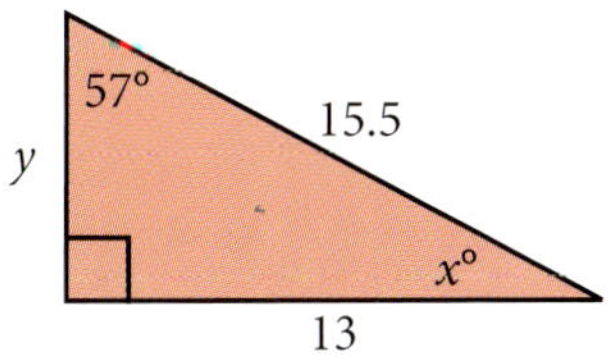

24.

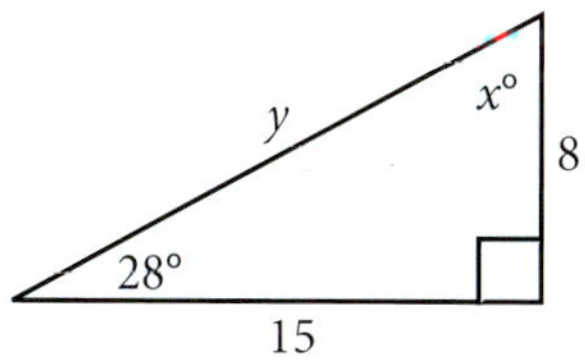

Tic-Tac-Toe ~ Creating Triples

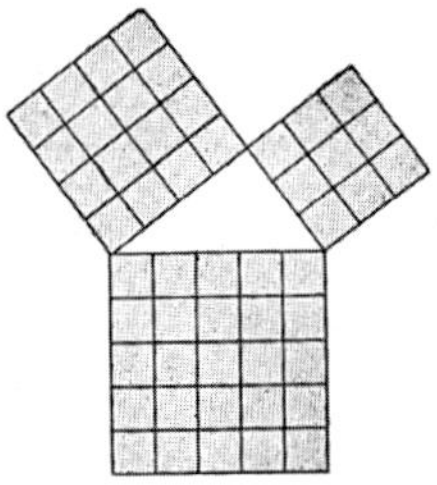

There are two kinds of Pythagorean triples: primitive and multiple. Look at examples of each.

Primitive:	3, 4, 5		
Multiples:	Multiplied by 2	–	6, 8, 10
	Multiplied by 3	–	9, 12, 15
	Multiplied by 10	–	30, 40, 50

The primitive Pythagorean triple is the "original". The three numbers have no common factors. The multiple Pythagorean triples are made by multiplying each number in the primitive Pythagorean triple by a constant.

You can create your own primitive Pythagorean triple by following the steps outlined below. Read the steps and follow along with the example provided.

Step 1: Choose an odd, positive integer. 21

Step 2: Square it. $21^2 = 441$

Step 3: Break the square into two numbers. The two numbers need to sum to the square in **Step 2** and have a difference of 1. $441 = 220 + 221$

Step 4: Write the Pythagorean triple. 21, 220, 221

Step 5: Verify it.

$$21^2 + 220^2 = 221^2$$
$$441 + 48400 = 48841$$
$$48841 = 48841$$

Make a poster to present primitive Pythagorean triples and their multiples. Create five of your own primitive Pythagorean triples using the process outlined above. Find two multiples for each of your primitive Pythagorean triples.

APPLYING THE PYTHAGOREAN THEOREM

LESSON 2.5

Apply the Pythagorean Theorem to solve problems in two and three dimensions.

The Pythagorean Theorem is used in many careers on a regular basis. Construction workers and cabinet makers use the Pythagorean Theorem to determine lengths of materials. Engineers use the theorem to design buildings. Pilots and ship captains use the Pythagorean Theorem to plan routes.

When a real-world situation requires finding a missing measure, it is helpful to draw a diagram. Label the diagram with the known information. Then solve for the missing measure.

SOLVING APPLICATION PROBLEMS

1. Draw a diagram to represent the situation.
2. Label the diagram with the given measures.
3. Solve for the missing measure. Label the answer.

EXAMPLE 1

An 8 foot ladder is placed 3.5 feet from the base of a wall. How high up the wall will the ladder reach? Round to the nearest tenth.

SOLUTION

Draw a diagram.

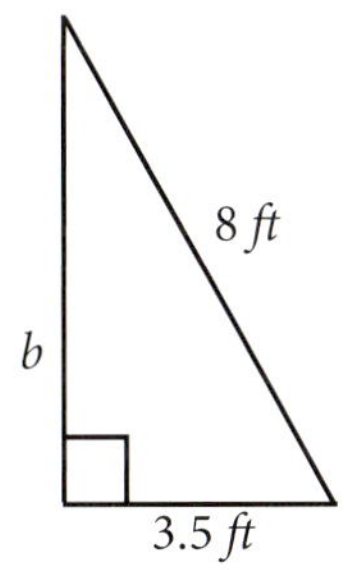

Substitute known values into the Pythagorean Theorem. $3.5^2 + b^2 = 8^2$

Simplify by squaring. $12.25 + b^2 = 64$

Subtract 12.25 from both sides of the equation. $-12.25 \quad -12.25$

$b^2 = 51.75$

Square root both sides of the equation. $\sqrt{b^2} = \sqrt{51.75}$

$b \approx 7.19$

Round to the nearest tenth. $b \approx 7.2$

The ladder will reach approximately 7.2 feet up the wall.

EXAMPLE 2 **A ship travels 320 miles due north and then makes a turn due east. It travels 200 miles east. How far is the ship from its starting point? Round to the nearest mile.**

SOLUTION Draw a diagram.

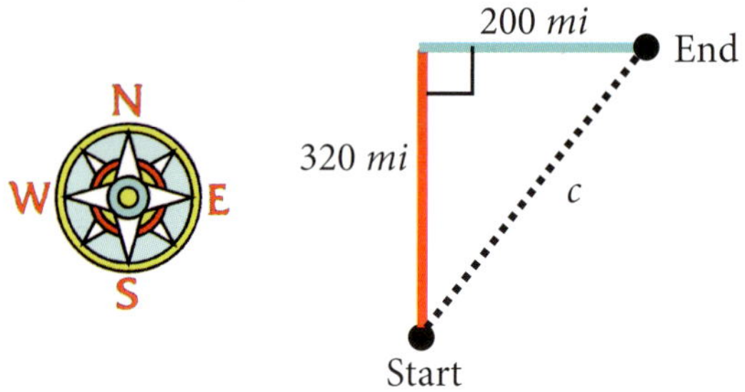

Substitute known values into the Pythagorean Theorem.	$320^2 + 200^2 = c^2$
Simplify by squaring.	$102400 + 40000 = c^2$
Add.	$142400 = c^2$
Square root both sides of the equation.	$\sqrt{142400} = \sqrt{c^2}$
Round to the nearest whole number.	$377 \approx c$

The ship is approximately 377 miles from its starting point.

Most things in the world are three dimensional. It is important to be able to solve problems using the Pythagorean Theorem in both two and three dimensions.

EXPLORE! **3D PT**

Renee plays softball for her college team. She went home to visit her parents for the weekend and took her favorite bat to practice at the batting cages. After returning to college, she realized she left her bat at her parent's house.

Step 1: The largest rectangular box that Renee's father can find has a base that is 28 inches by 18 inches. Use the Pythagorean Theorem to determine if Renee's 34-inch bat will lie across the diagonal of the bottom of the box.

Step 2: The box is 8 inches tall. Renee's father is hoping to put the bat at an angle in the box as shown by the green line in the diagram. Use the length of the base diagonal and the height of the box as the two legs of a new right triangle. Use the Pythagorean Theorem with this triangle to determine if the bat can fit in the box shown by the green line.

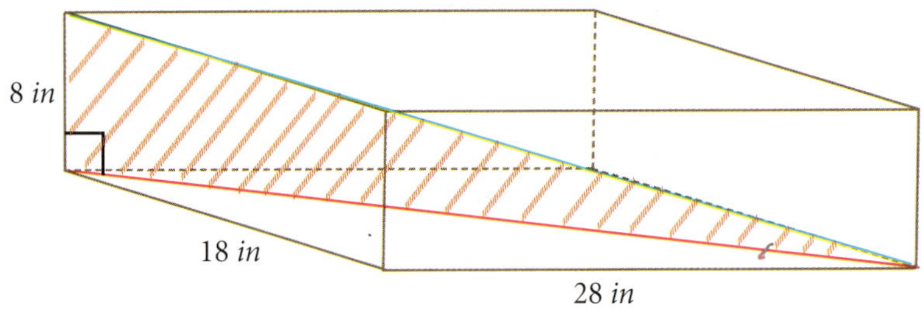

Step 3: The Pythagorean Theorem works in three dimensions with the formula:

$$a^2 + b^2 + c^2 = d^2$$

where a, b and c are the length, width and height of a rectangular prism. Use this formula to find the length of the longest diagonal in the box. Does it match your answer to **Step 2**?

EXPLORE! **(CONTINUED)**

Step 4: As this **Explore!** shows, you can find the length of the longest diagonal using the Pythagorean Theorem twice or using the three-dimensional Pythagorean Theorem formula. Which method do you prefer? Why?

Step 5: Find a rectangular box in your classroom. Measure and record the length, width and height of the box to the nearest tenth of a centimeter.

Step 6: Find the length of the longest stick that can fit in the box using either method in this **Explore!**.

THE PYTHAGOREAN THEOREM IN THREE DIMENSIONS

In a rectangular prism, the length of the longest diagonal d squared is equal to the sum of the squares of the length a, width b and height c of the prism.

$$a^2 + b^2 + c^2 = d^2$$

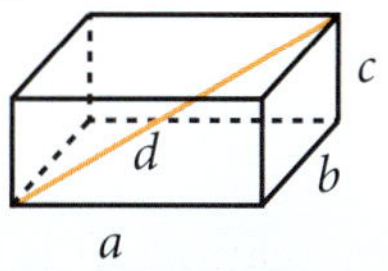

EXAMPLE 3

What is the longest object that Simone can put in a rectangular box that is 10 inches wide, 12 inches long and 20 inches tall? Round to the nearest tenth of an inch.

SOLUTION

Write the three-dimensional formula.	$a^2 + b^2 + c^2 = d^2$
Substitute known values into the formula.	$10^2 + 12^2 + 20^2 = d^2$
Simplify by squaring.	$100 + 144 + 400 = d^2$
Add.	$644 = d^2$
Square root both sides of the equation.	$\sqrt{644} = \sqrt{d^2}$
Round to the nearest tenth.	$25.4 \approx d$

The longest object that can fit in the box is about 25.4 inches.

EXERCISES

In this Exercise set, round to the nearest tenth, when necessary.

1. Jeff just bought a house on a triangular lot. The sides measure 85 feet, 132 feet and 157 feet. Is his lot a right triangle?

2. Paul is locked out of his house. An 18-foot ladder is outside and an upstairs window is open. Paul read the safety warning on the ladder recommending it be 6 feet away from the wall. He placed the ladder according to the warning and it reached the base of the window. How high up is the base of the window from the ground?

3. A ship traveled 140 miles due north, then made a turn due east. It traveled 180 miles east. How far is the ship from its starting point?

4. Jamar and Peggy live on opposite sides of a park. Peggy counted how many steps it takes her to get from her house to Jamar's house. She walks 52 steps west and 81 steps south.
   a. If she could just walk on a path directly from her house to Jamar's house, how many steps would it take?
   b. Approximately how many steps shorter would the direct route be?

5. A 60-foot cable is stretched from the top of a pole to an anchor on the ground. It is anchored on the ground 19 feet away from the base of the pole. How tall is the pole?

6. The diagonal of a rectangle measures 18.2 inches. The width of the rectangle is 6.7 inches.
   a. Find the length of the rectangle.
   b. What is the perimeter of the rectangle?

7. A square has an area of 225 m^2.
   a. Find the length of one side of the square.
   b. Determine the perimeter of the square.
   c. What is the length of the square's diagonal?

8. Jessica's cat is stuck in a tree. The fire department no longer assists in getting cats out of trees. Jessica's dad knows the cat is approximately 22 feet high. He has a 25-foot ladder and the directions say to be safe he must keep the base of the ladder 10 feet from the base of the tree. Will the ladder reach the cat so he can safely get it out of the tree?

9. Maria walked 8 miles south and then 3 miles east. Find her distance from her original starting point.

10. Jake drives a truck with a slide-out ramp for loading motorcycles. The tailgate of his truck is 1.6 meters above the ground. The ramp is 3.7 meters long. What is the horizontal distance the ramp can reach?

11. A rectangular prism is 3 feet long, 4 feet wide and 2 feet tall. What is the length of its longest diagonal?

12. Petra is sending her brother a giant candy cane stick for a gift. She will use a box that measures 10 inches by 6 inches by 2 inches. What is the maximum length the candy cane stick can be to fit in the box?

13. Elena needs to ship a 61 *cm* concert flute to a customer. She has two rectangular boxes. One is 25 *cm* by 25 *cm* by 50 *cm*. The other box is 10 *cm* by 12 *cm* by 58 *cm*.
   a. What is the longest object that will fit in the 25 *cm* × 25 *cm* × 50 *cm* rectangular box?
   b. What is the longest object that will fit in the 10 *cm* × 12 *cm* × 58 *cm* rectangular box?
   c. In which box will the flute best fit?

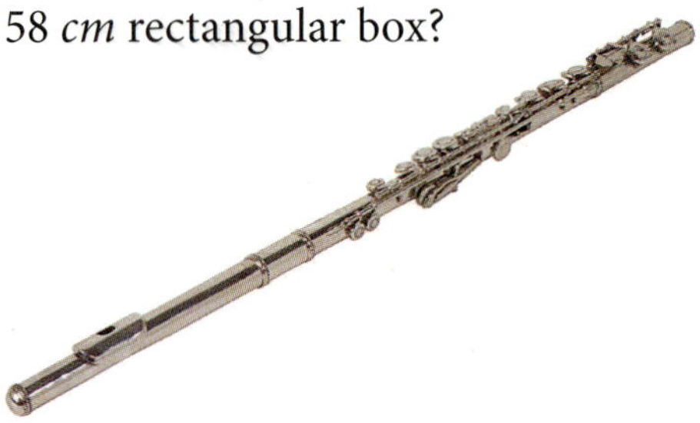

14. A steel box measures 9 inches by 6 inches by 6 inches. What is the measure of the longest diagonal in the box?

15. A cube has a surface area of 600 square meters.

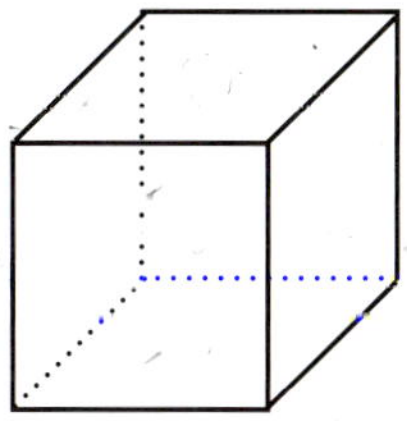

- **a.** How many faces does a cube have?
- **b.** What is the area of one face of the cube?
- **c.** Find the length of one side of the cube.
- **d.** What is the length of the cube's longest diagonal?

REVIEW

Determine the two positive integers that each square root is between.

16. $\sqrt{3}$ **17.** $\sqrt{87}$ **18.** $\sqrt{150}$

Solve for *x*. Round answers to the nearest tenth.

19. $x^2 + 5 = 37$ **20.** $3x^2 - 4 = 35$ **21.** $-5 - 4x^2 = -45$

Use the given Pythagorean triple to create another Pythagorean triple.

22. 5, 12, 13 **23.** 3, 4, 5

Tic-Tac-Toe ~ Applications

The Pythagorean Theorem is used in many types of real-world situations. Create a worksheet with ten real-world math problems that must be solved using the Pythagorean Theorem.

- At least two of the problems must include diagrams.
- At least two of the problems must require the use of the three-dimensional Pythagorean Theorem.
- Include an answer key for your worksheet.

DISTANCE ON THE COORDINATE PLANE

LESSON 2.6

Find the distance between two points on a coordinate plane using the Pythagorean Theorem.

Horizontal and vertical distances between points on a coordinate plane can be counted. The Pythagorean Theorem makes it possible to find diagonal distance on a coordinate plane.

EXPLORE! **PERIMETER ON THE COORDINATE PLANE**

Step 1: On a sheet of grid paper, draw a coordinate plane like the one shown below.

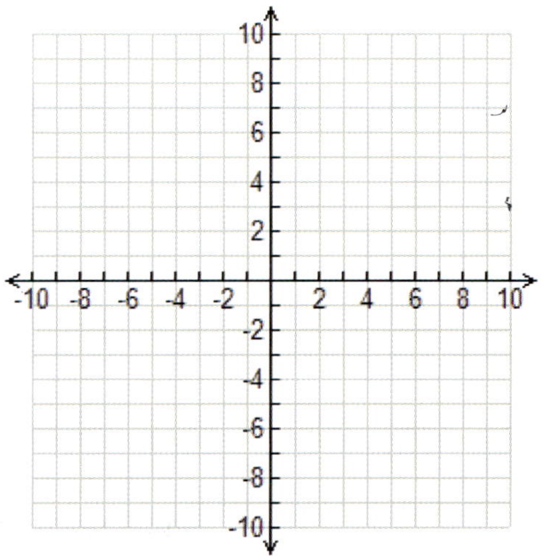

Step 2: Plot the ordered pairs below on the coordinate plane and connect them in the order given. Connect the last point to the first point.

A(3, 7) B(−8, 7) C(−8, −1) D(3, −1)

Step 3: What is the name of the quadrilateral formed?

Step 4: Find the length of each side of the quadrilateral. What is the perimeter of the figure?

Step 5: Draw another coordinate plane. Plot the points and connect them in the order given. Connect the last point to the first point.

X(−3, −2) Y(−3, 4) Z(5, 4)

Step 6: Classify the triangle based on its angles.

Step 7: Find the length of the legs of the triangle.

Step 8: Use the Pythagorean Theorem to find the length of the hypotenuse of the triangle.

Step 9: What is the perimeter of the triangle?

Step 10: On a new coordinate plane, sketch a right triangle with integer coordinates. Find the perimeter of the right triangle. Show all work.

You can find the distance between two points on a coordinate plane using the Pythagorean Theorem. Graph the two points. Use the segment between the two points as the hypotenuse of the right triangle. Draw the legs of the triangle and find their lengths. Use the Pythagorean Theorem to find the length of the hypotenuse.

Finding Distance on a Coordinate Plane

1. Graph the two ordered pairs. Connect the points and label the segment *c*.
2. Using *c* as the hypotenuse of a right triangle, draw the legs of the triangle.
3. Find the lengths of the legs.
4. Use the Pythagorean Theorem to find the length of the hypotenuse *c*. This length is the distance between the two points on the coordinate plane.

EXAMPLE 1 **Find the distance between (1, 3) and (4, −2). Round to the nearest tenth.**

Solution

Graph the points. Connect with a line segment and label it *c*.

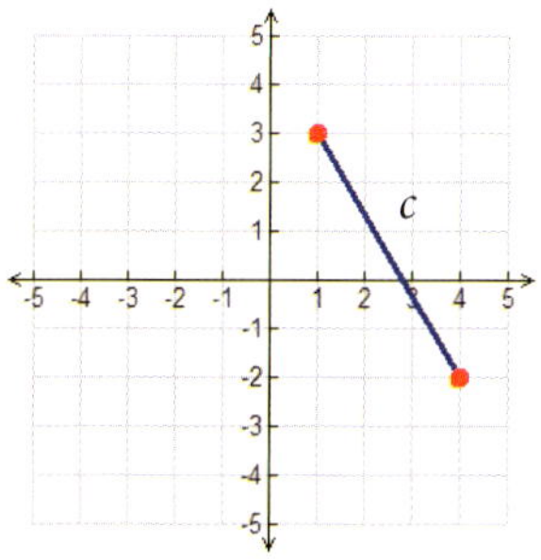

Using the line segment as the hypotenuse, draw the legs of the triangle. Record their lengths on the graph.

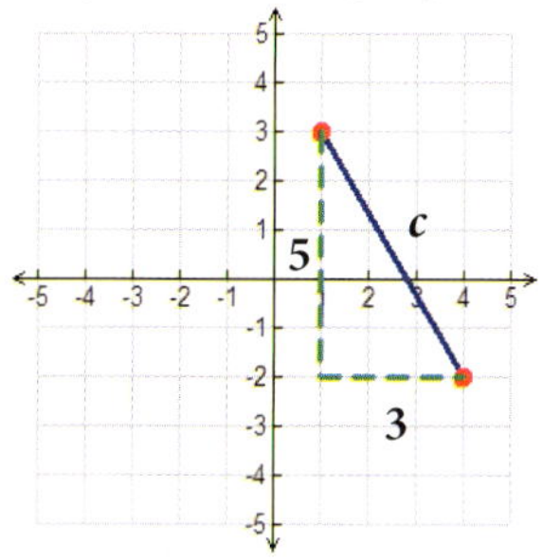

The legs are horizontal and vertical line segments on the coordinate plane.

Write the Pythagorean Theorem. $a^2 + b^2 = c^2$

Substitute the known values into the equation. $3^2 + 5^2 = c^2$

Simplify by squaring. $9 + 25 = c^2$

Add. $34 = c^2$

Square root both sides of the equation. $\sqrt{34} = \sqrt{c^2}$

Round to the nearest tenth. $5.8 \approx c$

The distance between (1, 3) and (4, −2) is about 5.8 units.

Use the positive square root of both sides of the equation since distance is a positive number.

EXERCISES

Find the length of *c* on each graph. When necessary, round to the nearest tenth.

1.

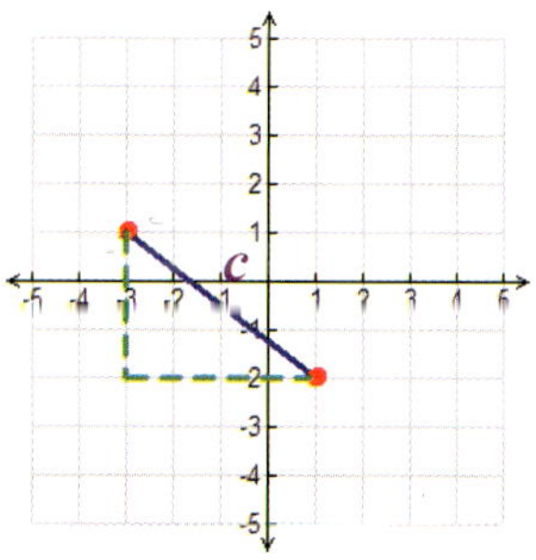

2.

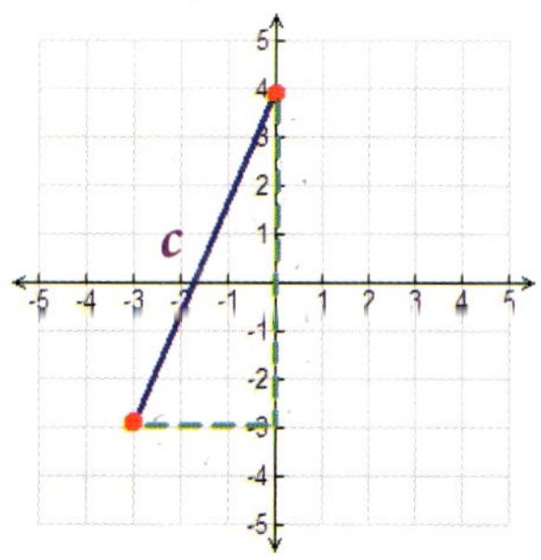

3.

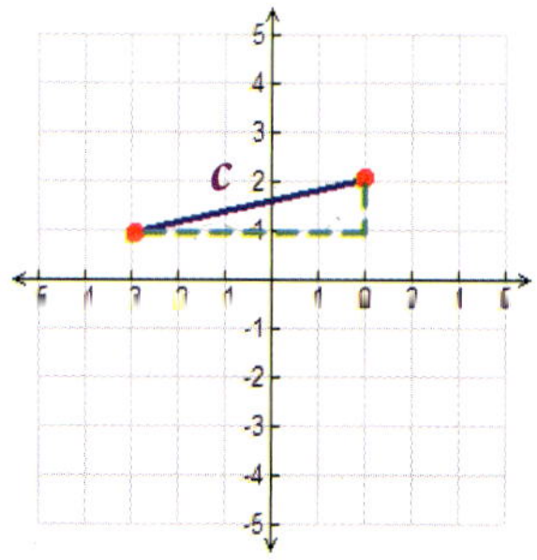

4. Use the ordered pairs (5, 2) and (1, 5).

a. Graph the two points on a coordinate plane.

b. Connect the two points with a line segment and label it c.

c. Use c as the hypotenuse and draw two legs to form a right triangle.

d. Find the lengths of the legs and write the lengths on the graph.

e. Use the Pythagorean Theorem to find the length of c.

Find the distance between each set of points. When necessary, round to the nearest tenth.

5. (2, 6) and (5, 1)

6. (0, −3) and (2, 0)

7. (−1, 4) and (2, −2)

8. (−3, −4) and (2, 3)

9. (0, 0) and (−7, 2)

10. (3.5, 2) and (0, −1)

11. (4, 2) and (−4, −5)

12. (3, 4) and (−3, −4)

13. (1, 3) and (−2, 2)

Find the perimeter of each triangle formed by the given set of points. When necessary, round to the nearest tenth.

14. (0, 0), (0, 3) and (4, 0)

15. (2, 5), (2, 8) and (−3, 5)

16. (−1, 4), (−8, 4) and (−1, −2)

17. (−7, 2), (3, 2) and (−7, −1)

18. On a map, the town of Bandon is located at (1, 3). The town of Union is located at (10, 8) on the map.

a. Find the distance between the two towns on the map.

b. Each unit on the map is equal to 45 miles. Approximately how far apart are Union and Bandon?

REVIEW

Determine if each set of numbers is a Pythagorean triple.

19. 1, 1, 2

20. 8, 15, 17

21. 33, 44, 55

Draw a diagram and solve for the missing measure. When necessary, round to the nearest tenth.

22. Two ships leave from the same port. One travels west 84 miles and the other travels 62 miles north. Determine the distance between the two ships.

23. Jackson and Shatiqua both leave from school at the same time. Jackson walks east 1.7 *km* and Shatiqua walks north 2.7 *km*. Find the shortest distance between Shatiqua and Jackson.

24. A freight train leaves the station and travels north 128 miles. A passenger train leaves the same station and travels east 349 miles. How far apart are the trains?

25. Peter leans a ladder against a wall. It reaches 8 feet off the ground. The ladder is 9 feet long. How far is the base of the ladder from the base of the building?

Tic-Tac-Toe ~ Percent Error

When estimating, it is important to know how close an estimate is to the actual answer. Percent error is used to determine how accurate an estimate is. In this activity, you will estimate the value of square roots to the nearest tenth. You will use your calculator to find the approximate value and then calculate your percent error. The percent error formula is given below.

$$\text{Percent error} = \frac{\text{estimated value} - \text{true value}}{\text{true value}} \times 100$$

Step 1: Estimate the value of $\sqrt{22}$ to the nearest tenth.

Step 2: Use your calculator to find the "true value" of $\sqrt{22}$ rounded to the nearest tenth.

Step 3: Insert your estimated value and the true value into the percent error formula above. Calculate your percent error.

Step 4: Repeat **Step 1 through Step 3** for the following values.

a. $\sqrt{11}$ **b.** $\sqrt{34}$ **c.** $\sqrt{45}$

d. $\sqrt{71}$ **e.** $\sqrt{150}$ **f.** $\sqrt{250}$

Step 5: What does it mean if your percent error is negative? What does it mean if it is positive?

Step 6: For which square root in **Step 4** did you get the greatest percent error? Why do you think that is?

THE DISTANCE FORMULA

LESSON 2.7

Find the distance between two points on a coordinate plane using the Distance Formula.

In the last lesson, you learned how to find the distance between two points on a coordinate plane using the Pythagorean Theorem. It required graphing two points, connecting them to see the hypotenuse and then drawing two legs to form a right triangle.

The horizontal leg of the right triangle was always the difference between the x-coordinates in the ordered pair.

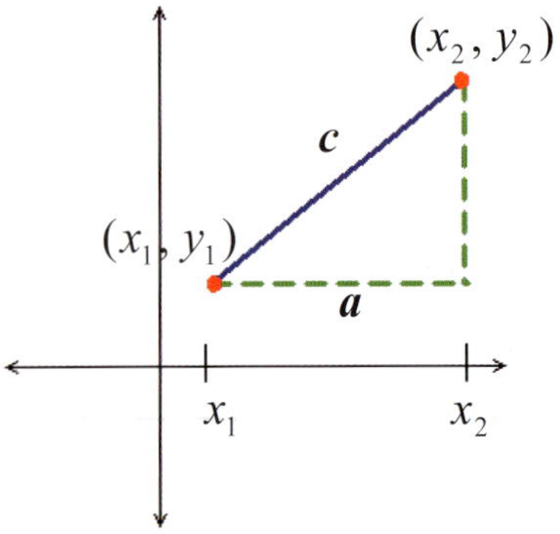

$$a = x_2 - x_1$$

The vertical leg of the right triangle was always the difference between the y-coordinates in the ordered pair.

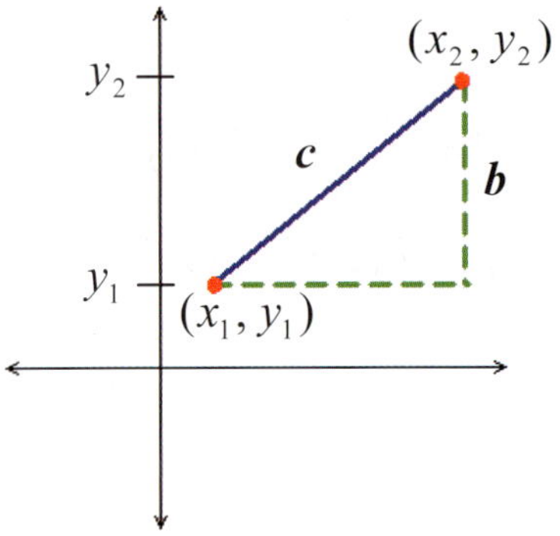

$$b = y_2 - y_1$$

Write the Pythagorean Theorem. $a^2 + b^2 = c^2$

Substitute $(x_2 - x_1)$ for a and $(y_2 - y_1)$ for b. $(x_2 - x_1)^2 + (y_2 - y_1)^2 = c^2$

Square root both sides of the equation. $\sqrt{(x_2 - x_1)^2 + (y_2 - y_1)^2} = c$

This is a formula that allows you to find the distance between any two points on a coordinate plane without graphing first. It is called the **Distance Formula**. The Distance Formula uses a d for the distance between the two points rather than a c.

The Distance Formula

The distance, d, between two points, (x_1, y_1) and (x_2, y_2), is found by:

$$d = \sqrt{(x_2 - x_1)^2 + (y_2 - y_1)^2}$$

EXAMPLE 1 **Find the distance between (1, 2) and (4, 6).**

Solution

Using the Pythagorean Theorem	Using the Distance Formula
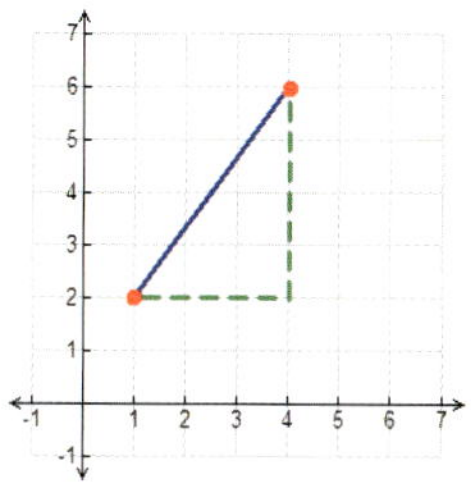 $3^2 + 4^2 = c^2$ $9 + 16 = c^2$ $25 = c^2$ $5 = c$	Let $(x_1, y_1) = (1, 2)$ and $(x_2, y_2) = (4, 6)$. $d = \sqrt{(x_2 - x_1)^2 + (y_2 - y_1)^2}$ $d = \sqrt{(4 - 1)^2 + (6 - 2)^2}$ $d = \sqrt{(3)^2 + (4)^2}$ $d = \sqrt{9 + 16}$ $d = \sqrt{25}$ $d = 5$

The distance between (1, 2) and (4, 6) is 5 units.

EXAMPLE 2 **Find the distance between (0, 8) and (3, 2). Round to the nearest tenth.**

Solution

Let (x_1, y_1) be (0, 8) and (x_2, y_2) be (3, 2).

Write the Distance Formula. $d = \sqrt{(x_2 - x_1)^2 + (y_2 - y_1)^2}$

Substitute each value from the ordered pairs into the formula. $d = \sqrt{(3 - 0)^2 + (2 - 8)^2}$

According to the order of operations, simplify the inside of the parentheses first. $d = \sqrt{(3)^2 + (-6)^2}$

Simplify by squaring. $d = \sqrt{9 + 36}$

Add. $d = \sqrt{45}$

Simplify. Round to the nearest tenth. $d \approx 6.7$

The distance between (3, 2) and (0, 8) is about 6.7 units.

EXAMPLE 3 **Find the perimeter of ΔMAT to the nearest tenth.**

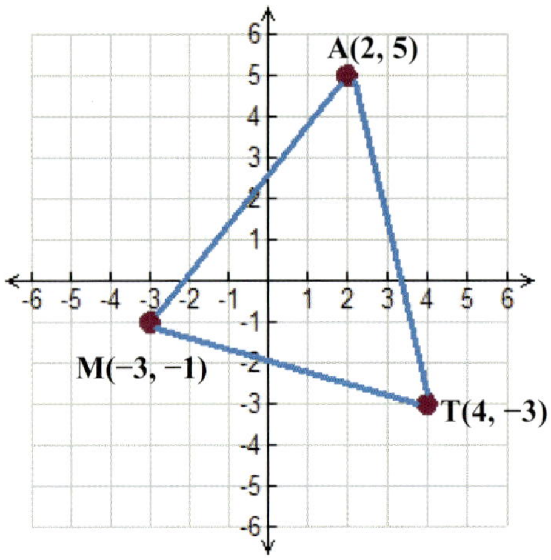

SOLUTION Find the length of each side of the triangle using the Distance Formula.

$\overline{AM}$: (2, 5) to (−3, −1)

$AM = d = \sqrt{(-3-2)^2 + (-1-5)^2}$

$AM = d = \sqrt{(-5)^2 + (-6)^2}$

$AM = d = \sqrt{25 + 36}$

$AM = d = \sqrt{61} \approx 7.8$

Remember that a negative number squared becomes positive.

$\overline{TA}$: (4, −3) to (2, 5)

$TA = d = \sqrt{(2-4)^2 + (5-(-3))^2}$

$TA = d = \sqrt{(-2)^2 + (8)^2}$

$TA = d = \sqrt{4 + 64}$

$TA = d = \sqrt{68} \approx 8.2$

$\overline{MT}$: (−3, −1) to (4, −3)

$MT = d = \sqrt{(4-(-3))^2 + (-3-(-1))^2}$

$MT = d = \sqrt{(7)^2 + (-2)^2}$

$MT = d = \sqrt{49 + 4}$

$MT = d = \sqrt{53} \approx 7.3$

Add the lengths of the three sides of the triangle. 7.8 + 8.2 + 7.3 = 23.3

The perimeter of ΔMAT is approximately 23.3 units.

EXERCISES

Use the Distance Formula to calculate the length of each segment. Round to the nearest tenth, if necessary.

1.

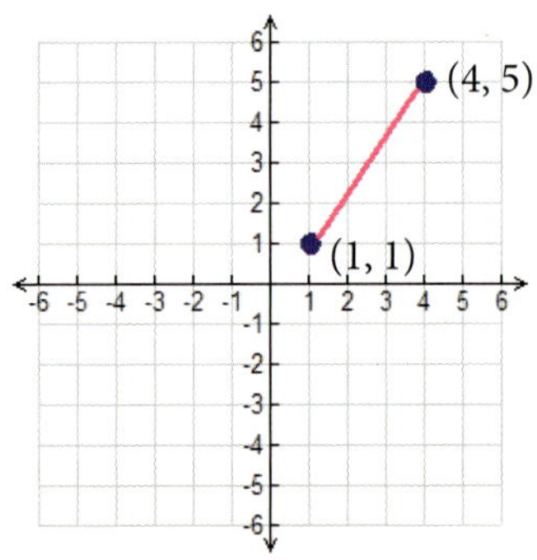

2.

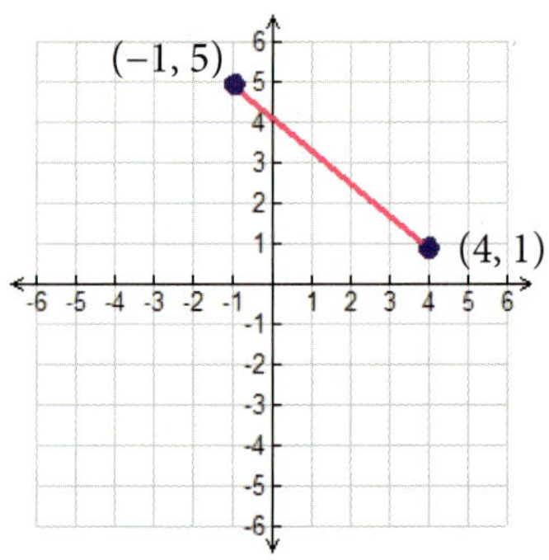

Use the Distance Formula to calculate the length of each segment. Round to the nearest tenth, if necessary.

3.

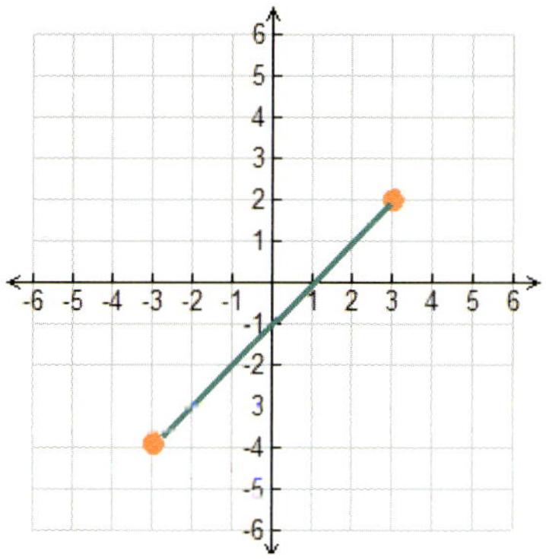

4.

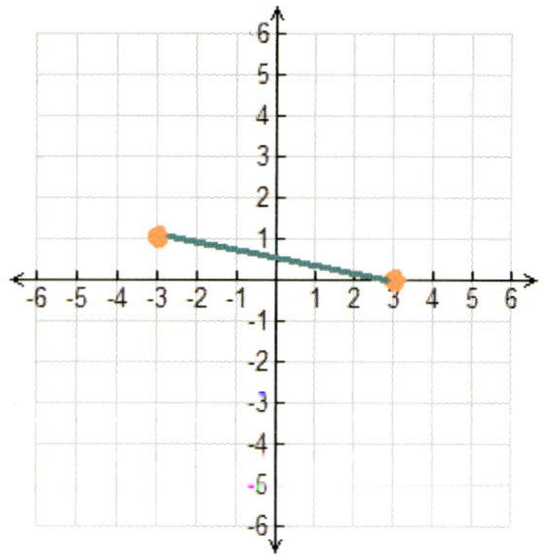

Find the distance between each pair of points. Round to the nearest tenth, if necessary.

5. (7, 9) and (4, 6)

6. (6, 2) and (1, 14)

7. (−8, 7) and (0, 1)

8. (9, −3) and (2, −4)

9. (0, −11) and (−4, −8)

10. (5, 8.5) and (3, 7.5)

11. ΔABC is formed by the points A(9, 2), B(−1, −3) and C(−5, 7).
  a. Find the length of $\overline{AB}$. Round to the nearest tenth.
  b. Find the length of $\overline{BC}$. Round to the nearest tenth.
  c. Find the length of $\overline{CA}$. Round to the nearest tenth.
  d. Find the approximate perimeter of the triangle.

12. ΔTUF is formed by the points T(1, 9), U(7, 1) and F(7, 7).
  a. Calculate the length of each side.
  b. What is the perimeter of the triangle?
  c. Classify the triangle based on its sides.

13. A circle has a center at (0, 0) and a point on the edge of the circle is located at (5, −12).
  a. Find the length of its radius.
  b. Calculate the circumference of the circle using the formula $C = 2\pi r$. Use 3.14 for π.
  c. Calculate the area of the circle using the formula $A = \pi r^2$. Use 3.14 for π.

14. A circle has a center at (5, −3) and a point on the edge of the circle is located at (8, 1).
  a. Find the length of its radius.
  b. Calculate the circumference of the circle using the formula $C = 2\pi r$. Use 3.14 for π.
  c. Calculate the area of the circle using the formula $A = \pi r^2$. Use 3.14 for π.

15. Show that ΔABC is isosceles based on the lengths of its sides.
  A(−1, 4) B(5, 2) C(3, 8)

16. Kayton is playing disc golf at the local park. He is standing at (3, 14) on the park grid. The first hole of the course is at (8, 2). Each unit on the grid is equal to 15 feet. How far is Kayton from the first hole in feet?

REVIEW

Solve for x.

17.

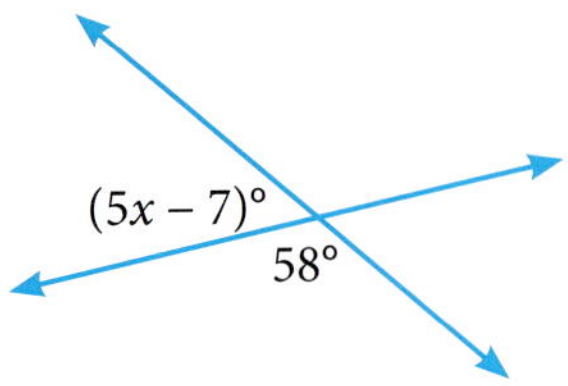

18.

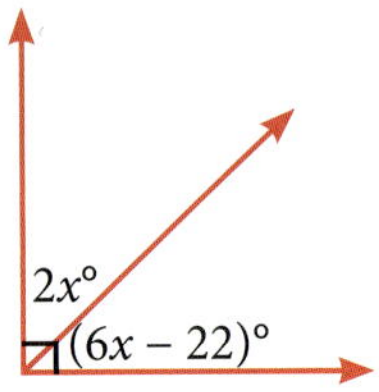

19.

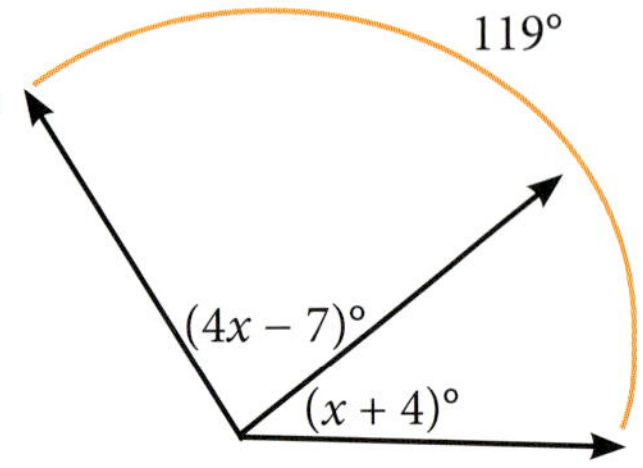

Determine if each triangle is possible based on its side lengths. If not, change one measurement to create a feasible triangle.

20.

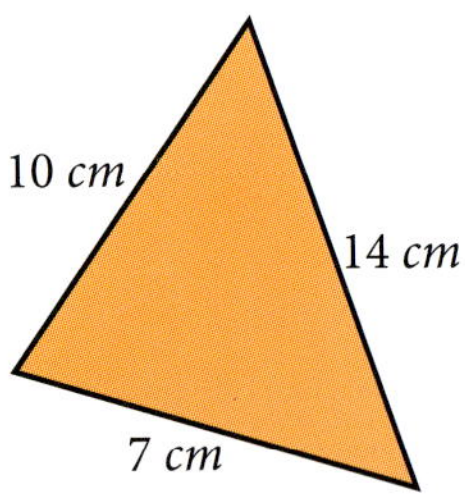

21.

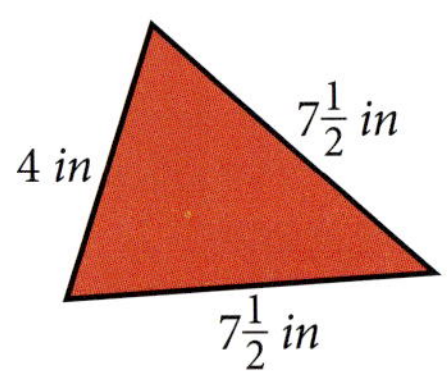

22.

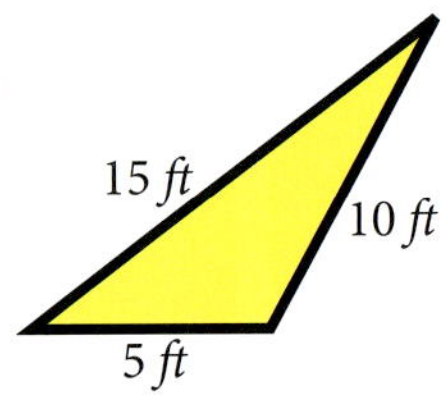

Use a protractor to draw an angle with the given measure. Then classify the angle.

23. 75°

24. 160°

25. 90°

26. 180°

Tic-Tac-Toe ~ Vocab Booklet

Geometry has many vocabulary terms and formulas. Create a small booklet of the vocabulary and formulas from this textbook. Organize it by topics in a way that makes sense to you. The topic of each section in your booklet should be clearly marked. Include diagrams and definitions. Include helpful hints for topics that have been difficult for you or other classmates. Create a cover page and table of contents.

Tic-Tac-Toe ~ 3D Coordinate Plane

So far, coordinate planes have been two-dimensional with two axes, x and y. There are three-dimensional coordinate planes with three axes: x, y and z. Points in a two-dimensional graph are written (x, y). In a three-dimensional graph they are written (x, y, z).

The Distance Formula and Pythagorean Theorem can be combined to form an equation that can find the distance between any two points in a three-dimensional coordinate plane.

$$d = \sqrt{(x_2 - x_1)^2 + (y_2 - y_1)^2 + (z_2 - z_1)^2}$$

Example: Below is a set of points on a three-dimensional coordinate plane. The distance found is the distance between the two points.

(1, 4, 6) and (2, −3, 8)

Write the three-dimensional Distance Formula.	$d = \sqrt{(x_2 - x_1)^2 + (y_2 - y_1)^2 + (z_2 - z_1)^2}$
Substitute the coordinates into the formula.	$d = \sqrt{(2 - 1)^2 + (-3 - 4)^2 + (8 - 6)^2}$
Simplify the inside of the parentheses.	$d = \sqrt{(1)^2 + (-7)^2 + (2)^2}$
Simplify by squaring.	$d = \sqrt{1 + 49 + 4}$
Simplify by adding.	$d = \sqrt{54}$
Square root and round to the nearest tenth.	$d \approx 7.3$

Find the distance between each pair of points. Round to the nearest tenth.

1. (4, 1, 5) and (8, 7, 3)

2. (0, −3, 14) and (−5, 2, −7)

3. (1, 8, −3) and (4, −2, 15)

4. (3.7, 2.5, 8) and (6.1, 0, 5.9)

5. A sphere's center is located at (0, 0, 0). The point (4, 9, −5) is located on the surface of the sphere. Find the length of the radius of the sphere. Round to the nearest tenth.

6. A sphere's center is at (9, −5, −5) and (−1, −2, −7) is a point on the surface of the sphere.
- **a.** Find the radius of the sphere. Round to the nearest tenth.
- **b.** Find the surface area of the sphere using the formula $SA = 4\pi r^2$. Use 3.14 for π.
- **c.** Find the volume of the sphere using the formula $V = \frac{4}{3}\pi r^3$. Use 3.14 for π.

REVIEW

BLOCK 2

Vocabulary

Distance Formula
hypotenuse
irrational numbers
leg
perfect square
rational number
square root
theorem

Recognize and find the values of perfect squares.
Estimate the values of square roots.
Use the Pythagorean Theorem to find missing side lengths in right triangles.
Determine if three side lengths create a right triangle.
Apply the Pythagorean Theorem to solve problems in two and three dimensions.
Find the distance between two points on a coordinate plane using the Pythagorean Theorem.
Find the distance between two points on a coordinate plane using the Distance Formula.

Lesson 2.1 ~ Perfect Squares

Find the value of each square root.

1. $\sqrt{25}$

2. $\sqrt{1}$

3. $\sqrt{100}$

4. $\sqrt{49}$

5. $\sqrt{64}$

6. $\sqrt{144}$

7. The Ransom family owns a square-based tent. The area of the base of the tent is 121 square feet. What is the length of each side of the base of the tent?

8. A square has an area of 36 meters. Find the perimeter of the square.

Solve each equation. Check each answer.

9. $x^2 + 5 = 86$

10. $4x^2 = 64$

11. $2x^2 + 12 = 20$

Lesson 2.2 ~ Estimating Square Roots

Determine the two positive integers each value is between.

12. $\sqrt{18}$

13. $\sqrt{70}$

14. $\sqrt{37}$

Estimate each square root to the nearest tenth.

15. $\sqrt{8}$ **16.** $\sqrt{15}$ **17.** $\sqrt{52}$

Use a calculator to approximate each square root. Round to the nearest tenth.

18. $\sqrt{43}$ **19.** $\sqrt{93}$ **20.** $\sqrt{231}$

21. Deidre bought a square plot of land. The total area of her land was 12,000 square feet. What is the length of one side of her plot? Round to the nearest foot.

Lesson 2.3 ~ The Pythagorean Theorem

Find the value of x in each right triangle. If necessary, round to the nearest tenth.

22.

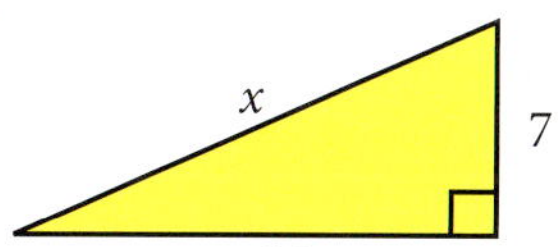

23.

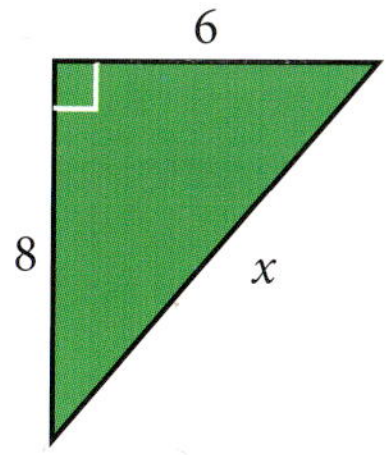

24.

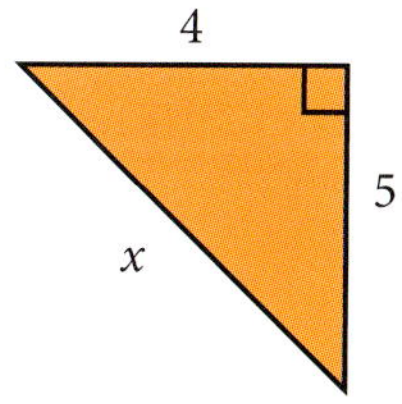

25.

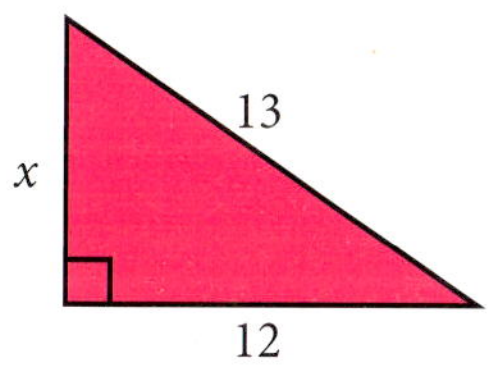

26.

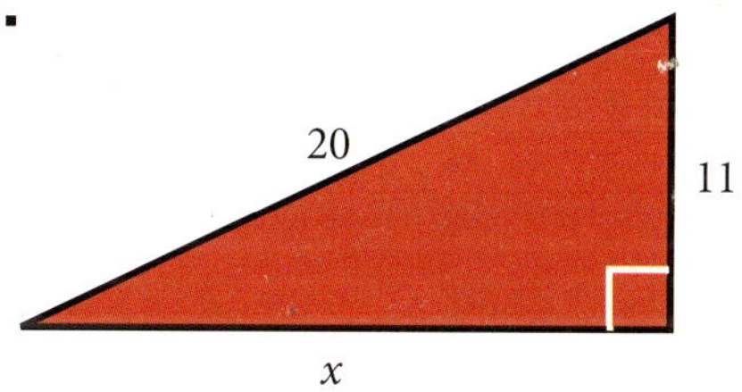

27.

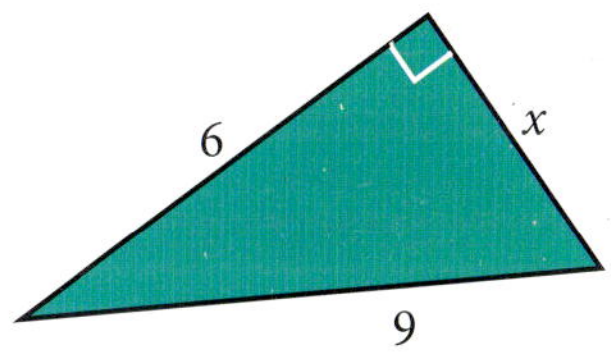

28. A rectangular gift card is 4 inches by 3 inches. Find the length of the card's diagonal.

29. The area of a square is 25 square meters.
  a. What is the length of each side of the square?
  b. What is the measure of the square's diagonal?

Lesson 2.4 ~ Converse of the Pythagorean Theorem

Determine if the given lengths form a right triangle. Support your answer with words and/or symbols.

30.

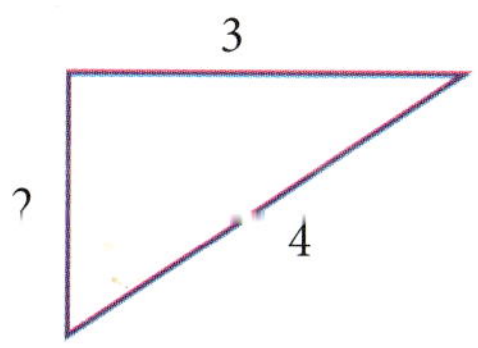

31.

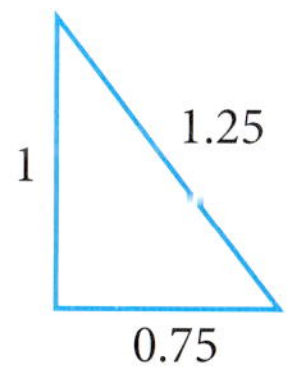

32. 16, 30, 34

Determine if each set of numbers is a Pythagorean triple.

33. 7, 24, 25

34. 10, 24, 23

35. 2.5, 6, 6.5

36. Use the Pythagorean triple with the integers 3, 4, 5.
- **a.** Multiply each number in the set by 2.
- **b.** Verify that the new set of numbers from **part a** form a right triangle.
- **c.** Create two more multiples of the Pythagorean triple 3, 4, 5.

Lesson 2.5 ~ Applying the Pythagorean Theorem

Draw a diagram and solve for the missing measure. When necessary, round to the nearest tenth.

37. Kyle cut a 3 inch by 5 inch card into two equal-sized right triangles. What is the length of the hypotenuse of each triangle?

38. Kirstina is painting her bedroom. Her ladder is 10 feet long. She wants it to reach 8 feet high on the wall. How far away from the base of the wall should she place the ladder?

39. A ship travels 120 miles due north then makes a turn due west. It travels 250 miles west. How far is the ship from its starting point?

40. Owen walks to school each day. He travels 12 blocks east and then 5 blocks south.
- **a.** If he could walk on a path directly from his house, what would the length of his walk be in blocks?
- **b.** How many blocks shorter would the direct route be?

41. A rectangular prism is 5 meters long, 6 meters wide and 8 meters tall. What is the length of its longest diagonal?

42. Bobby is sending his cousin a baseball bat for a gift. He needs to fit the bat into a box that measures 30 inches by 10 inches by 12 inches. What is the maximum length the baseball bat can be to fit in the box?

Lesson 2.6 ~ Distance on the Coordinate Plane

Find the length of c on each graph.

43.

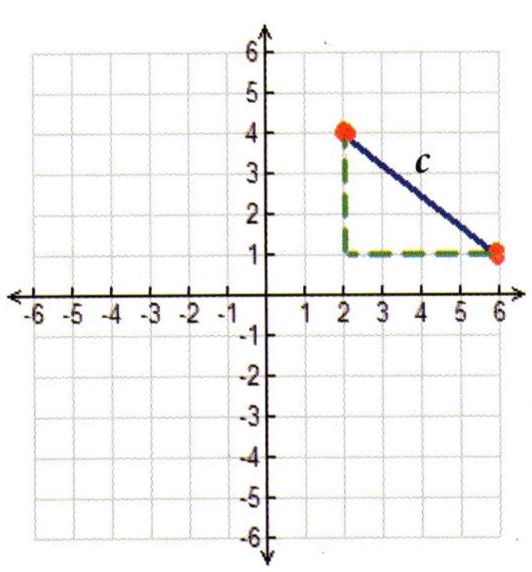

44.

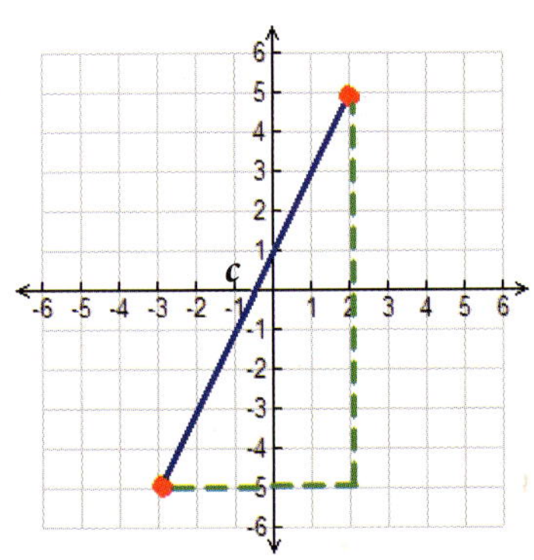

Find the distance between each set of points using the Pythagorean Theorem. When necessary, round to the nearest tenth.

45. (7, −1) and (1, 7) **46.** (0, −6) and (1, 0) **47.** (−3, 2) and (5, −2)

Find the perimeter of each triangle formed by the given set of points. When necessary, round to the nearest tenth.

48. (0, 0), (0, 6) and (8, 0) **49.** (−1, 3), (−1, −2) and (6, 3)

Lesson 2.7 ~ The Distance Formula

Use the Distance Formula to calculate the length of each segment. Round to the nearest tenth, if necessary.

50.

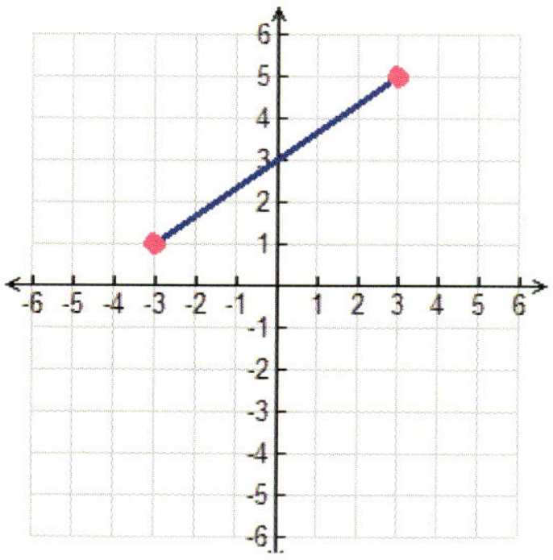

51.

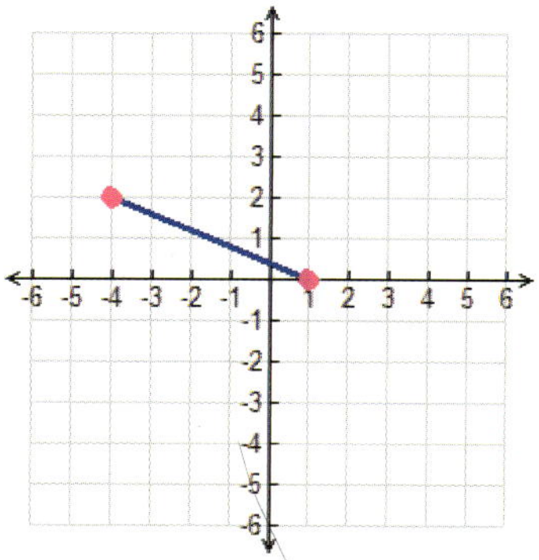

Use the Distance Formula to find the distance between each pair of points. Round to the nearest tenth, if necessary.

52. (1, 19) and (9, 2) **53.** (8, 3) and (4, −2) **54.** (−5, 0) and (−4, 1)

55. A circle has a center at (0, 0) and a point on the edge of the circle is located at (−4, 3).

a. Find the length of its radius.

b. Calculate the circumference of the circle using the formula $C = 2\pi r$. Use 3.14 for π.

c. Calculate the area of the circle using the formula $A = \pi r^2$. Use 3.14 for π.

CAREER FOCUS

CHARITY

CREDIT UNION ASSISTANT BRANCH MANAGER

I am an assistant branch manager for a Credit Union. My job includes many tasks each day. Part of being a manager is overseeing what other people who work for the Credit Union are doing. One of the main things that a Credit Union does is lend money to people. Loan officers help members apply for loans. Part of my job is to oversee the loan officers and to approve the loans that people want to take out from our Credit Union. I also work with members to help them find the best rate for their loan and keep their accounts with our Credit Union up to date.

There are many steps that go into a person being able to get a loan. One of these steps is called underwriting. Underwriting means to review all of a member's information to see if they qualify for a loan. When members of our Credit Union make loan requests, the loan officer will bring it to me and I begin the underwriting process. I use ratios to compare how much money a person is asking to borrow with how much money they make. I also look at things like a member's employment history and credit score. There are many factors to look at when deciding if somebody can borrow large amounts of money. It is my goal to put the member and the Credit Union in the best position so all parties involved are satisfied with the end result. Math helps me to make sound decisions towards reaching that goal.

A person can take a position in lending with a high school diploma. Each position has specific requirements, some of which need more education than others. Many lenders find it helpful to have a college degree in a financial subject area like economics or business.

In lending you can make from $19,200 per year as a full-time teller to $75,000 per year as a branch manager. Salaries vary depending on what position you have and your years of experience.

I love motivating my team to give excellent member service. Our goal is to provide a range of products and services that will appeal to our membership while making sound financial decisions.

CORE FOCUS ON GEOMETRY

BLOCK 3 ~ TRANSFORMATIONS

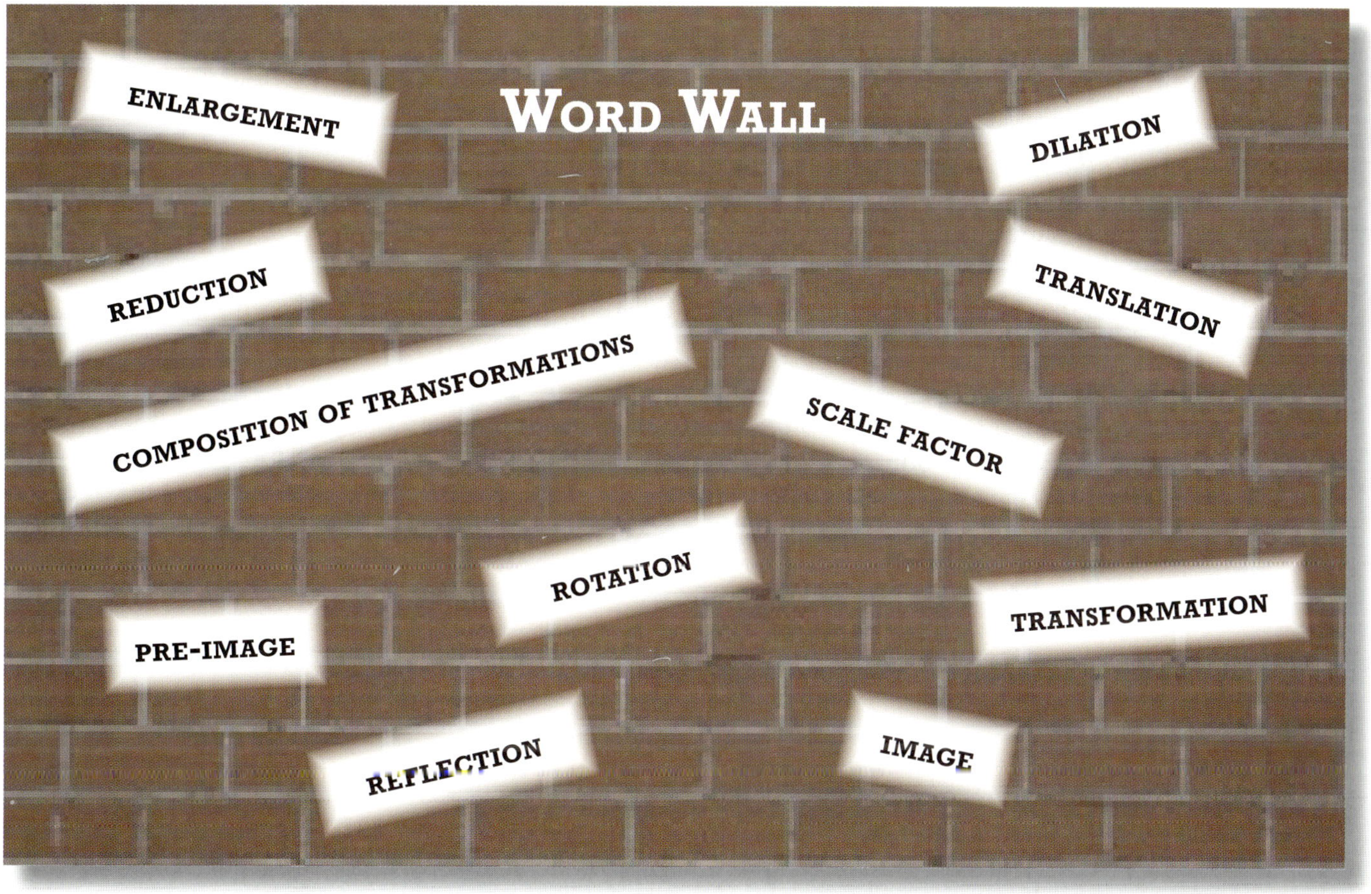

BLOCK 3 ~ TRANSFORMATIONS

TIC-TAC-TOE

Transformation Recap

Create a booklet and "Quick Check" summarizing the different types of transformations.

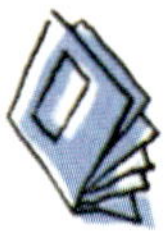

See page 110 for details.

Escher Art

Learn about M.C. Escher and create a tessellation of your own.

See page 116 for details.

Making It Large

Enlarge a small picture and determine the scale factor.

See page 116 for details.

Wallpaper Border

Create a wallpaper border using glide reflections.

See page 98 for details.

From Here to There

Create a series of transformations to map one point to another.

See page 120 for details.

Flashcards

Design a set of transformation flashcards and gather data on other strengths and weaknesses.

See page 106 for details.

Miniature Golf

Design three holes of a miniature golf course.

See page 94 for details.

Kaleidoscope

Research how transformations are used in the creation of a kaleidoscope.

See page 110 for details.

State Translation

Create a map of your state and translation rules from your town to other towns.

See page 106 for details.

REFLECTIONS

LESSON 3.1

Reflect an image on a coordinate plane.

A **transformation** is the movement of a point or figure that changes its size or position. The original figure is called the **pre-image** and the resulting figure is the **image**. A **reflection** is a type of transformation that flips a figure over a line. In a reflection, the pre-image and image are mirror images of each other. This means the pre-image and its image are equal distances from the line of reflection.

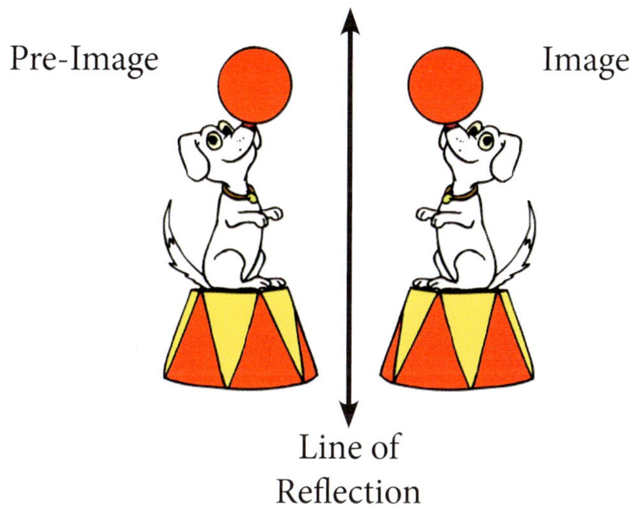

EXPLORE! **MIRROR, MIRROR**

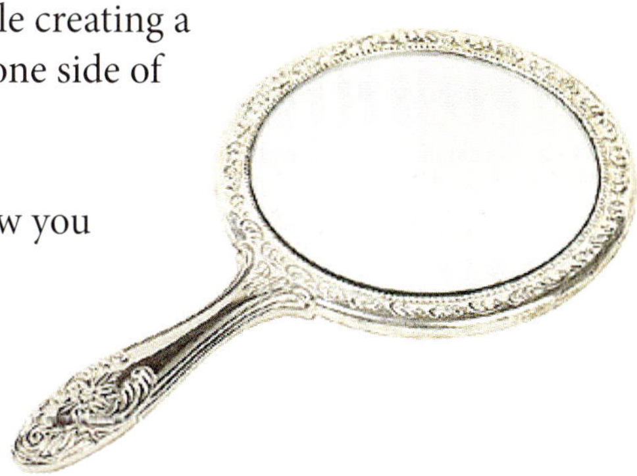

Step 1: Fold a piece of patty paper or tracing paper down the middle creating a vertical line. Open the paper and write your first name on one side of the line in uppercase letters.

Step 2: Create a reflection of your name. Explain to a classmate how you created your reflection.

Step 3: On a sheet of grid paper, graph a triangle at the top. Your vertices should have integer coordinates. This is your pre-image.

Step 4: Draw a horizontal line on a grid line below your pre-image. This is your line of reflection. Create the reflection of your triangle below the line WITHOUT folding the grid paper. Explain to a classmate how you created your image. Check your solution by folding the grid paper on the line of reflection.

Step 5: Are the pre-image and the image congruent? When working with polygons, do reflections always create congruent figures? Explain your reasoning.

The vertices of a pre-image are often labeled with letters. After a transformation occurs, the new image has vertices that are labeled with the same letter but an apostrophe is added. For example, if ΔABC is reflected over a line, then the image is labeled ΔA'B'C'. This is read "Triangle A prime, B prime, C prime".

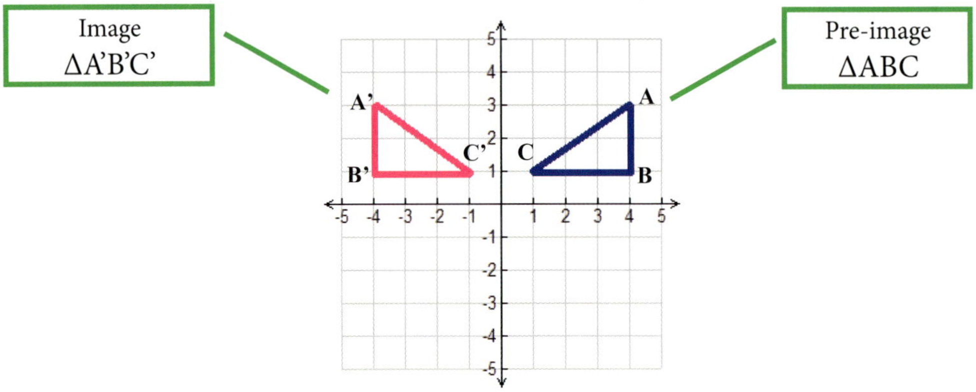

In mathematics, most transformations occur on a coordinate plane. The most common lines of reflection on a coordinate plane are the *x*- and *y*-axes. In the graph above, the line of reflection is the *y*-axis.

EXAMPLE 1 **The point D(−3, 4) is reflected over the *x*-axis. What are the coordinates of its image?**

SOLUTION Graph the point on a coordinate plane.

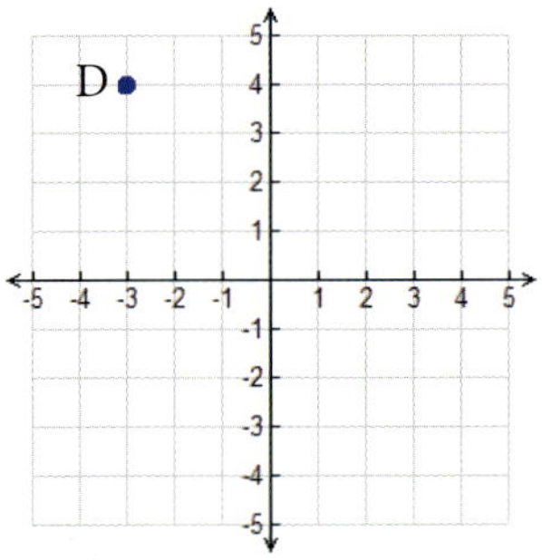

Reflect the point over the *x*-axis.

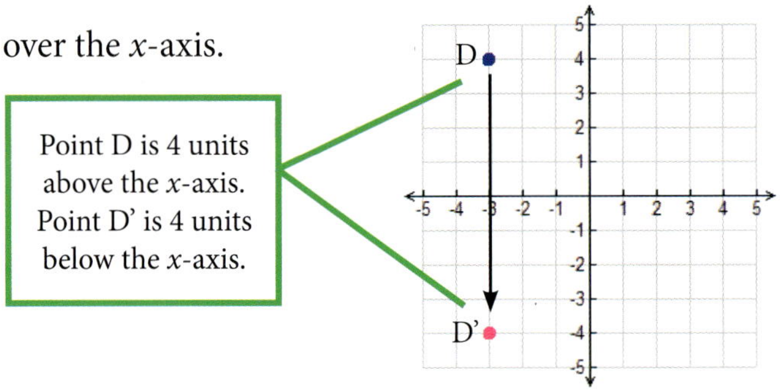

The coordinates of the image are D'(−3, −4).

Transformations, such as reflections, can be described using transformation rules. Transformation rules describe the effect of the transformation on an ordered pair. Transformation rules exist for reflections over the *x*- and *y*-axes. When reflecting over the *y*-axis, the *x*-coordinate changes to its opposite. When reflecting over the *x*-axis, the *y*-coordinate changes to its opposite.

y-axis reflection

When a figure is reflected over the y-axis, the x-coordinate changes sign.

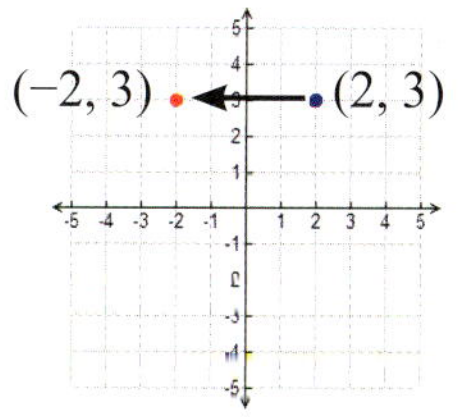

$(x, y) \longrightarrow (-x, y)$

x-axis reflection

When a figure is reflected over the x-axis, the y-coordinate changes sign.

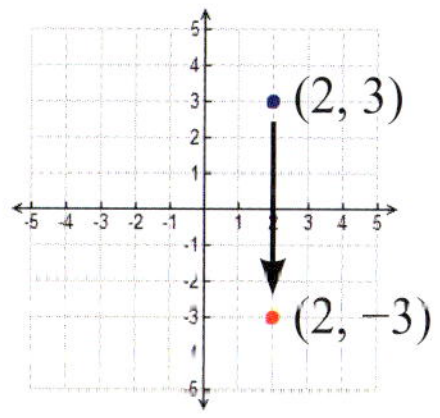

$(x, y) \longrightarrow (x, -y)$

EXAMPLE 2

Graph the image of ΔRST under the transformation rule $(x, y) \rightarrow (x, -y)$. Which axis is the image reflected over?

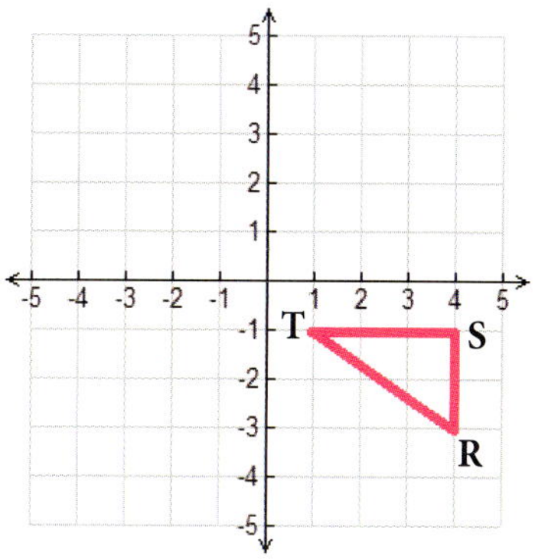

SOLUTION

Record the ordered pairs.

R(4, −3)
S(4, −1)
T(1, −1)

Change each of the y-coordinates to its opposite.

R'(4, 3)
S'(4, 1)
T'(1, 1)

Graph the ordered pairs of the image.

The x-axis is the line of reflection.

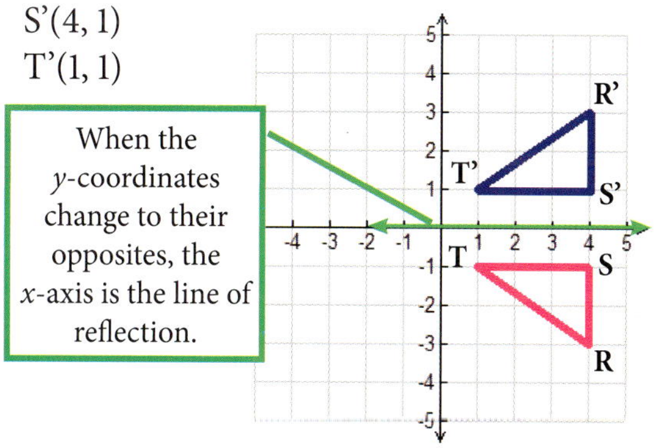

Reflections over the x- and y-Axes

When a point is reflected over the x-axis, the x-coordinate stays the same and the y-coordinate is changed to its opposite. $(x, y) \rightarrow (x, -y)$

When a point is reflected over the y-axis, the y-coordinate stays the same and the x-coordinate is changed to its opposite. $(x, y) \rightarrow (-x, y)$

EXAMPLE 3 **A square has coordinates M(−4, 2), N(−1, 2), P(−1, −1) and Q(−4, −1). The square is reflected over the y-axis. What are the coordinates of the image?**

SOLUTION A reflection over the y-axis causes the y-coordinates to stay the same. The x-coordinates change to their opposites.

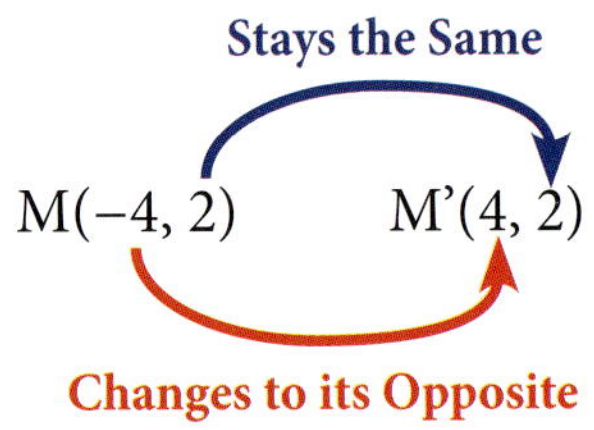

The coordinates of the image are M'(4, 2), N'(1, 2), P'(1, −1) and Q'(4, −1).

When using transformation rules, you can always check your answer by graphing the pre-image and image. Visually check the graph to confirm that the desired transformation occurred.

EXERCISES

1. The point (5, 2) is reflected over the x-axis. What are the coordinates of the new point?

2. The point (−7, 4) is reflected over the y-axis. What are the coordinates of the new point?

3. ΔFGH has vertices at F(1, 0), G(5, 2) and H(4, 4).
   a. Graph ΔFGH. This is the pre-image.
   b. Graph ΔF'G'H' after a reflection over the y-axis. List the coordinates of the vertices.

4. How can you draw a reflection of a figure if the figure is not on a coordinate plane? Draw an example to support your answer.

5. A pre-image has coordinates Y(9, 2), Z(3, −2) and W(6, 4). The image has coordinates Y'(−9, 2), Z'(−3, −2) and W'(−6, 4). What type of reflection occurred? Explain how you know.

6. A pre-image has coordinates A(−1, 8), B(5, −3) and C(−5, −1). The image has coordinates A'(−1, −8), B'(5, 3) and C'(−5, 1). What type of reflection occurred? Explain how you know.

Graph the pre-image and the image under the given reflection. Label the vertices correctly.

7. Triangle E(4, 1), F(−3, 5) and G(0, 3) — A reflection over the x-axis

8. Rectangle R(−5, 2), S(−4, 2), T(−4, 5) and U(−5, 5) — A reflection over the y-axis

9. Triangle M(3, 4), N(−1, 2) and P(0, 5) — A reflection over the y-axis

10. Square A(−1, 2), B(1, 2), C(1, 4) and D(−1, 4) — A reflection over the x-axis

11. Describe the reflection that maps ΔHJK onto H'J'K'.

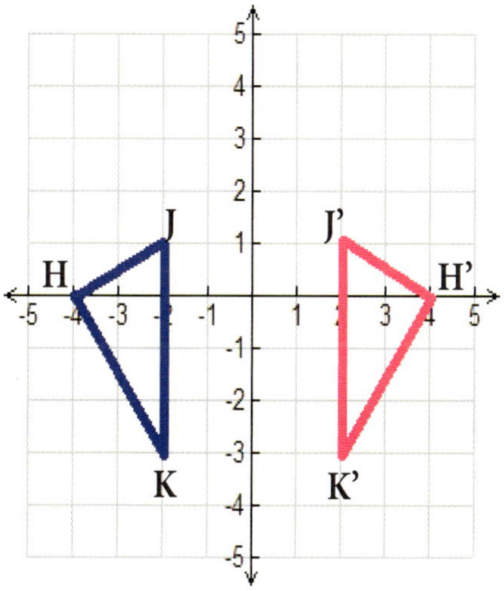

12. The point (0, −6) is reflected over the y-axis. What do you notice about the coordinates of the reflected point? Why is this?

13. A parabola can be graphed by finding a set of points that satisfy the equation $y = x^2$.

a. Copy and complete the table using the equation $y = x^2$.

x	y
−2	
−1	
0	
1	
2	

b. Graph the parabola. Connect the points with a smooth curve.

c. Reflect the parabola over the x-axis. What are the coordinates of the reflection? Graph the reflection on the same coordinate plane.

d. What would happen if you reflected the original parabola over the y-axis? How would the image compare to the pre-image?

14. The point A(−1, 8) is reflected over the line $x = 5$. What are the coordinates of the reflected point?

15. The point B(3, 2) is reflected over the line $y = 1$. What are the coordinates of the reflected point?

16. The point (x, y) is reflected over the x-axis. Its image is reflected over the y-axis. What ordered pair represents the final image?

REVIEW

Determine if the given lengths form a right triangle.

17. 2, 3, 5

18. 6, 8, 10

19. 10, 26, 24

20. 4, 7, 10

21. 8.5, 4, 7.5

22. $\frac{2}{3}, \frac{3}{4}, \frac{5}{7}$

23. A triangle has two angles which are complementary. What is the value of the third angle?

24. Two angles in a triangle are complementary. What is the measure of the third angle?

25. In ΔABC, angle A is twice the sum of angles B and C. Angle B is 24°. What is the measure of Angle A?

Tic-Tac-Toe ~ Miniature Golf

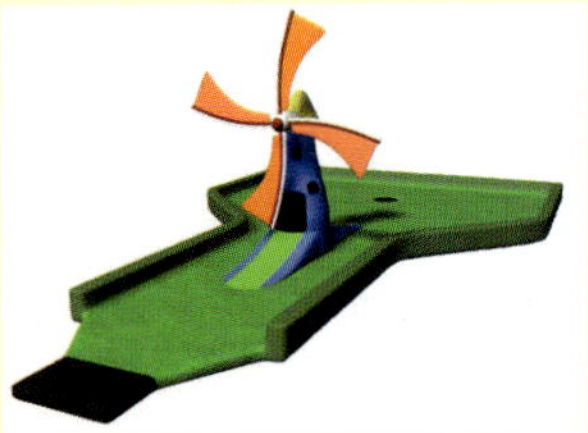

When a ball is hit without any spin against a wall, it will bounce off the wall at an angle equal to the angle at which it hit the wall. The path of the ball is reflected over a line perpendicular to where the ball hit the wall.

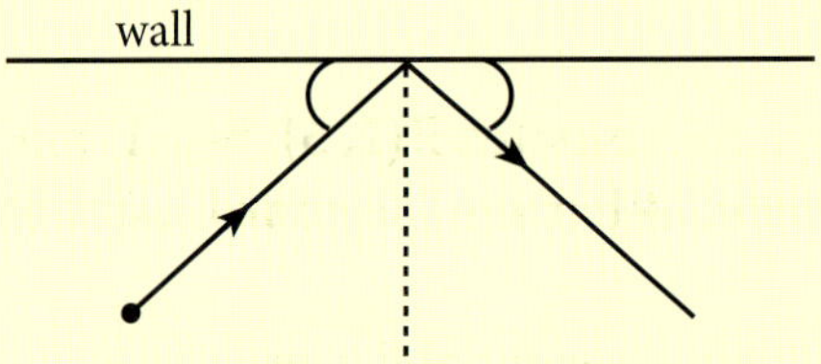

Sometimes when you play miniature golf, you need to hit the ball against the wall using reflections in order to get a hole-in-one.

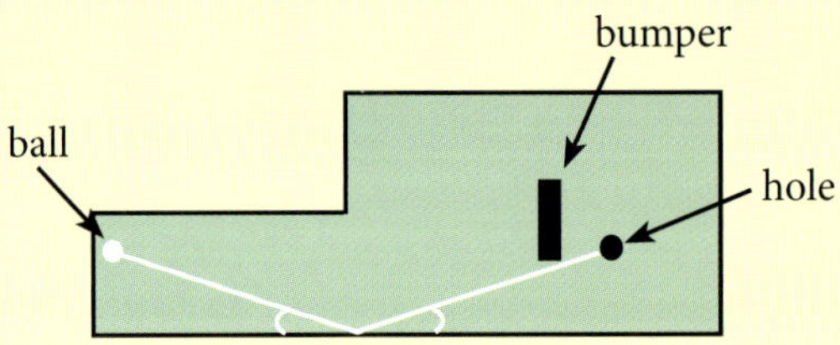

Suppose you are asked to design three miniature golf holes. The owner would like the first hole to require one reflection for a hole-in-one, the second hole to require two reflections for a hole-in-one and the third hole to require three reflections for a hole-in-one. You can use bumpers or the layout of the hole to meet these requirements.

Clearly draw your holes and show the path the ball will need to follow in order for the player to hit a hole-in-one. Use a straight edge. Be sure the reflections of the ball against each wall or bumper are clearly marked with congruent angles.

TRANSLATIONS

LESSON 3.2

Translate an image on a coordinate plane.

A **translation** is a transformation that moves a figure to a new position without turning it. The figure below has been translated both right and up from its original position. Notice that the figure below still has the same orientation (it is not turned or flipped) but is in a different location.

Translations can be performed on a coordinate plane when the move is defined by a specific number of units left or right and/or up or down.

EXAMPLE 1

A triangle with coordinates R(2, 5), S(0, 1) and T(3, –1) is translated 2 units right and 3 units down. What are the coordinates of ΔR'S'T'?

SOLUTION

Graph the pre-image (original figure).

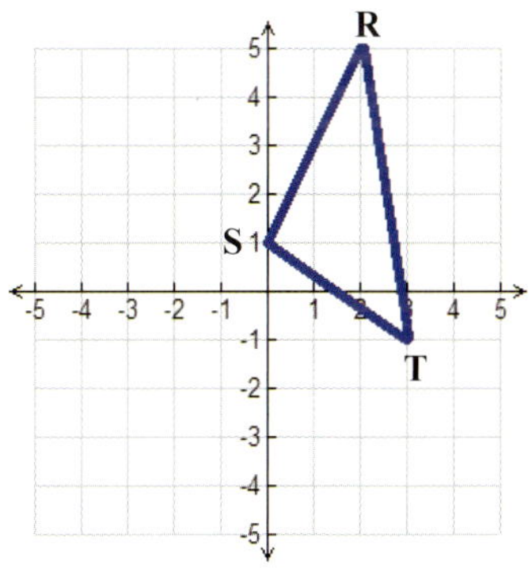

Graph the image by translating each point. Translate each point 2 units right (add 2 to the x-value) and 3 units down (subtract 3 from the y-value).

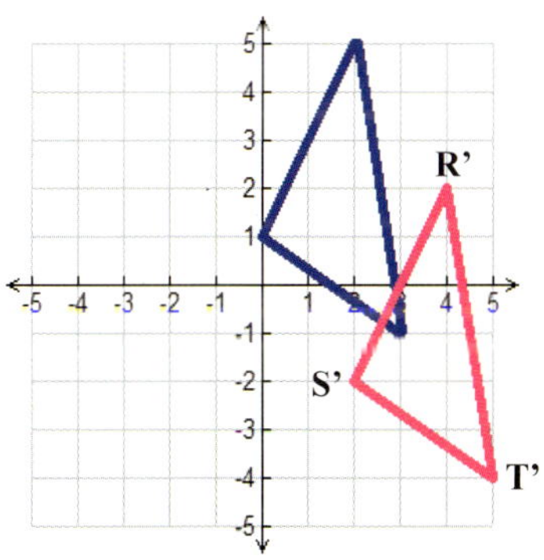

Write the ordered pairs for the coordinates of ΔR'S'T'.

R'(4, 2), S'(2, –2), T'(5, –4)

Translations can be described using transformation rules. A transformation rule for a translation is shown by stating the amount that is added or subtracted to the x- and y-coordinates in the pre-image.

TRANSLATIONS ON A COORDINATE PLANE

A point in a pre-image is translated a units horizontally and b units vertically using the transformation rule: $(x, y) \rightarrow (x + a, y + b)$.

If a > 0, the point is translated right.	If b > 0, the point is translated up.
If a < 0, the point is translated left.	If b < 0, the point is translated down.

EXAMPLE 2

Describe the horizontal and/or vertical path described by each transformation rule.
a. $(x, y) \rightarrow (x + 1, y + 4)$
b. $(x, y) \rightarrow (x - 2, y)$

SOLUTIONS

a. When the value being added to the x-coordinate is positive, the point moves right.

$(x, y) \rightarrow (x + 1, y + 4)$
Translated 1 unit right.

When the value being added to the y-coordinate is positive, the point moves up.

$(x, y) \rightarrow (x + 1, y + 4)$
Translated 4 units up.

b. When the value being added to the x-coordinate is negative, the point moves left.

$(x, y) \rightarrow (x - 2, y)$
Translated 2 units left.

Since there is no change to the y-coordinate, the graph is not translated vertically.

$(x, y) \rightarrow (x - 2, y)$
No vertical translation.

EXAMPLE 3

A transformation rule of $(x, y) \rightarrow (x + 6, y - 1)$ is applied to the triangle formed by the points D(−5, 2), A(−2, 2) and B(−2, 4). Graph ΔDAB and ΔD'A'B'.

SOLUTION

The transformation rule states that ΔDAB is translated right 6 units $(x + 6)$ and down 1 unit $(y - 1)$.

Add 6 to each x-coordinate and subtract 1 from each y-coordinate.

D(−5, 2) → (−5 + 6, 2 − 1) → D'(1, 1)
A(−2, 2) → (−2 + 6, 2 − 1) → A'(4, 1)
B(−2, 4) → (−2 + 6, 4 − 1) → B'(4, 3)

Graph the pre-image and image.

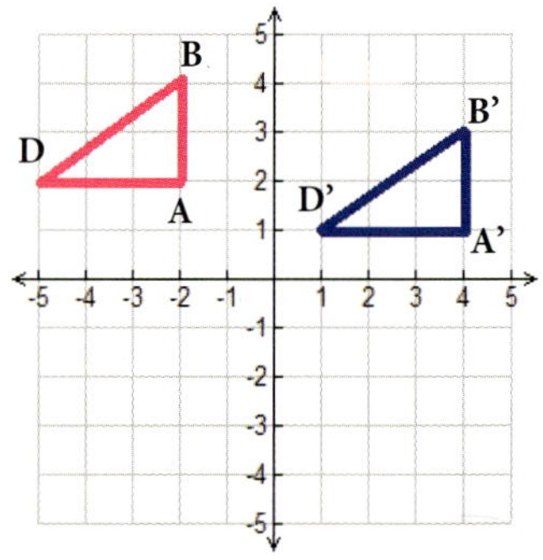

EXERCISES

Determine the coordinates of the image of F(−1, 9) under the following translations.

1. $(x, y) \rightarrow (x + 3, y)$

2. Translated down 5 units

3. Translated left 2 units and up 1 unit

4. $(x, y) \rightarrow (x - 4, y - 8)$

5. $(x, y) \rightarrow (x + 6, y + 11)$

6. Translated right 7 units and down 3 units

7. ΔPBM has vertices at P(1, 0), B(5, 2) and M(4, 4).
   a. Graph ΔPBM. This is the pre-image.
   b. Graph ΔP'B'M' after a translation 2 units left and 3 units down. List the coordinates of the vertices.

8. A pre-image has coordinates A(7, 1), B(1, −2) and C(5, −3). The image has coordinates A'(7, 6), B'(1, 3) and C'(5, 2). Write a transformation rule to describe the path the pre-image made to arrive at the image.

9. A pre-image has coordinates L(−1, 4), A(−5, 1) and K(−3, 7). The image has coordinates L'(−2, 2), A'(−6, −1) and K'(−4, 5). Write a transformation rule to describe the path the figure took to arrive at the image.

Graph the pre-image and the image under the given translation. Label the vertices correctly.

10. Triangle S(2, 1), A(−3, 2) and M(0, 0) $\quad (x, y) \rightarrow (x + 3, y + 5)$

11. Rectangle D(5, 4), I(1, 4), R(1, 5) and T(5, 5) $\quad (x, y) \rightarrow (x - 4, y - 6)$

12. Triangle W(−3, 4), I(−1, 2) and T(−1, 0) $\quad$ Translated right 2 units and down 4 units

13. Use the graph at the right.
   a. Describe the translation of ΔUTA to ΔU'T'A' in words.
   b. Write the transformation rule.

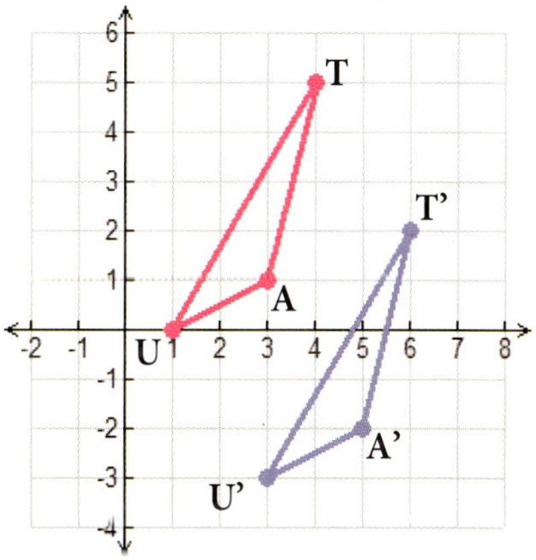

14. What are the coordinates of the point (x, y) after being translated p units left and m units up?

15. Graph a triangle in the third quadrant of the coordinate plane (all coordinates should have negative values). Translate your image using the transformation rule $(x, y) \rightarrow (x + 6, y + 5)$. Label your pre-image and image using proper notation.

16. Jeremy said that if you connect every point from a pre-image to its corresponding image point with a line segment, the line segments drawn will all be parallel. Do you agree or disagree? Support your answer with evidence.

17. A pre-image was translated twice. The first translation moved the triangle right 7 units and down 2 units. The second translation followed the transformation rule $(x, y) \rightarrow (x - 3, y + 8)$. Write one transformation rule that combines the two translations into one.

REVIEW

18. Carlos biked 4 miles north and 3 miles east. Then Carlos biked the shortest distance to return to his starting spot. How many miles did Carlos bike in all?

19. A 12-foot ladder is leaning against a wall. The ladder reaches 8 feet up the wall. To the nearest tenth of a foot, how far is the base of the ladder from the base of the wall?

20. A square photo frame has a diagonal of 20 centimeters. What is the side length of the photo frame to the nearest tenth of a centimeter?

21. The hypotenuse of a triangle is 15 inches long. The side lengths are whole numbers. What are the lengths of the two sides?

Tic-Tac-Toe ~ Wallpaper Border

A wallpaper border is a decorative strip of wallpaper that usually is placed along the top of a wall and goes around all the walls of a room. Research how "glide reflections" are often used in wallpaper borders. Create a wallpaper border of a design of your choosing. Follow the guidelines below to create your border: Start with a strip of paper that is at least 11.5" in height and 5 feet in length. You can tape pieces of paper together, if necessary.

- Draw your line of reflection in pencil down the midline of the strip of paper.
- Use a horizontal translation distance (measure and mark the distances in pencil) of your choosing to space out the images along the strip of paper. There should be at least 3 of your images on each side of the line of reflection (6 total images).
- Be sure to color your design appropriately.

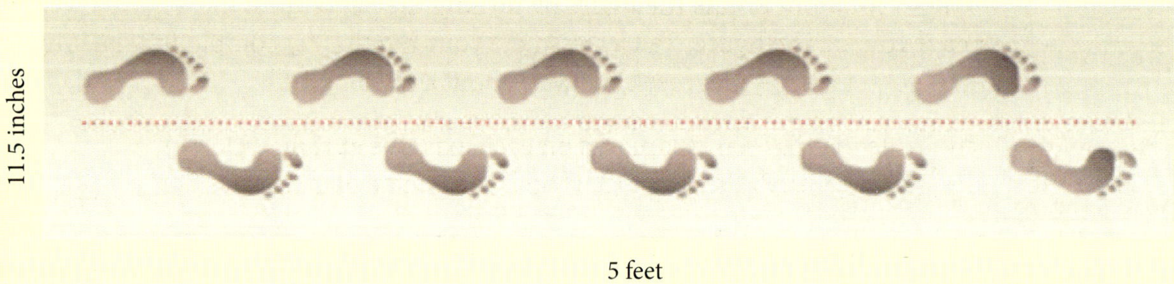

Write a half-page summary of your design. Describe how it fits the definition for a "glide reflection" and the steps involved in creating your border.

ROTATIONS

LESSON 3.3

Rotate an image on a coordinate plane.

A **rotation** is a transformation that turns a figure a specific number of degrees about a fixed point called the center of rotation. When working on a coordinate plane, the center of rotation is often the origin. Rotations are commonly done in increments of 90° either clockwise or counterclockwise.

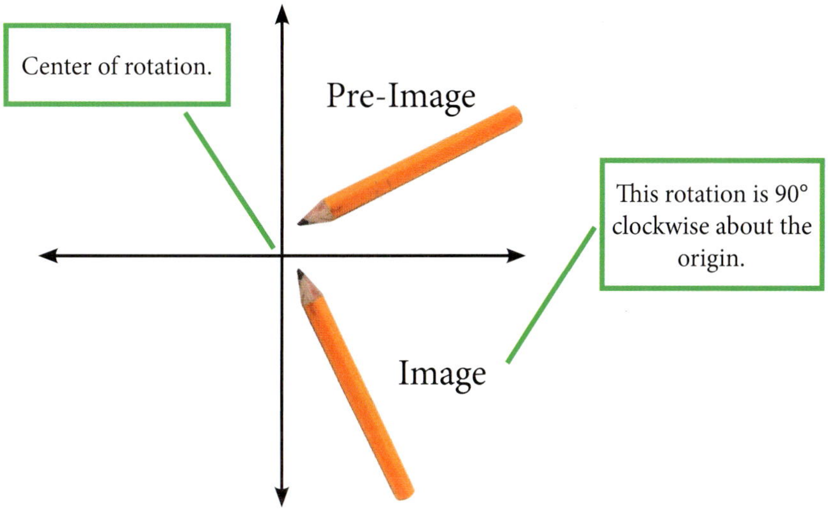

EXPLORE! **AROUND AND AROUND**

Step 1: Draw a simple happy face on the left side of a sheet of patty paper or tracing paper. Draw a point in the bottom left hand corner as the center of rotation. Draw a ray that goes through the left eye of the happy face. Draw the other side of the ray as shown. This forms the rotation angle.

Step 2: Measure your angle with a protractor. This will be your angle of rotation.

Step 3: Use a second piece of patty paper. Lay it on top of the happy face and trace the happy face. Put your pencil on the center of rotation. Rotate the figure by turning the paper until the left eye is on the second ray. You have rotated the figure. Tell a classmate your rotation by completing the sentence, "My happy face was rotated _____ degrees clockwise about a fixed point."

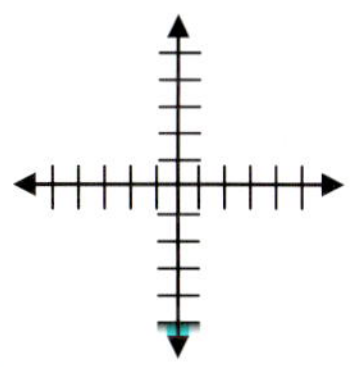

Step 4: Draw a coordinate plane that has ten tick marks on each axis (see at right). Do not write numbers on the axes. Graph the point (3, 1) on a coordinate plane.

Step 5: Turn your graph 90° clockwise. What are the coordinates of the point now? How do the coordinates compare to the original point? Complete the rotation transformation rule for a 90° clockwise rotation: $(x, y) \rightarrow$ (_____ , _____).

Step 6: Turn your graph another 90° clockwise. This is a 180° rotation from the original point (3, 1). What are the coordinates of the point now? How do the coordinates compare to the original point? Complete the rotation transformation rule for a 180° clockwise rotation: $(x, y) \rightarrow$ (_____ , _____).

EXPLORE! **(CONTINUED)**

Step 7: Turn your graph one more 90° turn clockwise. This is a 270° clockwise rotation or a 90° counterclockwise rotation from the original point (3, 1). What are the coordinates of the point now? How do the coordinates compare to the original point? Complete the rotation transformation rule for a 270° clockwise rotation: $(x, y) \rightarrow$ (____ , ____).

Step 8: Why do you think the most common rotation angles are 90°, 180° and 270°?

ROTATIONS ON A COORDINATE PLANE

Rotations about the origin in increments of 90° follow these transformation rules:

Clockwise ↷	Counterclockwise ↶	Transformation rule
90°	270°	$(x, y) \rightarrow (y, -x)$
180°	180°	$(x, y) \rightarrow (-x, -y)$
270°	90°	$(x, y) \rightarrow (-y, x)$

EXAMPLE 1

ΔRUS is rotated 180° about the origin. Graph ΔR'U'S'.

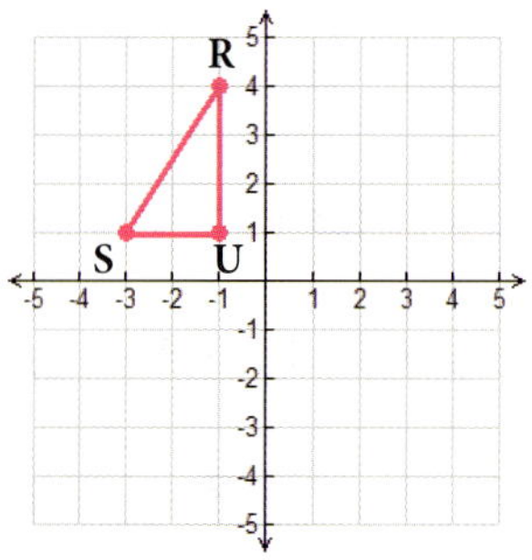

SOLUTION

Record the coordinates of ΔRUS.

R(−1, 4)
U(−1, 1)
S(−3, 1)

Use the transformation rule for 180° rotations about the origin.

$(x, y) \rightarrow (-x, -y)$
R(−1, 4) → R'(1, −4)
U(−1, 1) → U'(1, −1)
S(−3, 1) → S'(3, −1)

Graph the image.

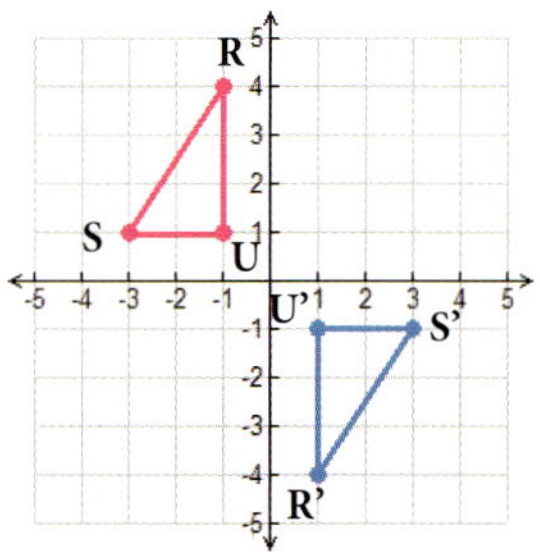

EXAMPLE 2 **The line segment is rotated about the origin. The pre-image, $\overline{PT}$, is located at P(2, −1) and T(4, −5). The image, $\overline{P'T'}$, is located at P'(1, 2) and T'(5, 4). How many degrees was the line segment rotated clockwise about the origin?**

SOLUTION Graph the pre-image and image.

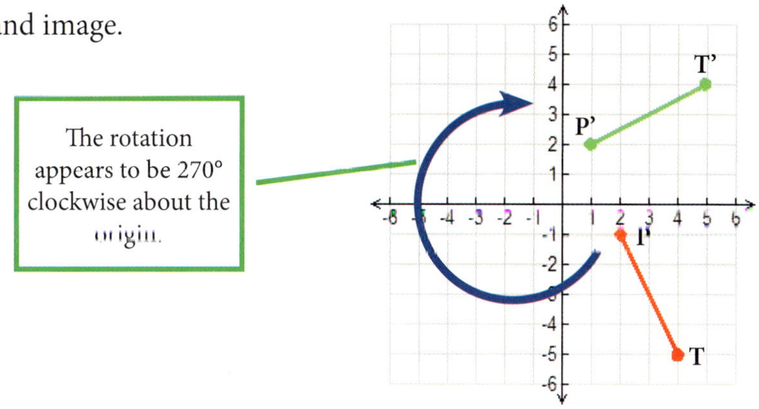

The transformation rule for a rotation 270° clockwise about the origin is $(x, y) \rightarrow (-y, x)$.

Examine the pre-image and image ordered pairs to confirm the rotation.

$(x, y) \rightarrow (-y, x)$
P(2, −1) → P'(1, 2)
T(4, −5) → T'(5, 4)

The line segment was rotated clockwise 270° about the origin.

EXERCISES

Determine the coordinates of the image of K(1, 4) under the following rotations.

1. $(x, y) \rightarrow (-x, -y)$

2. Rotation 90° clockwise about the origin

3. Rotation 180° counterclockwise about the origin

4. $(x, y) \rightarrow (-y, x)$

5. Which of **Exercises 1–4** result in the same image coordinates?

6. The point (4, −7) is rotated to (−7, −4). How many degrees was this clockwise rotation?

7. ΔFIG has vertices at F(2, 4), I(5, 4) and G(3, 2).
   a. Graph ΔFIG. This is the pre-image.
   b. ΔFIG is rotated 90° clockwise about the origin. Graph ΔF'I'G'.

8. A pre-image has coordinates W(−5, 2), A(−2, 0) and M(−3, −3). The image has coordinates W'(−2, −5), A'(0, −2) and M'(3, −3). Describe the counterclockwise rotational path of the figure.

9. A pre-image has coordinates U(−1, 4) and V(−5, 8). The image has coordinates U'(1, −4), and V'(5, −8). Describe the clockwise rotational path of the line segment.

Graph the pre image and the image under the given rotation. Label the vertices correctly.

10. Triangle A(1, 1), B(1, 4) and C(5, 1) $(x, y) \rightarrow (y, -x)$

11. Rectangle M(2, −3), I(2, −1), N(5, −1) and D(5, −3) 270° clockwise about the origin

12. Triangle G(−5, −4), A(−1, −2) and L(−3, 0) 180° about the origin

13. What are the coordinates of the point (x, y) after it is rotated 90° counterclockwise about the origin?

14. Explain how you can use patty paper or tracing paper to rotate a figure 45° clockwise about a point.

15. Xavier predicted that every image formed by a rotation will be similar but not congruent to its pre-image. Do you agree or disagree? Support your answer with your reasoning.

16. Kirsten wanted to write a series of two rotations that would cause her final image to be at the same location as her pre-image. Write three different pairs of rotations that would cause this to occur.

17. Ron said that a reflection over the x-axis followed by a reflection over the y-axis is the same as a 180° rotation about the origin. Do you agree or disagree? Support your answer by explaining your reasoning.

18. Tim rotated a figure 360° about the origin. How does the pre-image compare to the image?

REVIEW

Determine if each statement is a translation or a reflection. Determine the coordinates of the image of P(1, −3) under each transformation.

19. $(x, y) \rightarrow (-x, y)$

20. Moved right 4 units

21. Moved left 1 unit and down 2 units

22. Flipped over the x-axis

23. $(x, y) \rightarrow (x + 4, y + 9)$

24. $(x, y) \rightarrow (x, -y)$

Determine the two positive integers each value is between.

25. $\sqrt{71}$

26. $\sqrt{52}$

27. $\sqrt{13}$

28. $\sqrt{120}$

29. $\sqrt{33}$

30. $\sqrt{84}$

DILATIONS

LESSON 3.4

Dilate an image on a coordinate plane.

A **dilation** is a transformation that changes the size of a figure, but does not change the shape. The image is either an enlarged or reduced version of the pre-image. Similar to rotations, dilations have a point which serves as the center for the dilation. In this lesson, the origin will be used as the center of dilation. Every dilation has a **scale factor**. The scale factor is the ratio of two corresponding sides of the pre-image and image.

DILATIONS ON A COORDINATE PLANE

A figure is dilated using scale factor b when each x- and y-coordinate is multiplied by b ($b \neq 0$). The dilation transformation rule for scale factor b is: $(x, y) \rightarrow (bx, by)$.

EXAMPLE 1

ΔABC is drawn below. Find its image under a dilation centered at the origin with a scale factor of 2. Graph the image.

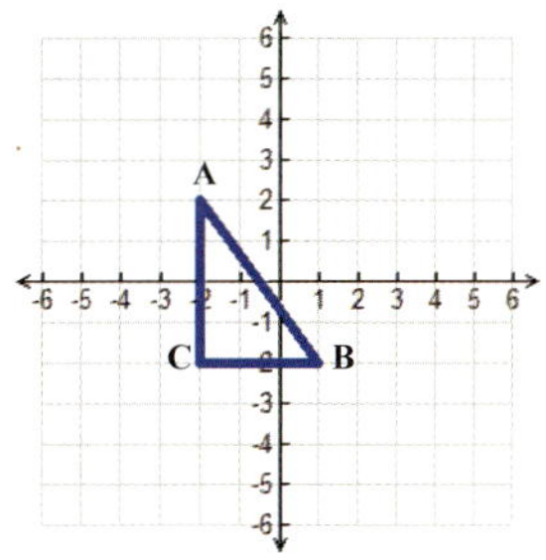

SOLUTION

Multiply the coordinates of each point by the scale factor of 2.

A(−2, 2) → (2 · −2, 2 · 2) → A'(−4, 4)
B(1, −2) → (2 · 1, 2 · −2) → B'(2, −4)
C(−2, −2) → (2 · −2, 2 · −2) → C'(−4, −4)

Graph the image.

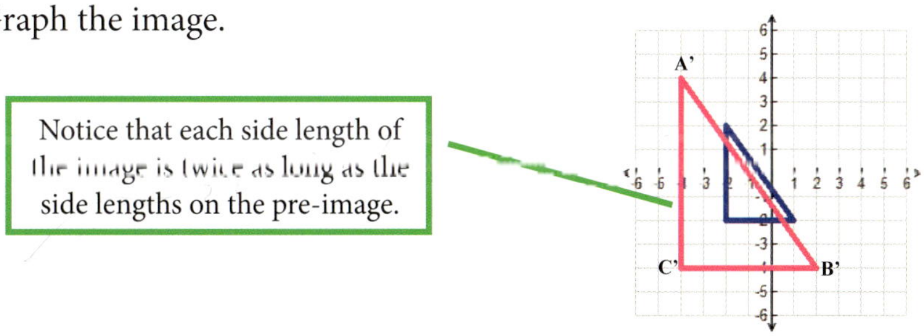

EXPLORE! **BIGGER OR SMALLER?**

A dilation that creates an image larger than the pre-image is called an **enlargement**. A dilation that creates an image smaller than the pre-image is a **reduction**. Certain scale factor values create enlarged images compared to the pre-image. Other values create images that are reduced in size. In this **Explore!**, you will determine which values of the scale factor create each of these situations.

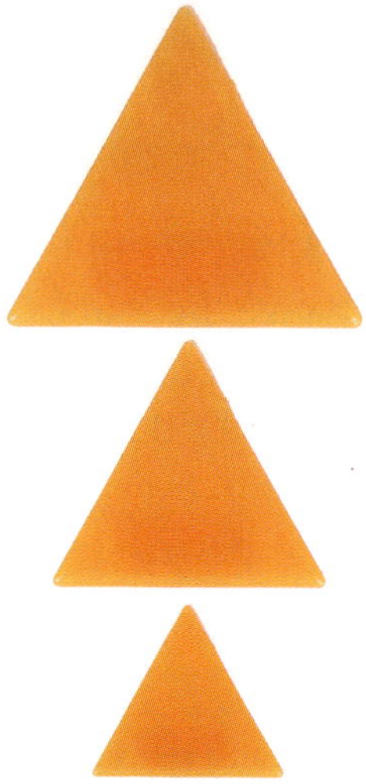

Step 1: Graph H(2, 2), J(4, 2) and K(4, 4) on a coordinate plane.

Step 2: Using a colored pencil, graph the image ΔH'J'K' after a dilation with a scale factor of 0.5 on the same coordinate plane.

Step 3: Using a different colored pencil, graph the image ΔH"J"K" after a dilation with a scale factor of 2 on the same coordinate plane.

Step 4: If b represents the scale factor, write an inequality that describes scale factors that create enlargements. Write another inequality that describes scale factors that create reductions. How did you arrive at your answer?

Step 5: What type of image does a scale factor of 1 create? Support your answer with evidence.

TYPES OF DILATIONS

$(x, y) \rightarrow (bx, by)$

When the absolute value of the scale factor, b, is greater than 1, the image created by the dilation is an enlargement.

When the absolute value of the scale factor, b, is between 0 and 1, the image created by the dilation is a reduction.

EXAMPLE 2

Give the scale factor for each dilation. Determine if the scale factor creates an enlargement or a reduction in size compared to the pre-image.

a. $(x, y) \rightarrow (0.3x, 0.3y)$

b. $(x, y) \rightarrow (\frac{4}{3}x, \frac{4}{3}y)$

SOLUTIONS

a. The scale factor is the value that each coordinate is multiplied by. The scale factor is 0.3. — Scale Factor = 0.3

Compare the scale factor to 1. — $0.3 < 1$

Since the scale factor is less than 1, the dilation is a reduction.

b. The scale factor is the value that each coordinate is multiplied by. The scale factor is $\frac{4}{3}$. — Scale Factor = $\frac{4}{3}$

Compare the scale factor to 1. — $\frac{4}{3} > 1$

Since the scale factor is larger than 1, the dilation is an enlargement.

EXERCISES

1. ΔHAT has vertices at H(1, 1), A(3, 0) and T(2, 2).

a. Graph ΔHAT. This is the pre-image.
b. Graph ΔH'A'T' after a dilation with a scale factor of 3.
c. Describe this dilation as a reduction or enlargement.

Give the scale factor for each dilation centered at the origin. Determine if the scale factor creates an enlargement or a reduction in size compared to the pre-image.

2. $(x, y) \to (6x, 6y)$

3. $(x, y) \to (0.25x, 0.25y)$

4. $(x, y) \to \left(\frac{3}{5}x, \frac{3}{5}y\right)$

5. $(x, y) \to \left(\frac{5}{2}x, \frac{5}{2}y\right)$

6. $(x, y) \to (3.8x, 3.8y)$

7. $(x, y) \to \left(\frac{9}{10}x, \frac{9}{10}y\right)$

8. A pre-image has coordinates P(−2, 2), L(−4, 3) and C(−3, 4). The image has coordinates A'(−5, 5), B'(−10, 7.5) and C'(−7.5, 10). Write the transformation rule used to create this image.

9. A pre-image has coordinates A(9, 3), B(3, 3) and C(6, 12). The image has coordinates A'(3, 1), B'(1, 1) and C'(2, 4). Write the transformation rule used to create this image.

Graph the pre-image and the image under the given dilation centered at the origin. Label the vertices correctly.

10. Triangle M(2, −1), A(4, −3) and T(2, −6) $(x, y) \to \left(\frac{1}{2}x, \frac{1}{2}y\right)$

11. Triangle R(−4, 5), A(−3, 2) and T(0, 4) $(x, y) \to (1.5x, 1.5y)$

12. Rectangle G(−1, 2), R(−1, 1), I(1, 1) and M(1, 2) $(x, y) \to (2x, 2y)$

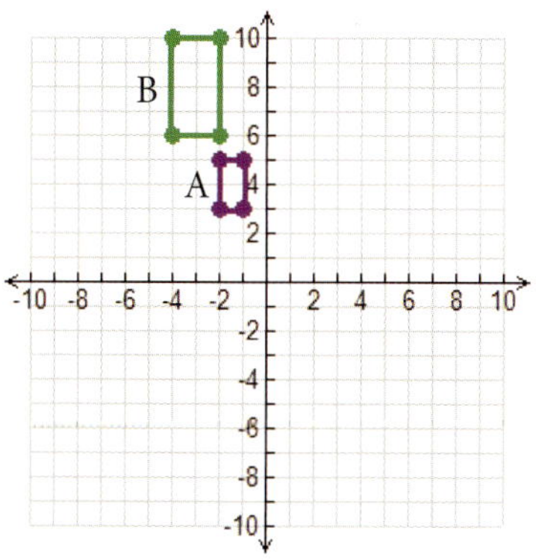

13. Rectangle B at left is the image of Rectangle A under a dilation with a scale factor of 2 centered at the origin.

a. How does the perimeter of Rectangle A compare to the perimeter of Rectangle B?
b. How does the area of Rectangle A compare to the area of Rectangle B?

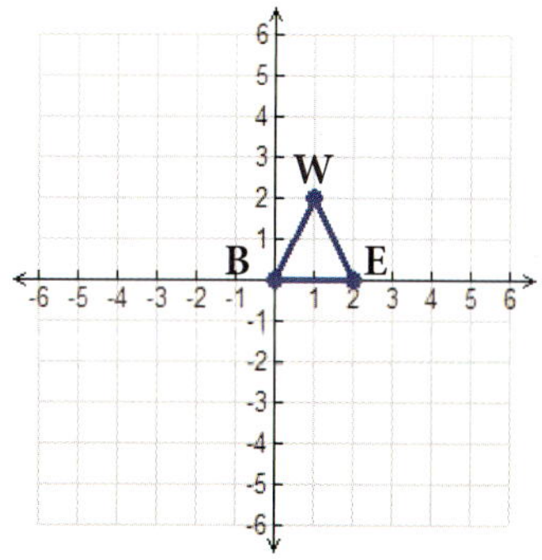

14. Use the graph at the right.

a. Graph an enlargement of ΔWEB using a dilation with a scale factor of 3 centered at the orign. Label the image ΔW'E'B'.
b. How many times greater is the area of ΔW'E'B' than ΔWEB?

15. Shanequa graphed a triangle with the coordinates R(−2, 1), S(2, 1) and T(0, −3). She wanted to graph a dilation of the figure with a scale factor of 2 centered at the origin so she graphed the points R'(−2, 2), S'(2, 2) and T'(0, −6).

  a. Is ΔR'S'T' a dilation of ΔRST? Why or why not?

  b. If Shanequa was wrong, correct her answer by giving the correct coordinates for ΔR'S'T'.

  c. Shanequa's teacher said Shanequa had made a common mistake. Explain her mistake.

REVIEW

Write the transformation rule for each transformation.

16. A reflection over the x-axis

17. A rotation of 180° about the origin

18. A translation down 3 units

19. A translation right 1 unit and up 5 units

20. A reflection over the y-axis

21. A 90° rotation clockwise about the origin

22. Winston found one angle in a triangle measured 76°. What is the sum of the other two angles in the triangle?

Tic-Tac-Toe ~ Flashcards

Create a set of twenty transformation flashcards. The set should include five cards of each type of transformation. On one side of the flashcard, graph the pre-image and image on a coordinate plane. On the opposite side, write a single transformation rule that represents the transformation.

Quiz at least 5 different peers each with eight randomly selected flashcards from your set. Show the graph and have them give the transformation rule. Collect data on which type of transformation rules your peers struggle with the most and what type they struggle with the least.

Write a one-paragraph summary of your findings to turn in with your flashcards.

Tic-Tac-Toe ~ State Translations

Find or print a map of your state that is no larger than 8.5" by 11". Draw a coordinate plane over the top of your map where the center of the paper is the origin. Each axis should extend from −10 to 10. Locate the city or town you live in and record the approximate ordered pair for this location. Write translation rules for the path from your city to ten other towns in your state. Include towns from all four quadrants of your graph.

TRANSFORMATIONS AND CONGRUENCE

LESSON 3.5

Understand the relationship between the pre-image and the image of a transformation.

Two figures are congruent if they are the exact same shape and the exact same size. Figures are similar if they are the same shape but different sizes.

In the **Explore!** you will draw the image of a triangle given a transformation and then determine if the image is congruent or similar to the original (or pre-image) triangle.

EXPLORE! **CONGRUENT OR SIMILAR?**

Step 1: List four types of transformations on four separate lines of your paper.

Step 2: Each type of transformation forms an image that is either similar or congruent to the original figure. Make a conjecture (an educated guess) for each transformation you listed in **Step 1** about whether the image will be congruent or similar to the original figure.

Step 3: Use ΔABC as the pre-image for **Steps 4–7**: A(2, 5), B(4, 3) and C(3, 0). Use graph paper to make five coordinate planes (each from −10 to 10 on both the *x*- and *y*-axes). Graph ΔABC on the first coordinate plane.

Step 4: One type of transformation is a translation. On the second coordinate plane, graph ΔA'B'C' by shifting ΔABC 3 units left and 2 units up. What do you notice about the image compared to the pre-image? Use the word "similar" or "congruent" in your observation.

Step 5: Another type of transformation is a reflection. On the third coordinate plane, graph ΔA'B'C' by reflecting ΔABC over the *x*-axis. What do you notice about the image compared to the pre-image? Use the word "similar" or "congruent" in your observation.

Step 6: A dilation is another type of transformation. On the fourth coordinate plane, graph ΔA'B'C' using a dilation with a scale factor of 2. What do you notice about the image compared to the pre-image? Use the word "similar" or "congruent" in your observation.

Step 7: The last type of transformation is a rotation. On the last coordinate plane, graph ΔA'B'C' by rotating ΔABC 180° clockwise about the origin. What do you notice about the image compared to the pre-image? Use the word "similar" or "congruent" in your observation.

Step 8: Compare your observations with your original conjectures. Did your work in **Steps 4–7** support your conjectures? If not, rewrite your conjecture(s) to match your work.

Congruency and Similarity with Transformations

Under a translation, reflection or rotation, the original figure and its image are congruent.

Under a dilation, the original figure and its image are similar.

EXAMPLE 1

Name the type of transformation given by each rule. Then state if the image will be congruent or similar to its pre-image.

a. $(x, y) \rightarrow (x - 3, y + 4)$ **b.** $(x, y) \rightarrow (4x, 4y)$

SOLUTIONS

a. $(x - 3)$ means translate 3 units left and $(y + 4)$ means translate 4 units up. This is a translation. Translations form congruent figures.

b. A transformation that multiplies the coordinates of a figure by a scale factor (in this case, 4), is a dilation. Dilations form similar figures.

EXAMPLE 2

ΔEFG was formed by a single transformation of ΔMNP.

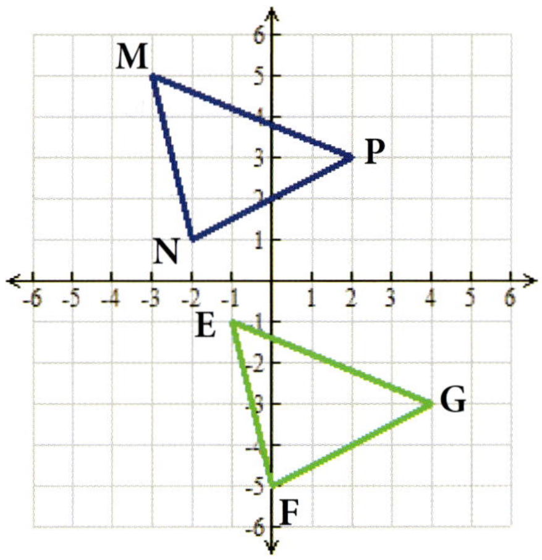

a. Are ΔEFG and ΔMNP congruent or similar?
b. Write a translation rule that maps ΔMNP onto ΔEFG.

SOLUTIONS

a. ΔEFG and ΔMNP are congruent because they are the same shape and same size.
b. ΔMNP is moved 2 units to the right and 6 units down to make ΔEFG. This transformation rule can be written $(x, y) \rightarrow (x + 2, y - 6)$.

EXERCISES

Write the transformation rule for each transformation. State whether the transformation creates similar or congruent figures.

1. A reflection over the y-axis

2. A translation up 4 units

3. A dilation with a scale factor of 3

4. A reflection over the x-axis

5. A translation 2 units left and 8 units down

6. A dilation with a scale factor of 0.25

Name the type of transformation given by each rule. State whether the transformation creates similar or congruent figures.

7. $(x, y) \rightarrow (7x, 7y)$

8. $(x, y) \rightarrow (-x, -y)$

9. $(x, y) \rightarrow (-x, y)$

10. $(x, y) \rightarrow (x + 5, y - 4)$

11. Which transformation rule does not always create a congruent figure?

12. Jacob shifted a figure five units down and two units left. He predicted that the resulting figure would be similar to, but not congruent to, his original shape. Do you agree or disagree? Explain your answer.

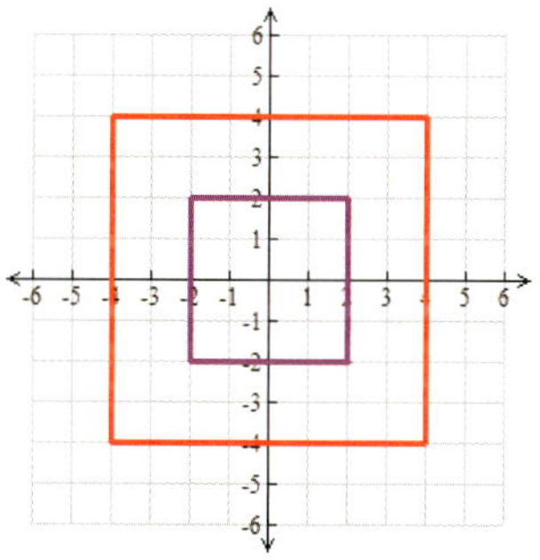

13. The red square at the left was graphed first. A transformation was performed on the red square to form the purple square.

a. Write a transformation rule that maps the red square onto the purple square.

b. What type of transformation is this?

c. Are the figures similar or congruent?

14. Write a transformation rule that creates an image congruent to its pre-image.

15. Write a transformation rule that creates a figure similar to its original.

16. Use the triangle formed by the vertices P(−3, 4), A(0, 5) and N(2, 1).

a. ΔP'A'N' is a triangle formed by using the transformation rule $(x, y) \rightarrow (x, -y)$. Graph ΔP'A'N'.

b. What type of transformation is ΔPAN → ΔP'A'N'?

c. Are ΔPAN and ΔP'A'N' congruent or similar?

17. Nachelle wrote the transformation rule $(x, y) \rightarrow (2x, 3y)$. She says it will produce figures that are similar to one another but not congruent. Do you agree or disagree with Nachelle's statement? Support your answer by graphing an original figure and its image using the transformation rule.

18. Jina stated that there is one dilation that creates congruent figures. Is she correct? If so, what scale factor creates congruent images?

19. Rectangle ABCD is transformed to form rectangle EFGH. The coordinates for ABCD are A(2, 3), B(2, 5), C(5, 5) and D(5, 3). The coordinates for EFGH are E(−2, 3), F(−2, 5), G(−5, 5) and H(−5, 3). Describe how the original figure and the new figure are related (transformation type and congruence/similarity).

20. Jeff said he could reflect a figure and then translate the reflected figure. He said the final image would be similar, but not congruent, to the pre-image. Is he correct? Support your answer with reasoning and/or an example.

REVIEW

Find the missing side length in each right triangle. If necessary, round to the nearest tenth.

21.
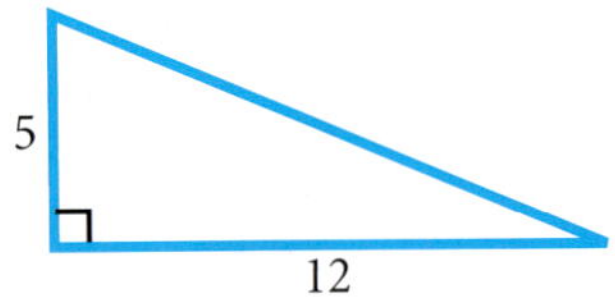

22.
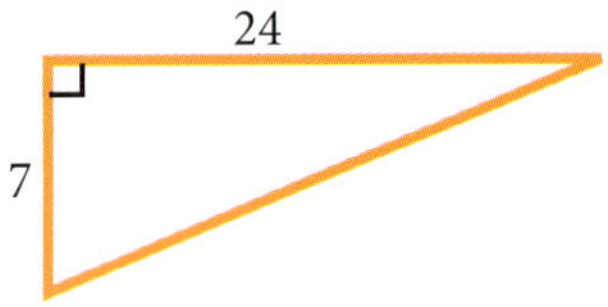

23.
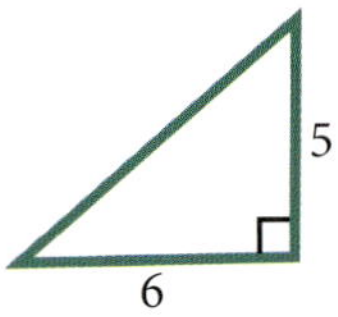

24.
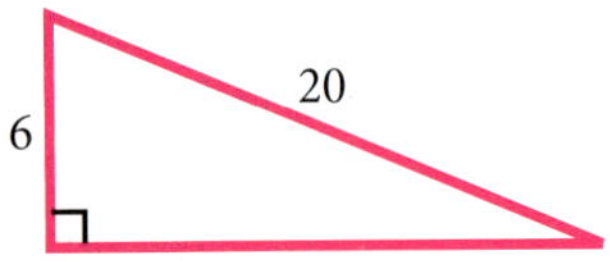

25.
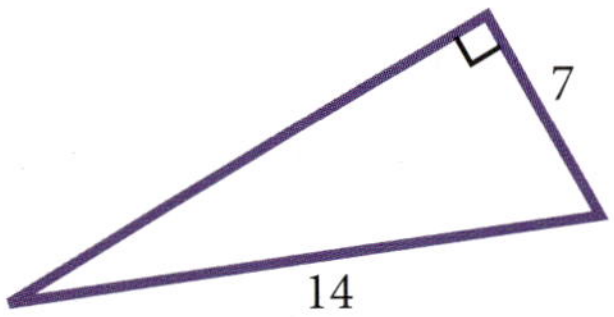

26.
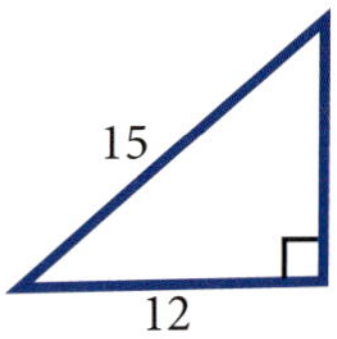

Tic-Tac-Toe ~ Kaleidoscope

The word "kaleidoscope" means to examine beautiful forms. Sir David Brewster invented the kaleidoscope in 1815. When it was manufactured, 200,000 kaleidsoscopes were sold in London and Paris. Research how a kaleidoscope is made and works. Write a booklet explaining how to make a kaleidoscope and explain how it works.

Tic-Tac-Toe ~ Transformation Recap

Create a booklet which gives the highlights of each of the four types of transformations. Include definitions, examples, transformation rules and information about whether the image would be congruent or similar to the pre-image for each type of transformation.

In the last page of your booklet, write a five-to ten-problem "Quick Check" that others could use to test their knowledge of the different transformations. Include the answers on the back cover of your booklet.

COMPOSITION OF TRANSFORMATIONS

LESSON 3.6

Perform multiple transformations on a figure including a combination of reflections, rotations, translations and dilations.

EXPLORE! **ON A TRIP**

Step 1: Graph the point A(3, 3) on a coordinate plane.

Step 2: Let point B be the image of point A under the transformation $(x, y) \rightarrow (x + 1, y - 5)$. Graph point B and give the ordered pair for its location.

Step 3: Let point C be a reflection of point B over the y-axis. Graph point C and give its ordered pair.

Step 4: Let point D be a rotation of point C 90° clockwise about the origin. Graph point D and give its ordered pair.

Step 5: Let point E be the image of point D under the transformation $(x, y) \rightarrow \left(\frac{1}{2}x, \frac{1}{2}y\right)$. Graph point E and give its ordered pair.

Step 6: Write a translation rule that moves point E back to the original location (point A).

Step 7: Create your own "trip" around the coordinate plane for someone else to follow. Start by giving the ordered pair for a point you choose to be point S. Give four transformation directions that will move the points to four different locations around the coordinate plane $S \rightarrow T \rightarrow U \rightarrow V \rightarrow W$.

Step 8: Create an answer key for your path by giving an ordered pair for points T, U, V and W.

A series of transformations is called a **composition of transformations**. The transformations that are included in the sequence will determine whether or not the image is similar or congruent to the original figure.

Transformations that create congruent figures include translations, reflections and rotations. Any combination of these transformations on a figure create a congruent image. If a dilation occurs on a figure during a transformation sequence, that image will be similar to the pre-image.

On a coordinate plane, you can prove two figures are similar or congruent if you can write a sequence of transformations that maps one figure onto the other. If the sequence does not include a dilation, the figures are congruent.

You can show that two figures are congruent if you can map one onto another using transformation rules that include reflections, rotations and/or translations.

EXAMPLE 1

Use the triangle formed by the points H(−2, 4), K(1, 2) and J(0, −3).
a. Graph ΔH'K'J' by reflecting ΔHKJ over the *x*-axis and then following the translation rule $(x, y) \rightarrow (x - 4, y + 3)$.
b. Are the figures congruent or similar to each other? How do you know?

SOLUTIONS

a. Graph the original figure, ΔHKJ, on a coordinate plane.

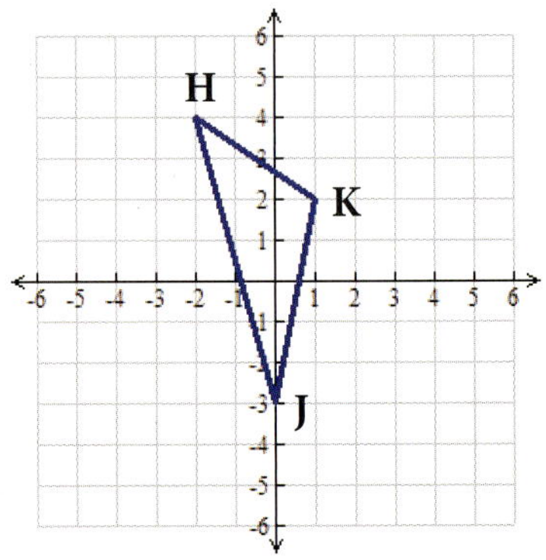

Reflection over the *x*-axis.
$(x, y) \rightarrow (x, -y)$

H(−2, 4) → (−2, −4)
K(1, 2) → (1, −2)
J(0, −3) → (0, 3)

Translation left 4 units and up 3 units.
$(x, y) \rightarrow (x - 4, y + 3)$

(−2, −4) → H'(−6, −1)
(1, −2) → K'(−3, 1)
(0, 3) → J'(−4, 6)

Graph the resulting figure, ΔH'K'J', on the coordinate plane.

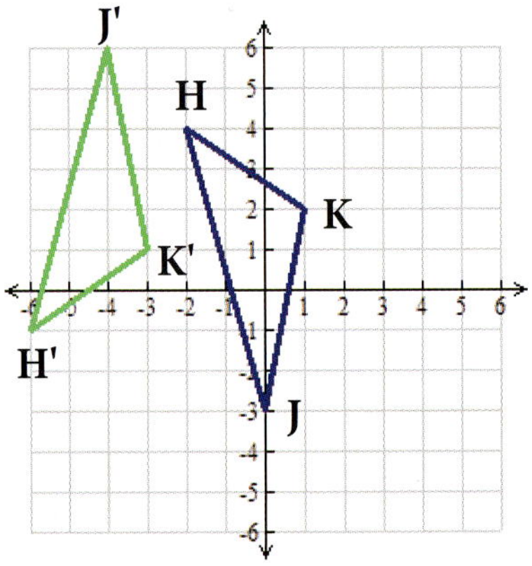

b. The figures are congruent because ΔH'K'J' is a result of two transformations (reflection and translation) on ΔHKJ. Each of these transformations creates congruent images from its pre-images.

You can show that two triangles are similar if you can map one onto another using transformation rules that include at least one dilation.

EXAMPLE 2 **Write a series of two transformations that shows ΔMNP is similar to ΔM'N'P'.**

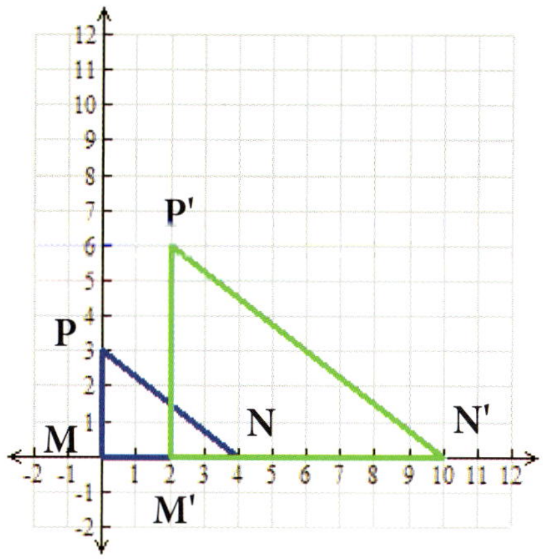

SOLUTION The final triangle appears to be longer and moved to the right. This can be done using a dilation and a translation.

Record the original coordinates. M(0, 0), N(4, 0), P(0, 3)

The image appears larger. Find the scale factor by looking at the ratio of one side of ΔM'N'P' and the corresponding side of ΔMNP.

$$\frac{M'N'}{MN} = \frac{8}{4} = 2$$

Multiply each vertex of ΔMNP by a scale factor of 2.

$(x, y) \rightarrow (2x, 2y)$
M(0, 0) → (0, 0)
N(4, 0) → (8, 0)
P(0, 3) → (0, 6)

The point at the origin is shifted 2 units to the right. Add 2 units to each x-coordinate for a translation rule of $(x, y) \rightarrow (x + 2, y)$.

$(x, y) \rightarrow (x + 2, y)$
(0, 0) → M'(2, 0)
(8, 0) → N'(10, 0)
(0, 6) → P'(2, 6)

Since these points match ΔM'N'P' on the graph above, this series of transformations (dilation $(x, y) \rightarrow (2x, 2y)$ and translation $(x, y) \rightarrow (x + 2, y)$) maps ΔMNP onto ΔM'N'P'.

The two figures are similar (not congruent) because a dilation was part of the transformation sequence that made the triangle larger.

EXERCISES

Start with the point Q(1, 5) for Exercises 1–5. Write the ordered pair for the final location of the given point after completing the series of transformations in the order listed.

1. $(x, y) \rightarrow (x - 1, y + 3)$, reflection over the x-axis

2. A 180° rotation about the origin, $(x, y) \rightarrow (-x, y)$

3. $(x, y) \rightarrow (x + 3, y - 7)$, $(x, y) \rightarrow (x - 2, y + 4)$, $(x, y) \rightarrow (2x, 2y)$

4. Reflection over the y-axis, reflection over the x-axis, $(x, y) \rightarrow (x, y + 6)$

5. $(x, y) \rightarrow (4x, 4y)$, $(x, y) \rightarrow \left(\frac{1}{2}x, \frac{1}{2}y\right)$, 90° rotation clockwise about the origin

6. Which composition of transformations in **Exercises 1–5** form figures that are congruent to their original figures? Explain how you know.

7. Write a series of two transformations that would form similar (but not congruent) figures. Explain how you can know the figures will be similar without graphing.

Determine if each series of transformations forms a similar figure or congruent figure.

8. Translation, Reflection, Reflection

9. Reflection, Rotation, Dilation (scale factor of 3)

10. Rotation, Rotation, Translation

11. Dilation (scale factor of 0.5), Reflection, Translation

12. Write a series of transformations that includes a reflection followed by a translation to map A(−1, 3) onto A'(4, 5).

13. Write a series of transformations that includes a translation followed by a dilation to map B(0, 0) onto B'(4, −6).

14. What dilation scale factor can be used to create an image that is congruent to its pre-image?

15. Graph a triangle with any three points in the third quadrant (all coordinates should have negative values). Write a translation rule followed by a reflection and graph your new image. Are the figures similar or congruent? Explain your reasoning.

16. Larry graphed a figure in the first quadrant of a coordinate plane. He performed two transformations. The resulting figure was twice as large as the original figure and all the x-values were now negative. Write a possible series of transformations that he might have performed.

Show that each set of figures is either similar or congruent by writing a series of transformation rules that maps the blue figure onto the green figure.

17.

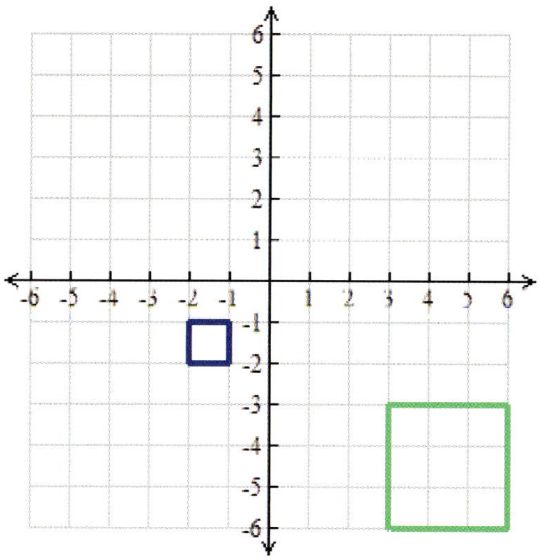

18.

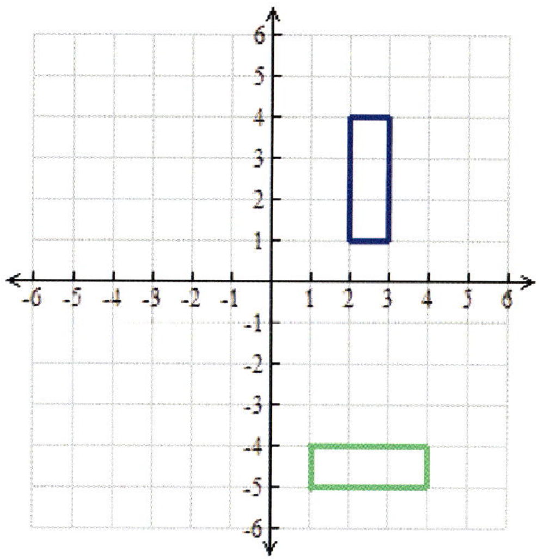

19.

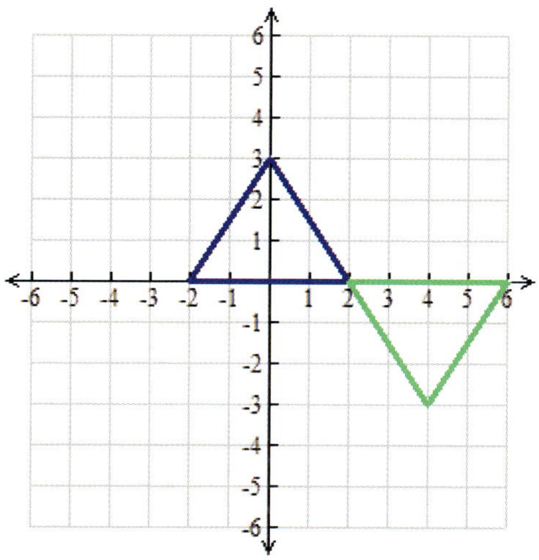

20.

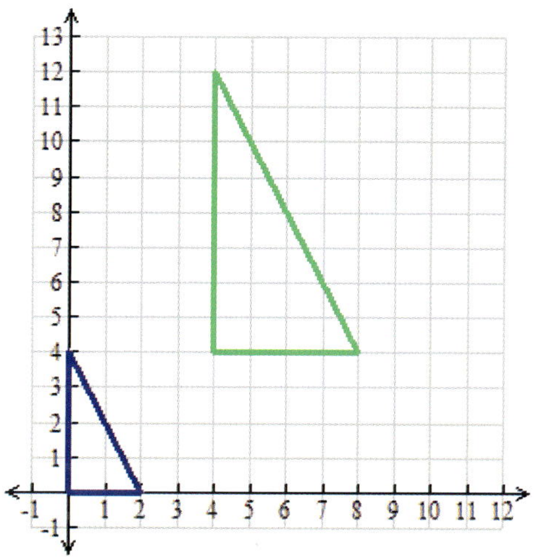

21. Graph two similar triangles on a coordinate plane. Label them and record the ordered pairs for each vertex. Write a series of three transformations that would map the original figure onto the resulting figure.

22. Is there always just one unique series of transformations that can be written to map a figure onto another? Explain your reasoning.

23. N(2, 4) was mapped onto N'(−3, −5). Evan said this was a 180° rotation followed by $(x, y) \rightarrow (x + 1, y - 1)$. Vicky said this was a reflection over the y-axis followed by $(x, y) \rightarrow (x - 1, y - 9)$. Who is correct? Support your answer with evidence.

24. Write two different series of at least two different transformations that move point A(1, 2) to B(−3, 4).

REVIEW

Write the transformation rule for each transformation.

25. A dilation with a scale factor of 7

26. A reflection over the *y*-axis

27. A translation right 1 unit

28. A translation left 3 units and down 2 units

29. A 270° rotation clockwise about the origin

30. A reflection over the *x*-axis

31. A line segment had endpoints A(4, −5) and P(3, −1). It was transformed and the new endpoints were A'(4, 5) and P'(3, 1). Write a transformation rule that created A'P'.

Tic-Tac-Toe ~ Making it Large

Step 1: Find a rectangular picture, drawing or panel from a cartoon that has a length less than or equal to 6 centimeters and a height less than or equal to 6 centimeters. The length and width do not need to be equal.

Step 2: Use a ruler to draw straight lines around the outside of the rectangle.

Step 3: Measure the length and separate the length into four equally spaced columns.

Step 4: Measure the height and separate the height into four equally spaced rows.

Step 5: On a blank 8.5 inch by 11 inch piece of paper, you will redraw your picture, drawing or cartoon panel. Determine how many times larger you can make your picture so it maximizes the space on the blank paper. Use centimeters for your measurements. Record this scale factor at the top of your paper.

Step 6: Multiply each length and width by the scale factor and redraw the enlarged border and grid lines from **Steps 2–4** on the blank piece of paper.

Step 7: Look at each rectangle in the grid separately. Draw an enlargement of that rectangle in the corresponding location on the blank paper.

Step 8: Continue until all rectangles have been enlarged. You have made a scale drawing which is like a dilation. Tape your picture to the bottom left corner of your drawing which is like the origin on a graph.

Tic-Tac-Toe ~ Escher Art

M.C. Escher was a Dutch artist known for his tessellation-inspired pieces of art. Research how to create an Escher tessellation. Create your own Escher-esque piece of art following the guidelines below:

- Start with a plain white piece of poster paper at least 18" by 18".
- Your tessellated image should be small enough that at least 12 of them fit on the poster paper.
- Use at least 2 different colors in your artwork.

Write a half-page biography of M.C. Escher. Also include printed images of two pieces of his tessellation artwork.

REVIEW

BLOCK 3

Vocabulary

composition of transformations
dilation
enlargement
image
pre-image
reduction
reflection
rotation
scale factor
transformation
translation

Reflect an image on a coordinate plane.
Translate an image on a coordinate plane.
Rotate an image on a coordinate plane.
Dilate an image on a coordinate plane.
Understand the relationship between the pre-image and the image of a transformation.
Perform multiple transformations on a figure including a combination of reflections, rotations, translations and dilations.

Lesson 3.1 ~ Reflections

1. The point (1, –6) is reflected over the *y*-axis. What are the coordinates of the new point?

2. The point (2, 3) is reflected over the *x*-axis. What are the coordinates of the new point?

3. ΔKBM has vertices at K(–4, 2), B(0, 6) and M(–2, 3).
- **a.** Graph ΔKBM. This is the pre-image.
- **b.** Graph ΔK'B'M' after a reflection over the *x*-axis. List the coordinates of the vertices.

4. A pre-image has coordinates A(3, –1), B(5, 2) and C(4, –1). The image has coordinates A'(–3, –1), B'(–5, 2) and C'(–4, –1). What type of reflection occurred? Explain how you know.

5. The transformation rule $(x, y) \rightarrow (x, -y)$ represents a reflection over which axis?

Graph the pre-image and the image under the given reflection. Label the vertices correctly.

6. Triangle P(–1, 4), I(1, 6) and E(4, 3) — A reflection over the *y*-axis

7. Rectangle W(2, –3), A(4, –3), R(4, 0) and M(2, 0) — A reflection over the *x*-axis

Lesson 3.2 ~ Translations

Determine the coordinates of the image of P(6, –5) under the following translations.

8. Translated up 4 units

9. $(x, y) \rightarrow (x - 1, y + 1)$

10. $(x, y) \rightarrow (x - 3, y + 10)$

11. Translated right 2 units and down 3 units

12. A pre-image has coordinates F(–3, 1), I(1, 4) and G(2, 1). The image has coordinates F'(0, 0), I'(4, 3) and G'(5, 0). Write a transformation rule to describe the path the figure took to arrive at the image.

Graph the pre-image and the image under the given translation. Label the vertices correctly.

13. Triangle D(0, 1), A(1, 7) and P(2, 0) $(x, y) \rightarrow (x + 3, y - 4)$

14. Rectangle R(–3, –5), I(–3, –1), C(0, –1) and E(0, –5) $(x, y) \rightarrow (x - 3, y + 7)$

Lesson 3.3 ~ Rotations

Determine the coordinates of the image of Q(–3, 8) under the following rotations.

15. Rotation 270° counterclockwise about the origin.

16. $(x, y) \rightarrow (y, -x)$

17. The point (5, 2) is rotated to (–2, 5). What degree was this clockwise rotation about the origin?

18. ΔLAM has vertices at L(–1, 1), A(2, 4) and M(4, 1).
- **a.** Graph ΔLAM. This is the pre-image.
- **b.** ΔLAM is rotated 180° about the origin. Graph ΔL'A'M'.

19. A pre-image has coordinates B(–4, 2) and W(–5, 6). The image has coordinates B'(2, 4) and W'(6, 5). Describe the <u>clockwise</u> rotational path taken by the figure.

Graph the pre-image and the image under the given rotation. Label the vertices correctly.

20. Triangle A(1, 1), B(1, 4) and C(5, 1) $(x, y) \rightarrow (y, -x)$

21. Rectangle M(2, –3), I(2, –1), N(5, –1) and D(5, –3) 270° clockwise about the origin

22. What are the coordinates of the point (x, y) after it is rotated 90° counterclockwise about the origin?

Lesson 3.4 ~ Dilations

Give the scale factor for each dilation centered at the origin. Determine if the scale factor creates an enlargement or a reduction in size compared to the pre-image.

23. $(x, y) \rightarrow (0.3x, 0.3y)$

24. $(x, y) \rightarrow (7x, 7y)$

25. $(x, y) \rightarrow \left(\frac{4}{3}x, \frac{4}{3}y\right)$

26. $(x, y) \rightarrow \left(\frac{1}{2}x, \frac{1}{2}y\right)$

27. A pre-image has coordinates Y(−3, 4), A(−4, 5) and B(−3, 2). The image has coordinates Y'(−12, 16), A'(−16, 20) and B'(−12, 8). Write the transformation rule used to create this image.

Graph the pre-image and the image under the given dilation centered at the origin. Label the vertices correctly.

28. Triangle P(−4, −1), I(−2, 0) and T(0, −1) $(x, y) \rightarrow (2x, 2y)$

29. Triangle B(2, 6), A(6, 2) and X(4, 5) $(x, y) \rightarrow (0.5x, 0.5y)$

Lesson 3.5 ~ Transformations and Congruence

Write the transformation rule for each transformation. State whether the transformation creates similar or congruent figures.

30. A reflection over the x-axis

31. A translation right 2 units and up 5 units

32. A dilation with a scale factor of 0.7

33. A 180° rotation about the origin

Name the type of transformation given by each rule. State whether the transformation creates similar or congruent figures.

34. $(x, y) \rightarrow (x, y + 6)$

35. $(x, y) \rightarrow (-y, x)$

36. $(x, y) \rightarrow (-x, y)$

37. $(x, y) \rightarrow (4x, 4y)$

38. Which type of transformation rule does not always create a congruent figure?

39. Write a transformation rule that creates an image congruent to its pre-image.

Use the point V(4, –1) as the pre-image in Exercises 40–44. Write the ordered pair for the final location of the given point after completing the series of transformations in the order listed.

40. $(x, y) \rightarrow (x - 1, y + 3)$, reflection over the x-axis

41. A 180° rotation about the origin, $(x, y) \rightarrow (-x, y)$

42. $(x, y) \rightarrow (x + 3, y - 7)$, $(x, y) \rightarrow (x - 2, y + 4)$, $(x, y) \rightarrow (2x, 2y)$

43. Reflection over the y-axis, reflection over the x-axis, $(x, y) \rightarrow (x, y + 6)$

44. Which composition of transformations in **Exercises 40–44** form figures that are similar to their original figure? Explain how you know.

Determine if each series of transformations forms a similar figure or congruent figure.

45. Translation, Rotation, Reflection

46. Reflection, Reflection, Translation

47. Reflection, Dilation (scale factor of 1.75), Translation

48. Write a series of two transformations that includes a reflection followed by a translation to map A(5, 2) onto A'(–2, 1).

Show that each set of figures is either similar or congruent by writing a series of transformation rules that maps the blue figure onto the red figure.

49.

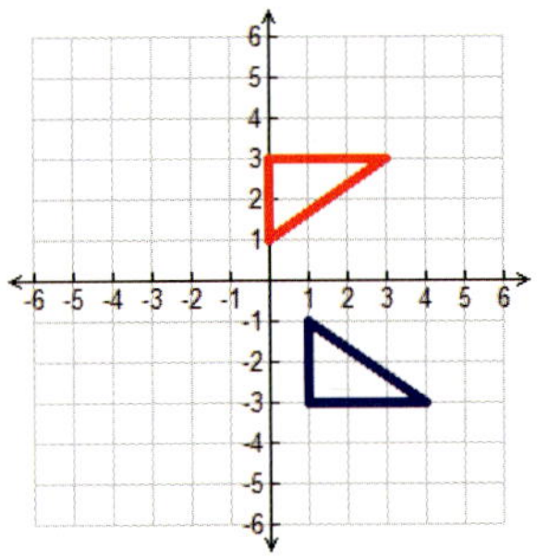

50.

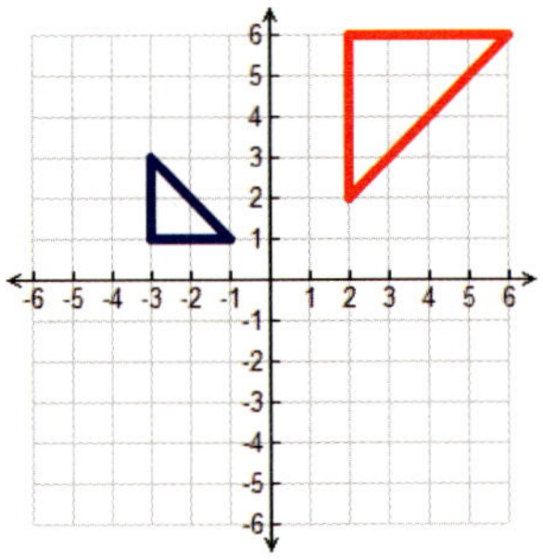

Tic-Tac-Toe ~ From Here to There

The point A(–4, 5) moved under a series of transformations to A'(2, –1). Write a series of transformation rules that would map A to A' for each of the following descriptions:

1. Three transformations: A reflection followed by a rotation and then a translation.
2. Four transformations: A rotation, dilation, reflection and translation in any order.
3. Three transformations: A translation followed by two reflections.
4. Three transformations: A translation followed by two rotations.
5. Three transformations: A rotation, dilation and reflection in any order.

CAREER FOCUS

Chris
Chemist

I am an analytical chemist. I test environmental samples such as drinking water or ground soils and look for pollutants and toxins that might be in each. In my job I use a combination of tools, from bench chemistry like distilling and extracting, to high tech instruments that help me determine what exactly is in each sample. There are many different chemicals that can be present in any sample. It is my job to figure out which chemical, and how much of the chemical, there is in the sample.

I am constantly using math in my profession. I use ratios to make sure that I have the right quantities of chemicals in my solutions. There is also a lot of Algebra that I use on a daily basis. Different formulas help me to calculate the right amounts and tell me how much error there might be in my instruments. I use the standard curve and statistics to make predictions and estimates about what may be in each sample. It is very important that my calculations are precise, so having a strong knowledge of math is very necessary.

In order to become a chemist, you must have a Bachelor's degree in some sort of science field. I have a degree in environmental science, but other chemists get degrees in different areas. The starting salary for a chemist is anywhere from $25,000-$40,000 per year, but the salary can rise to as high as $100,000 per year with experience.

I like my job as a chemist. I enjoy working with high tech instrumentation and learning how to operate and maintain it. I also like doing something that is a direct help to society, ensuring the safety of our drinking water and taking part in helping to keep our environment clean.

CORE FOCUS ON GEOMETRY

BLOCK 4 ~ EXPONENTS AND VOLUME

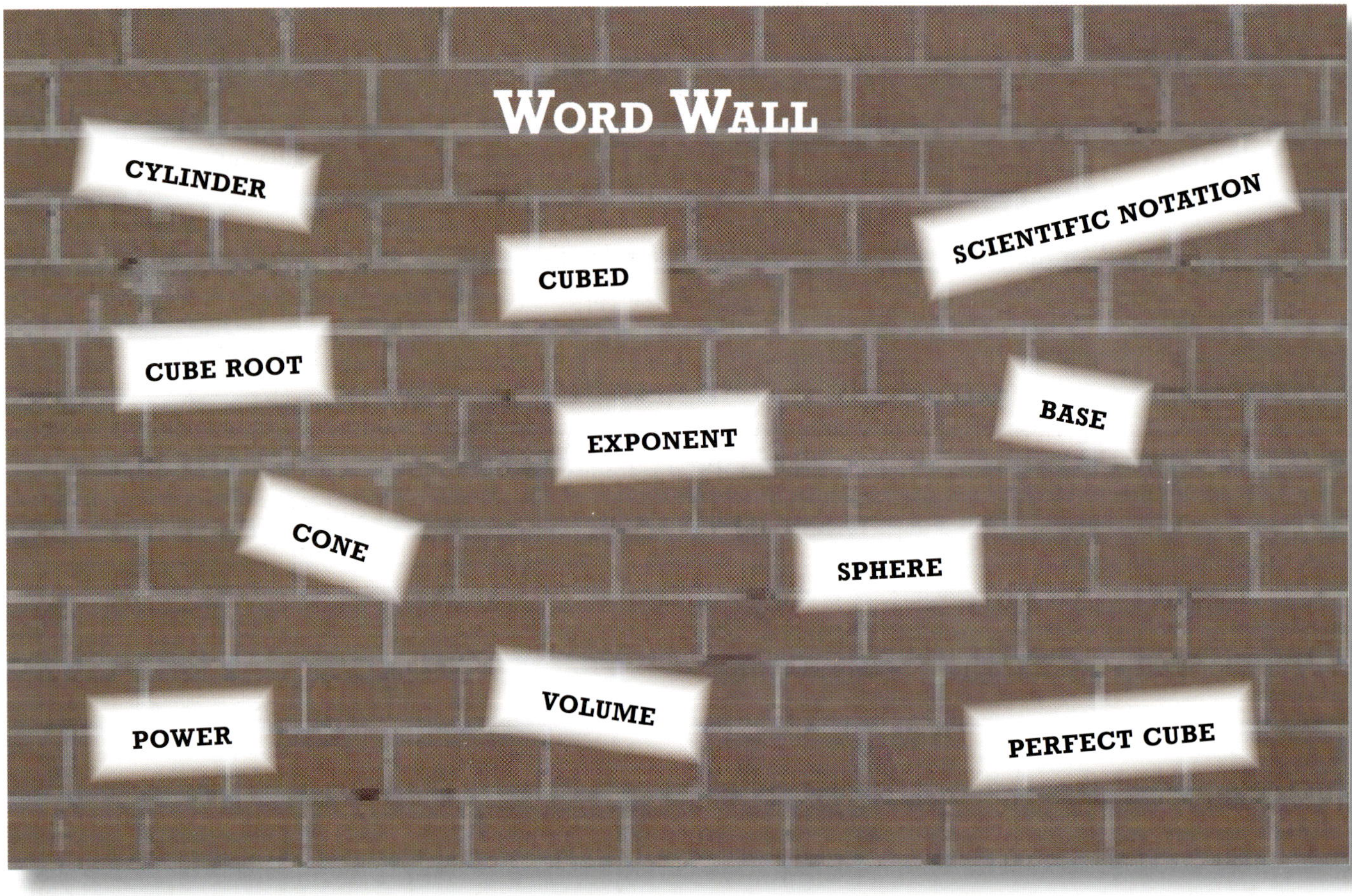

BLOCK 4 ~ EXPONENTS AND VOLUME

TIC-TAC-TOE

The Science of Large and Small

Research large and small numbers in science. Write values in scientific notation.

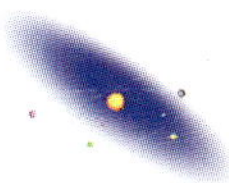

See page 138 for details.

The Solar System

Find the surface area of the planets in the solar system.

See page 128 for details.

Simplifying Solutions

Solve equations requiring square roots. Write answers in simplified radical form.

See page 148 for details.

Measure Of Money

Find the circumference, surface area and volume of various coins.

See page 153 for details.

Matching Game

Create a matching game using equivalent expressions that apply the exponent properties.

See page 133 for details.

Teaching Volume

Design an activity kit that teaches the concept of volume to an elementary class.

See page 158 for details.

Displacement

Find the volume of objects using displacement.

See page 159 for details.

Comparing Country Size

Find the sizes (in square miles) of different countries and compare their sizes.

See page 147 for details.

Field Test

Make three cylinders with the same volume. Find out which cylinder is more popular.

See page 153 for details.

MULTIPLICATION PROPERTIES OF EXPONENTS

LESSON 4.1

Simplify expressions involving multiplication using properties of exponents.

When a numerical expression is the product of a repeated factor, it can be written using a **power**. A power consists of two parts, the **base** and the **exponent**. The base of the power is the repeated factor. The exponent shows the number of times the factor is repeated.

POWERS, BASES AND EXPONENTS

exponent — expanded form

base → $3^4 = 3 \times 3 \times 3 \times 3$

power

It is important to know how to read powers correctly.

Power	Reading the Expression	Expanded Form	Value
5^2	"five to the second power" or "five squared"	5×5	25
6^3	"six to the third power" or "six **cubed**"	$6 \times 6 \times 6$	216
2^4	"two to the fourth power"	$2 \times 2 \times 2 \times 2$	16

EXPLORE! — EXPAND IT

Use expanded form to discover two different exponent multiplication properties.

Step 1: Write each of the following products in expanded form.

a. $5^3 \cdot 5^4$ **b.** $4^2 \cdot 4^8$ **c.** $x^3 \cdot x^2$

Step 2: Rewrite each of the products in **Step 1** as a single term with one base and one exponent.

Step 3: What relationship do you see between the original bases and the single term's base? What about the original exponents and the single term's exponent?

Step 4: Based on your findings, write a statement explaining how to find the product of two powers with the same base WITHOUT writing the terms in expanded form.

EXPLORE! **(CONTINUED)**

Step 5: Write each of the following powers in expanded form. Then rewrite the power as a single term. The first one is done for you.

a. $(3^2)^4 \rightarrow (3^2)(3^2)(3^2)(3^2) \rightarrow \underbrace{3 \cdot 3}\cdot\underbrace{3 \cdot 3}\cdot\underbrace{3 \cdot 3}\cdot\underbrace{3 \cdot 3} \rightarrow 3^8$

b. $(7^3)^2$

c. $(x^3)^5$

Step 6: What is the relationship between the final exponent and the power to a power? Based on your findings, write a statement explaining how to find the power of a power WITHOUT expanding the power.

You can also simplify a power of a product. Look at $(df)^3$.

Written in expanded form: $(df)(df)(df)$

Group like variables together: $(d \cdot d \cdot d)(f \cdot f \cdot f)$

Simplify: d^3f^3

MULTIPLICATION PROPERTIES OF EXPONENTS

Product of Powers

To multiply two powers with the same base, add the exponents.

$$a^m \cdot a^n = a^{m+n}$$

Power of a Power

To find the power of a power, multiply the exponents.

$$(a^m)^n = a^{mn}$$

Power of a Product

To find the power of a product, find the power of each factor and multiply.

$$(ab)^n = a^n b^n$$

EXAMPLE 1 **Simplify each of the following.**

a. $y^3x^2y^6x$ **b.** $(b^3w^2)^4$ **c.** $(5p^4)(2p^3)$

When a term has no exponent written, the exponent is 1.

SOLUTIONS

a. Group like variables together. $y^3x^2y^6x = y^3y^6 \cdot x^2x$

Add exponents with the same base. $y^{3+6}x^{2+1} = y^9x^3$

b. Distribute the exponent to each base. $(b^3w^2)^4 = (b^3)^4(w^2)^4$

Multiply exponents. $(b^3)^4(w^2)^4 = b^{12}w^8$

c. Group like terms together. $(5p^4)(2p^3) = (5 \cdot 2)(p^4p^3)$

Multiply coefficients. Add exponents with the same base. $(5 \cdot 2)(p^4p^3) = 10p^7$

A simplified expression should have:

- each base appear exactly once,
- no powers to powers,
- no numeric values with powers, and
- fractions written in simplest form.

EXAMPLE 2 **Simplify each of the following.**

a. $6x^2y^4z^3 \cdot 3x^5z^2$ **b.** $(4m^3w)^2(5m^2w^2)^3$

SOLUTIONS

a. Rewrite the expression. $6x^2y^4z^3 \cdot 3x^5z^2$

Group like terms together. $6 \cdot 3 \cdot x^2 \cdot x^5 \cdot y^4 \cdot z^3 \cdot z^2$

Multiply coefficients. Add exponents with the same base. $18x^7y^4z^5$

b. Rewrite the expression. $(4m^3w)^2(5m^2w^2)^3$

Distribute the exponent to each base. $4^2(m^3)^2(w)^2 \cdot 5^3(m^2)^3(w^2)^3$

Find the values of the coefficients with exponents. $16(m^3)^2(w)^2 \cdot 125(m^2)^3(w^2)^3$

Multiply exponents. $16m^6w^2 \cdot 125m^6w^6$

Multiply coefficients. Add exponents with the same base. $2000m^{12}w^8$

EXERCISES

Simplify.

1. x^3x^2

2. $(y^5)^2$

3. $(pq)^5$

4. $(4x^5)^3$

5. $(w^2y^4z^6)(w^5y^3z)$

6. $(2a^6b)(3a^3b^3)$

7. $(5gh^2)^2$

8. $(9x^4y^5)(-2x^2y^7)$

9. $(0.5f^2d^9)^3$

10. A farmer wants to fence in a square pasture for his sheep. The length of one side of the pasture is represented by kp^3. What term represents the area of the pasture?

11. Write and solve a problem that requires you to multiply exponents.

12. Write and solve a problem that requires you to add exponents.

13. Jake, Joe and Jenny attempted to rewrite $(5^2)(4^3)$ as a single power. Who got the correct answer? How do you know?

Jake	Joe	Jenny
$(5^2)(4^3) = 20^5$	$(5^2)(4^3)$ cannot be written as a single power.	$(5^2)(4^3) = 9^5$

Simplify.

14. $(3x^2)^3(4x^5)^2$

15. $(-4y^5w^2)^2(2y^4)^3$

16. $(5p^3)(5p^2)^3$

17. $(2x^2)^3(3x^4)^2(-2x^3)^3$

18. $(-2y^2)(-3xy^4)^2(5x^6)^2$

19. $(4a^2b)^3(5a^4b^5)^2$

Write the volume of each rectangular prism as a single term.

20.

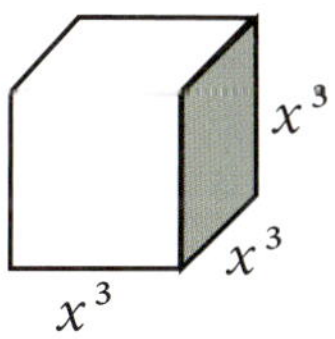

21.

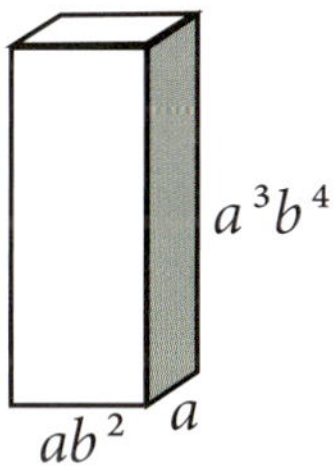

22.

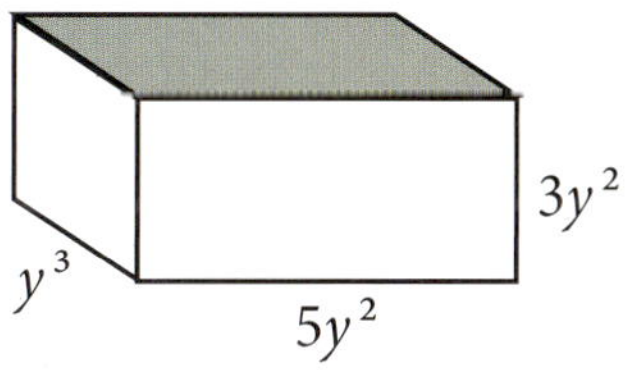

23. The length of a rectangle is five times its width. If its width is x^3 units, what is the area of the rectangle?

24. The width of a rectangular prism is four times its length. The height is three times its length. If its length is m units, what is the volume of the prism?

REVIEW

Graph the pre-image and the image under the given transformation. Label the vertices correctly.

25. Triangle S(4, 5), W(2, 4) and R(3, 1) $(x, y) \rightarrow (y, -x)$

26. Rectangle N(3, −3), I(−3, −3), C(−3, −1) and E(3, −1) Reflection over the x-axis

27. Triangle P(3, 1), A(6, −2) and L(4, 2) $(x, y) \rightarrow (x - 2, y + 5)$

28. Jill moved a figure 3 units down and 6 units right. She predicted that the resulting figure would be congruent to her original shape. Do you agree or disagree? Explain your answer.

29. What type of transformation creates a figure similar to its pre-image, but not congruent?

30. A figure is reflected over the x-axis and then reflected over the y-axis. Finally, the figure is dilated with a scale factor of 1.5. Is the image similar or congruent to the original figure? Explain your answer.

Tic-Tac-Toe ~ The Solar System

There are five basic solids: prisms, cylinders, pyramids, cones and spheres. The surface area of a sphere is not the sum of the lateral area and the base area. The formula is given below.

Surface Area of a Sphere

The surface area of a sphere is the product of 4, pi and the radius squared.

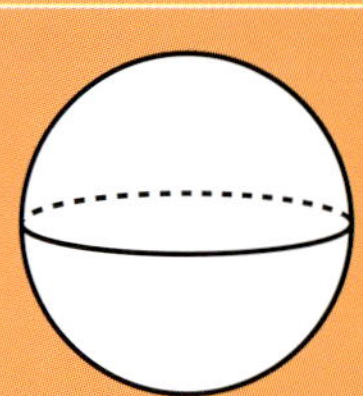

Surface Area = $4\pi r^2$

1. Use resources to find the radius of the eight planets in our solar system.

2. Assume each planet is a perfect sphere. Calculate the surface area of each planet in square miles. Use 3.14 for π.

3. Create a booklet showing the eight planets and their surface areas. Include at least one other interesting fact about each planet.

DIVISION PROPERTIES OF EXPONENTS

LESSON 4.2

Use properties of exponents to simplify expressions involving division.

You discovered in **Lesson 4.1** that when two powers with the same base are multiplied, the base remains the same and the exponents are added together. Examine the division problems below.

$$\frac{5^7}{5^3} = \frac{\cancel{5} \cdot \cancel{5} \cdot \cancel{5} \cdot 5 \cdot 5 \cdot 5 \cdot 5}{\cancel{5} \cdot \cancel{5} \cdot \cancel{5}} = 5^4 \qquad \frac{x^6}{x^4} = \frac{\cancel{x} \cdot \cancel{x} \cdot \cancel{x} \cdot \cancel{x} \cdot x \cdot x}{\cancel{x} \cdot \cancel{x} \cdot \cancel{x} \cdot \cancel{x}} = x^2$$

Simplify the fractions by cancelling common factors.

What is the relationship between the original exponents and the resulting exponent?

$$\frac{5^7}{5^3} = 5^4 \qquad \frac{x^6}{x^4} = x^2$$

When two powers with the same base are divided, the base remains the same and the exponents are subtracted.

You can also simplify a power of a quotient or fraction by "distributing" the power to both the numerator and denominator.

$$\left(\frac{3}{4}\right)^3 = \frac{3}{4} \cdot \frac{3}{4} \cdot \frac{3}{4} = \frac{3^3}{4^3} = \frac{27}{64}$$

Division Properties of Exponents

Quotient of Powers

To divide two powers with the same base, subtract the exponents.

$$\frac{a^m}{a^n} = a^{m-n}$$

Power of a Quotient

To find the power of a quotient, find the power of the numerator and denominator.

$$\left(\frac{a}{b}\right)^n = \frac{a^n}{b^n}$$

EXAMPLE 1 **Simplify each of the following.**

a. $\dfrac{x^5y^8}{x^2y}$ b. $\left(\dfrac{3m^4}{5}\right)^2$

SOLUTIONS

a. Group powers that have the same base. $\left(\dfrac{x^5}{x^2}\right)\left(\dfrac{y^8}{y}\right)$

Subtract exponents. $x^{5-2}y^{8-1} = x^3y^7$

b. Distribute the power to each base. $\left(\dfrac{3m^4}{5}\right)^2 = \dfrac{3^2 \cdot (m^4)^2}{5^2}$

Multiply powers of powers and evaluate coefficients. $\dfrac{9m^8}{25}$

EXPLORE! **GOING NEGATIVE**

Step 1: Copy the tables below. Find the value of each power with a calculator. If the value of the power is less than 1, write the power as a fraction.

Power	Value
2^4	
2^3	
2^2	
2^1	

Power	Value
2^{-4}	
2^{-3}	
2^{-2}	
2^{-1}	

Step 2: What do you notice about the powers with opposite exponents (i.e. 2^2 and 2^{-2})?

Step 3: Use your observation from **Step 2** to predict the value of each power below.

a. Given that $4^2 = 16$, what is the value of 4^{-2}?

b. Given that $3^5 = 243$, what is the value of 3^{-5}?

c. Given that $6^{-3} = \dfrac{1}{216}$, what is the value of 6^3?

Step 4: Look at the statements below. What is the value of each expression (written without an exponent)?

a. $\dfrac{5^2}{5^2} = \dfrac{25}{25} = ?$ b. $\dfrac{2^3}{2^3} = ?$ c. $\dfrac{3^4}{3^4} = ?$

Step 5: Notice that $\frac{5^2}{5^2} = 5^{2-2} = 5^0$, $\frac{2^3}{2^3} = 2^0$ and $\frac{3^4}{3^4} = 3^0$ using the Division Property of Exponents.

Based on your findings in **Step 4**, what is the value of 5^0, 2^0 and 3^0?

Step 6: Use your calculator to raise other numbers to a power of 0. Try whole numbers and decimal values. What can you conclude?

More Properties of Exponents

Negative Exponents

For any nonzero number a and integer n, the expression a^{-n} is the reciprocal of a^n. Also, a^n is the reciprocal of a^{-n}.

$$a^{-n} = \frac{1}{a^n} \text{ and } a^n = \frac{1}{a^{-n}}$$

Zero Exponents

Any nonzero number, a, with an exponent of 0 is 1.

$$a^0 = 1$$

An expression is in simplest form only if it has no negative or zero exponents. As shown in the blue box above, an exponent with its base can be moved to the opposite side of the fraction bar to change its sign.

EXAMPLE 2

Simplify each of the following.

a. $\left(\frac{4p^5k}{3}\right)^0$ **b.** $\frac{x^2y^{-4}}{z^{-3}}$ **c.** $\frac{6a^2b^{-6}c^8}{-2a^2b^{-5}c^{-1}}$

Solutions

a. Any term to the power of 0 equals 1. $\left(\frac{4p^5k}{3}\right)^0 = 1$

b. Write as separate factors. $\left(\frac{x^2y^{-4}}{z^{-3}}\right) = \left(\frac{x^2}{1}\right)\left(\frac{y^{-4}}{1}\right)\left(\frac{1}{z^{-3}}\right)$

Use the rules for negative exponents. $\left(\frac{x^2}{1}\right)\left(\frac{1}{y^4}\right)\left(\frac{z^3}{1}\right)$

Negative exponents become positive when a power is moved to the opposite side of the fraction bar.

Multiply factors. $\frac{x^2z^3}{y^4}$

c. Write as separate factors. Group like bases. $\frac{6a^2b^{-6}c^8}{-2a^2b^{-5}c^{-1}} = \left(\frac{6}{-2}\right)\left(\frac{a^2}{a^2}\right)\left(\frac{b^{-6}}{b^{-5}}\right)\left(\frac{c^8}{c^{-1}}\right)$

Subtract exponents on like bases. $\left(\frac{6}{-2}\right)(a^{2-2})(b^{-6-(-5)})(c^{8-(-1)})$

Simplify. $(-3)(a^0)(b^{-1})(c^9)$

Use rules for zero and negative exponents. $(-3)(1)\left(\frac{1}{b}\right)c^9$

Multiply. $\frac{-3c^9}{b}$

EXERCISES

Simplify.

1. $\dfrac{8^{12}}{8^5}$

2. $\dfrac{x^5}{x^2}$

3. $\dfrac{a^6b^9}{a^3b^5}$

4. $\dfrac{2w^5v^4}{10wv^2}$

5. $\left(\dfrac{d^2}{g^3}\right)^5$

6. $\left(\dfrac{2y^3}{3}\right)^3$

7. $(4yh^2)^0$

8. $\left(\dfrac{5x^{11}}{3w^{-4}}\right)^0$

9. 5^{-2}

10. 2^{-4}

11. $\dfrac{k^{-3}m^2}{n^{-7}}$

12. $7p^{-2}q^{-5}$

13. A contractor plans to build a rectangular office complex. The area of the first floor of the complex is represented by $15x^5y^9$. The length of the complex is represented by $3x^2y^5$. What expression represents the width?

14. Write and solve a problem that requires you to subtract exponents.

15. A truckload of cement weighs approximately 4^7 pounds. The driver of the truck weighs 4^4 pounds. The truckload of cement weighs how many times more than the driver?

Simplify.

16. $\dfrac{-3m^5}{m^{11}}$

17. $\dfrac{24x^7y^{-4}}{4x^{-3}y^2}$

18. $\dfrac{10p^2w^6}{6p^{-2}w^6}$

19. $\left(\dfrac{r^{-2}t^0}{n^{-5}}\right)^3$

20. $\left(\dfrac{6y^2}{z^{-3}}\right)^2$

21. $\left(\dfrac{2q^{-2}w^3}{3x^{-4}y^5}\right)^{-1}$

Find the missing dimension when given the volume. Write the missing dimension in simplest form.

22. Volume $= 10x^4y^{-5}z$

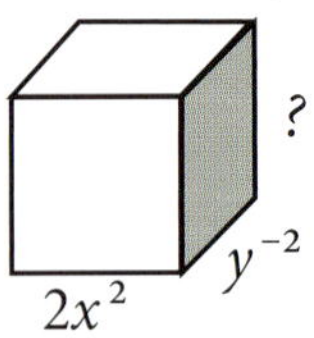

23. Volume $= 14a^{-3}m^8p^{-1}$

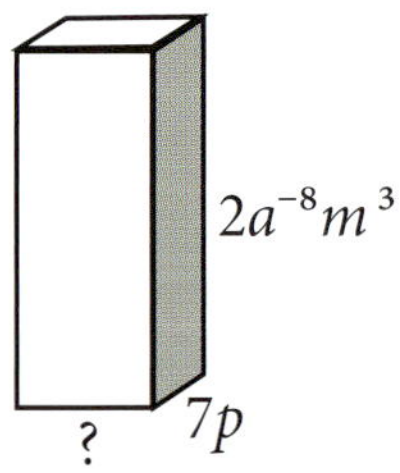

REVIEW

Simplify.

24. p^3p^9

25. $(x^5)^3$

26. $(2cd)^3$

27. $(2x^2)^2(3x^2)^3$

28. $(5y^3)^2(y^4)^3$

29. $(\frac{1}{2}m)^2(2m^2)^3$

30. The length of a rectangle is three times its width. If its width is $2x^2$ units, what is the area of the rectangle?

31. The width of a rectangular prism is twice its length. The height is three times its length. If its length is p units, what is the volume of the prism?

TIC-TAC-TOE ~ MATCHING GAME

Write 15 non-simplified variable expressions that contain exponents. Make sure you have a minimum of five expressions containing fraction bars. Cut thicker paper (such as cardstock, construction paper, index cards or poster board) into 30 equal-sized pieces. Write each expression on one card. Write the expression in simplest form on another card. Use these cards to play a memory game with a friend, classmate or family member. Record each pair of cards that each participant wins on a sheet of paper by listing the non-simplified and simplified expressions. Give the cards and the game sheet to your teacher.

SCIENTIFIC NOTATION

LESSON 4.3

Express numbers in scientific notation and standard notation.

The table below shows the approximate distance to the sun from each planet in our solar system.

Planet	Approximate Distance to the Sun (in miles)	Distance Written in Scientific Notation
Mercury	36,000,000	3.6×10^7
Venus	67,000,000	6.7×10^7
Earth	93,000,000	9.3×10^7
Mars	142,000,000	1.42×10^8
Jupiter	484,000,000	4.84×10^8
Saturn	888,000,000	8.88×10^8
Uranus	1,784,000,000	1.784×10^9
Neptune	2,799,000,000	2.799×10^9

The approximate distances are written in both standard notation and scientific notation. It is useful to write numbers in **scientific notation** when they are very large or very small. Examine the numbers in scientific notation. How would you describe a number in scientific notation?

Scientific notation uses powers of 10 since numbers are in a base-ten system. Each time you multiply a decimal value by 10, you move the decimal point one place to the right. Each time you divide by 10, the decimal point moves one place to the left. Look at the difference between 4.5×10^4 and 4.5×10^{-4}.

Multiplying by Tens

4.5
$\times 10$
45
$\times 10$
450
$\times 10$
4500
$\times 10$
45000

$45000. = 4.5 \times 10^4$

Dividing by Tens

4.5
$\div 10$
0.45
$\div 10$
0.045
$\div 10$
0.0045
$\div 10$
0.00045

$0.00045 = 4.5 \times 10^{-4}$

Dividing by 10 is the same as multiplying by 10^{-1}.

When 4.5 was multiplied by 10 four times, the decimal point moved four places to the right. When 4.5 was divided by 10 four times, the decimal point moved four places to the left.

Scientific Notation

Scientific notation is an exponential expression using a power of 10 where $1 \le |N| < 10$ and P is an integer.

$$N \times 10^P$$

EXAMPLE 1 **Convert 52,000 to scientific notation.**

SOLUTION

A decimal point is at the end of every whole number. Move the decimal point to the **left** until it creates a number whose absolute value is equal to or larger than 1 and less than 10.

$52{,}000 \rightarrow 5.2$

Count how many places the decimal point was moved in the number above.

52000. — The decimal point was moved 4 places to the left.

The decimal point was moved 4 places to the **left**. Since the decimal point was moved left, the P value is positive.

$52{,}000 = 5.2 \times 10^4$.

In **Example 1**, 5.2 was less than the original number (52,000) so the number was multiplied by 10^4 to make it larger.

EXAMPLE 2 **Convert −0.00492 to scientific notation.**

SOLUTION

Move the decimal point to the **right** until it creates an absolute value equal to or larger than 1 and less than 10.

$-0.00492 \rightarrow -4.92$ — $1 \le |-4.92| < 10$

Count how many places the decimal point was moved in the number above.

−0.00492 — The decimal point was moved 3 places to the right.

The decimal point was moved 3 places to the **right**. Since the decimal point was moved right, the P value is negative.

$-0.00492 = -4.92 \times 10^{-3}$.

There are times when numbers are written in scientific notation and you want to convert them to standard notation. You must expand the number by multiplying the decimal by the given power of ten. The exponent on the 10 gives the number of digits the decimal point is moved to the right or left.

EXAMPLE 3 **Write each of the following numbers in standard notation.**
a. 3.8×10^5 **b.** 6.12×10^{-3}

SOLUTIONS

a. METHOD 1 – Moving the Decimal Point
Move the decimal five places to the right. 3.8‿‿‿‿. →

Write the answer in standard notation. $3.8 \times 10^5 = 380{,}000$

METHOD 1 – Multiplying by a Power of Ten
Change the power of 10 to standard notation. $10^5 = 100{,}000$

Multiply the decimal value by the power of 10. $3.8(100000) = 380{,}000$

b. METHOD 1 – Moving the Decimal Point
Move the decimal three places to the left. .‿‿‿6.12 ←

Write the answer in standard notation. $6.12 \times 10^{-3} = 0.00612$

METHOD 1 – Multiplying by a Power of Ten
Change the power of 10 to standard notation. $10^{-3} = 0.001$

Multiply the decimal value by the power of 10. $6.12(0.001) = 0.00612$

You may need to compare numbers in scientific and standard notation. Convert each number to standard notation and then compare the numbers as asked.

EXAMPLE 4 **Four truck drivers kept track of their mileage for the year. The chart below shows the number of miles each one drove. List the drivers in order from the least mileage driven to the greatest mileage driven.**

Name	Mileage
Sam	5.41×10^4
Tom	3×10^4
Pete	25,000
Juan	1.1×10^5

SOLUTION Convert each number to standard notation.

Name	Mileage	Standard Notation
Sam	5.41×10^4	54,100
Tom	3×10^4	30,000
Pete	25,000	25,000
Juan	1.1×10^5	110,000

List the drivers from least mileage to greatest mileage. Pete, Tom, Sam, Juan

EXERCISES

Write each number in scientific notation.

1. 0.000058

2. 9,700

3. 2,000,000,000

4. −0.00921

5. 0.0000007

6. 81,400,000

7. 560

8. 0.001092

9. −750,000

Write each number in standard notation.

10. 2.3×10^7

11. 7.1×10^{-3}

12. -6.04×10^3

13. -8×10^{-4}

14. 9.45×10^{-2}

15. 4.317×10^5

16. According to most-expensive.net, *Spider-Man 3* cost more to produce than any other movie before it. The budget was \$258,000,000. Write this number in scientific notation.

17. Maryland's population in 2009 could be approximated by 5.7×10^6. Write this number in standard notation.

18. Jillian ate 6.1×10^5 milligrams of rice. Chris ate 4×10^6 milligrams of rice. Who ate more?

19. List the following numbers in order from least to greatest.

$0.091, 9.9 \times 10^{-3}, 0.06, 8 \times 10^{-2}$

20. The approximate heights of four Oregon mountains are listed in the table below. List the mountains in order from the shortest to the tallest.

Name	Height (feet)
South Sister	10,300
Mt. Hood	1.1×10^4
Mt. McLoughlin	9,300
Mt. Jefferson	1.02×10^4

21. The diameter of Jupiter (the largest planet) is approximately 1.4×10^5 kilometers. What is this distance in standard notation?

22. Owen weighed 3 different bugs. List the bugs in order from lightest to heaviest.

Bug	Weight (ounces)
Fly	2.5×10^{-4}
Spider	7.1×10^{-5}
Bee	4×10^{-3}

23. Jamal determined that he is approximately 1.04×10^{-3} miles tall. How many places, and in which direction, must you move the decimal to determine his height in standard notation?

24. In what types of jobs might people work with very large or very small numbers? Give examples of the values they may work with.

REVIEW

Simplify.

25. $(d^3f^2)(d^5f)$

26. $\left(\frac{p^8x^5}{p^2}\right)^0$

27. $\frac{3a^4b^2}{12a^2b^2}$

28. $(4w^2b^5)^3$

29. $\frac{5x^{-3}y^2}{y^6}$

30. $\left(\frac{3k^4}{5}\right)^2$

31. $4y^{-4}z^{-1}$

32. $\left(\frac{3t^6}{2t^{-5}}\right)^2$

33. $(4x^3y^4)(2x^3)^2$

Tic-Tac-Toe ~ The Science of Large and Small

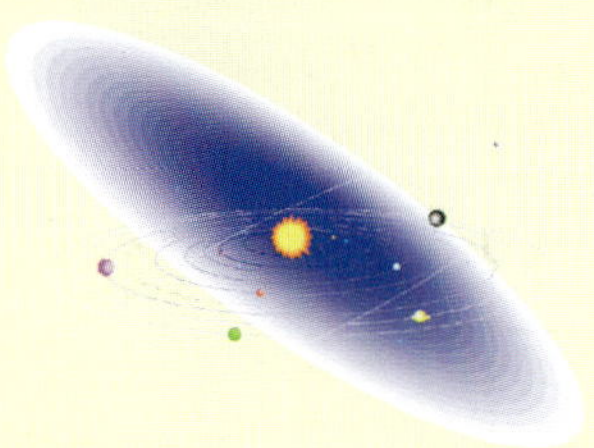

Scientific notation is commonly used by scientists, doctors, engineers and mathematicians. Scientists often encounter very large and very small numbers when they work with elements in the world around us. Scientific notation is used to represent these types of numbers.

Find ten real-world objects or situations where very large or very small numbers are used and are expressed in scientific notation. Record each number in scientific and standard notation. Describe what the number represents. Include pictures or diagrams.

APPLICATIONS OF SCIENTIFIC NOTATION

LESSON 4.4

Compute with numbers in scientific notation.

EXPLORE! **POPULATIONS**

The table below shows the estimated 2005 populations of ten countries based on the approximations of the U.N. World Population Prospects.

Country / Territory	Approximate Population July 2005
People's Republic of China	1.3×10^9
India	1.1×10^9
United States of America	3×10^8
Mexico	1.1×10^8
France	6.5×10^7
Kenya	3.4×10^7
Uzbekistan	2.6×10^7
Nicaragua	5.5×10^6
Iceland	3×10^5
Tonga	1×10^5

Step 1: Write the population of the People's Republic of China and the population of France in standard notation.

Step 2: How many times larger is the population of China than the population of France?

Step 3: Look at the equivalence statement: $\left(\frac{1.3 \times 10^9}{6.5 \times 10^7}\right) = \left(\frac{1.3}{6.5}\right) \times \left(\frac{10^9}{10^7}\right)$. Simplify each factor.

Write your answer as a decimal. — $\square \times \square$ — Remember to use the Division Property of Exponents.

Step 4: Is your answer in **Step 3** in scientific notation? If not, change it so that it is in scientific notation. How does your answer compare to your answer in **Step 2**?

Step 5: You have either: (1) converted to standard notation and then divided or (2) divided the factors while in scientific notation. Which of the two methods for dividing numbers in scientific notation do you like better. Why?

Step 6: Use either method in **Step 5** to answer the following questions. Write each answer in standard and scientific notation.

a. How many times larger is the population of India compared to Nicaragua?

b. How many times larger is the population of the USA compared to Iceland?

c. According to the US Department of Agriculture, the average American consumes 1.6×10^2 pounds of sugar each year. Approximately how many pounds of sugar are consumed in America in one year?

d. In 2005, if the land in Mexico was split evenly among the population, each person would have 1.9×10^5 square feet. Approximately how many square feet of land is there in Mexico? Write your answer in both standard notation and scientific notation.

Computing with numbers in scientific notation can be completed using different methods. One method involves converting numbers in scientific notation to standard notation and then completing the computation. Another method involves using exponent properties to simplify and compute. In **Examples 1 and 2**, you will see how a division problem and a multiplication problem can be solved using each method.

EXAMPLE 1

Find the value of $\frac{8.6 \times 10^4}{4.3 \times 10^{-3}}$. Write the answer in scientific notation and standard notation.

Solution

METHOD 1 – Working in Standard Notation

Convert both numbers to standard notation. $\frac{8.6 \times 10^4}{4.3 \times 10^{-3}} = \frac{86000}{0.0043}$

Divide. $\frac{86000}{0.0043} = 20{,}000{,}000$

Write the answer in scientific notation. $20{,}000{,}000 = 2 \times 10^7$

METHOD 2 – Using Exponent Properties

Group factors. $\frac{8.6 \times 10^4}{4.3 \times 10^{-3}} = \left(\frac{8.6}{4.3}\right)\left(\frac{10^4}{10^{-3}}\right)$

Divide each factor. Subtract exponents when dividing with like bases. $\left(\frac{8.6}{4.3}\right)\left(\frac{10^4}{10^{-3}}\right) = 2 \times 10^{4-(-3)} = 2 \times 10^7$

Write the answer in standard notation. $2 \times 10^7 = 20{,}000{,}000$

EXAMPLE 2 **Find the value of $(2.4 \times 10^5)(6 \times 10^9)$. Write the answer in scientific notation and standard notation.**

SOLUTION

METHOD 1 – Working in Standard Notation

Convert both numbers to standard notation.	$(2.4 \times 10^5)(6 \times 10^9)$ $= 240000 \times 6000000000$
Multiply.	240000×6000000000 $= 1{,}440{,}000{,}000{,}000{,}000$
Write the answer in scientific notation.	$1{,}440{,}000{,}000{,}000{,}000 = 1.44 \times 10^{15}$

METHOD 2 – Using Exponent Properties

Group like factors.	$(2.4 \times 10^5)(6 \times 10^9) = (2.4 \times 6)(10^5 \times 10^9)$
Multiply each factor. Add exponents when multiplying with like bases.	$(2.4 \times 6)(10^5 \times 10^9) = 14.4 \times 10^{14}$
Convert to scientific notation. In this case the decimal point must be moved so the leading value is less than 10.	$14.4 \times 10^{14} = 1.44 \times 10^{15}$
Write the answer in standard notation.	$1.44 \times 10^{15} = 1{,}440{,}000{,}000{,}000{,}000$

Computing with Numbers in Scientific Notation

Method 1 - Working in Standard Notation

1. Convert numbers to standard notation.
2. Perform calculations.
3. Convert answer back to scientific notation when necessary.

Method 2 - Using Exponent Properties

1. Group like factors.
2. Multiply or divide. Use properties of exponents when finding products or quotients of powers of 10.
3. If necessary, convert to scientific notation so the absolute value of the leading number is greater than or equal to 1 and less than 10.
4. Convert answer to standard notation when necessary.

EXERCISES

Find each product. Write each answer in scientific and standard notation.

1. $(3 \times 10^6)(2 \times 10^3)$

2. $(1.5 \times 10^{-5})(5 \times 10^{-3})$

3. $(5.4 \times 10^3)(3 \times 10^7)$

4. $(-7 \times 10^4)(2 \times 10^{-6})$

5. $(8.6 \times 10^5)(3.1 \times 10^3)$

6. $(2.2 \times 10^{-10})(8 \times 10^{-7})$

7. Which method(s) from this lesson did you use to find the products? Why did you choose this method for multiplying?

Find each quotient. Write each answer in scientific and standard notation.

8. $\frac{9 \times 10^9}{3 \times 10^4}$

9. $\frac{7.7 \times 10^9}{1.1 \times 10^1}$

10. $\frac{2.55 \times 10^{-2}}{5 \times 10^{-5}}$

11. $\frac{4.96 \times 10^3}{6.2 \times 10^7}$

12. $\frac{1.8 \times 10^2}{9 \times 10^{-8}}$

13. $\frac{1.6 \times 10^1}{6.4 \times 10^{-5}}$

14. Which method(s) from this lesson did you use to find the quotients? Why did you choose this method for dividing?

15. The population of Russia in 2005 was approximately 1.4×10^8 people. The population of Greenland was approximately 5.6×10^4 people. How many times more people were in Russia than Greenland? Write your answer in scientific and standard notation.

16. There were approximately 3.1×10^8 people in the United States of America in 2010. The average person drank 3.3×10^3 ounces of soda.

a. Approximately how many ounces of soda were drunk in the USA in 2010?

b. The average global citizen (outside of the USA) drank 6.8×10^2 ounces of soda in 2010. Approximately how many times more ounces did an American drink per year compared to a global citizen?

17. The sun has a mass of approximately 2×10^{30} *kg*. The planet Neptune has a mass of approximately 1×10^{26} *kg*. About how many times more mass does the sun have compared to Neptune? Write your answer in standard notation.

18. Both Liam and Micah performed the same calculation. Liam got an answer of 24×10^8. Micah arrived at an answer of 2.4×10^9.

a. Are their answers equal to each other? How do you know?

b. Which answer is in scientific notation? Why is the other answer not in scientific notation?

19. In 2008, the average American produced 4.5 pounds of garbage per day (including recycled garbage). The population of the United States at that time was approximately 3×10^8 people. How much garbage was produced in America during 2008? Assume there are 365 days in one year.

20. Mr. Knight's class wanted to determine how many people lived in Mexico, the United States and Canada combined. They discovered that in 2005 Canada had an approximate population of 3.2×10^7 people, Mexico had approximately 1.1×10^8 people and the USA had 3×10^8 people.

a. Frank said he added together all of the front numbers (3.2, 1.1 and 3) and then added the exponents to get a total of 7.3×10^{23} people. Does Frank's method work? Why or why not?

b. What was the approximate population of Canada, Mexico and the USA in 2005? How did you find this total?

REVIEW

Find the length of the hypotenuse given two leg lengths. Round to the nearest tenth, if necessary.

21. 7 inches and 24 inches

22. 5 meters and 9 meters

23. 10 *ft* and 24 *ft*

24. 4 *cm* and 12 *cm*

25. 30 *mm* and 40 *mm*

26. 1 mile and 2 miles

EXPONENTS AND ROOTS

LESSON 4.5

Use roots to solve equations with exponents.

In **Block 2** you used square roots to solve problems involving right triangles. In this lesson you will use squre roots as well as cube roots to solve problems. A **cube root** is one of the three equal factors of a number. The mathematical symbol for a cube root is $\sqrt[3]{\ }$. A **perfect cube** is a number whose cube root is an integer. The first ten perfect cubes are shown at the right.

Perfect Cubes	*Cube Root*
$1 \cdot 1 \cdot 1 = 1$	$\sqrt[3]{1} = 1$
$2 \cdot 2 \cdot 2 = 8$	$\sqrt[3]{8} = 2$
$3 \cdot 3 \cdot 3 = 27$	$\sqrt[3]{27} = 3$
$4 \cdot 4 \cdot 4 = 64$	$\sqrt[3]{64} = 4$
$5 \cdot 5 \cdot 5 = 125$	$\sqrt[3]{125} = 5$
$6 \cdot 6 \cdot 6 = 216$	$\sqrt[3]{216} = 6$
$7 \cdot 7 \cdot 7 = 343$	$\sqrt[3]{343} = 7$
$8 \cdot 8 \cdot 8 = 512$	$\sqrt[3]{512} = 8$
$9 \cdot 9 \cdot 9 = 729$	$\sqrt[3]{729} = 9$
$10 \cdot 10 \cdot 10 = 1000$	$\sqrt[3]{1000} = 10$

Roots are the inverse operation of the corresponding powers. When solving equations, you can "undo" an exponent by using a root. For example, to "undo" x^4 you can use a fourth root ($\sqrt[4]{\ }$).

When finding the root of a number, the answer may be positive, negative or both. For example, if $x^2 = 9$, then x can equal either 3 or -3 because $(3)(3) = 9$ and $(-3)(-3) = 9$. Look at the following rules below:

- When finding the root of a term with an even exponent, the answer will be both negative and positive (shown using the symbol ±).
- When finding the root of a term with an odd exponent, the answer will have the same sign as the term under the root.
- In real-world situations, negative answers may not always make sense. Look closely at the situation to determine if only a positive answer is needed.

USING ROOTS TO SOLVE EQUATIONS WITH EXPONENTS

1. Isolate the variable with the exponent.
2. Use inverse operations to undo the exponent with the corresponding root.
3. Determine if the answer should be positive, negative or both based on the original exponent and the application setting.

EXAMPLE 1 **Solve $\frac{x^2}{2} + 11 = 43$ for x.**

SOLUTION

Subtract 11 from both sides of the equation. $\frac{x^2}{2} + 11 - 11 = 43 - 11$

Multiply both sides of the equation by 2. $2 \cdot \frac{x^2}{2} = 32 \cdot 2$

Square root both sides of the equation. $\sqrt{x^2} = \pm\sqrt{64}$

Include both positive and negative answers. $x = \pm 8$

✓ $(-8)(-8) = 64$ and $(8)(8) = 64$

EXAMPLE 2 **Solve $2x^3 - 5 = -59$ for x.**

SOLUTION

Add 5 to both sides of the equation. $2x^3 - 5 + 5 = -59 + 5$

Divide both sides of the equation by 2. $\frac{2x^3}{2} = \frac{-54}{2}$

Cube root both sides of the equation. $\sqrt[3]{x^3} = \sqrt[3]{-27}$

$x = -3$

✓ $(-3)(-3)(-3) = (-27)$

When solving equations with exponents, answers may not always be integers. When this is the case, round to the specified place value.

EXAMPLE 3 **The volume of a sphere can be found using the formula $V = \frac{4}{3}\pi r^3$. Finley has a spherical water balloon filled with water. She knows there are 56 cubic inches of water in the balloon. What is the approximate radius, r, of the balloon? Round the answer to the nearest hundredth.**

SOLUTION

Write the formula for thevolume of a sphere. $\text{Volume} = \frac{4}{3}\pi r^3$

Substitute all known values for the variables. $56 \approx \frac{4}{3}(3.14)r^3$

Multiply both sides of the equation by the reciprocal of $\frac{4}{3}$. $\frac{3}{4} \cdot 56 \approx \frac{4}{3}(3.14)r^3 \cdot \frac{3}{4}$

Divide both sides of the equation by 3.14. $\frac{42}{3.14} \approx \frac{3.14r^3}{3.14}$

$13.38 \approx r^3$

Cube root both sides of the equation. $\sqrt[3]{13.38} \approx \sqrt[3]{r^3}$

$2.37 \approx r$

The radius of the balloon is approximately 2.37 inches.

EXERCISES

1. Match the inverse operations.

a. $(\ )^2$	**A.** $-$
b. $(\ )^5$	**B.** $\sqrt{\ }$
c. $\div$	**C.** $\sqrt[5]{\ }$
d. $\sqrt[3]{\ }$	**D.** $\times$
e. $+$	**E.** $(\ )^3$

2. Listed below are equations. Determine which sign the solution will have (+, − or ±).

a. $x^5 = 32$ **b.** $x^2 = 121$ **c.** $x^3 = -343$ **d.** $x^4 = 49$

3. Jennifer used a square root to solve a math problem that helped her find the distance between home plate and 2nd base on a baseball field. She said her approximate answer was ±127.3 feet. Her teacher said her answer was almost correct. What is wrong with her answer?

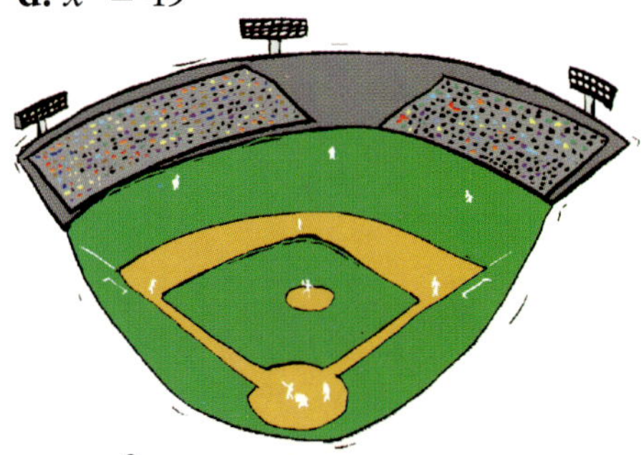

Solve each equation.

4. $x^3 = 512$ **5.** $x^2 + 4 = 53$ **6.** $\frac{x^2}{5} = 20$

7. $10 + 7x^3 = -46$ **8.** $4x^2 - 20 = 556$ **9.** $\frac{2}{3}x^3 - 1 = 17$

10. $\frac{x^3}{10} + 12 = 112$ **11.** $x^5 - 1 = 242$ **12.** $5x^2 - 30.5 = 149.5$

13. The surface area of a sphere is found using the formula $SA = 4\pi r^2$ where r is the radius. A playground ball has a surface area of 113.04 square inches. What is the radius of the ball? Use 3.14 for π.

14. The volume of a sphere is found using the formula $V = \frac{4}{3}\pi r^3$ where r is the radius. A large globe has a volume of 3,052.08 cubic inches. What is the radius of the globe? Use 3.14 for π.

15. In a previous block, you learned that the Pythagorean Theorem states that in a right triangle, the sum of the squares of the lengths of the legs is equal to the square of the length of the hypotenuse. Use this to solve for the missing side lengths.

$$a^2 + b^2 = c^2$$

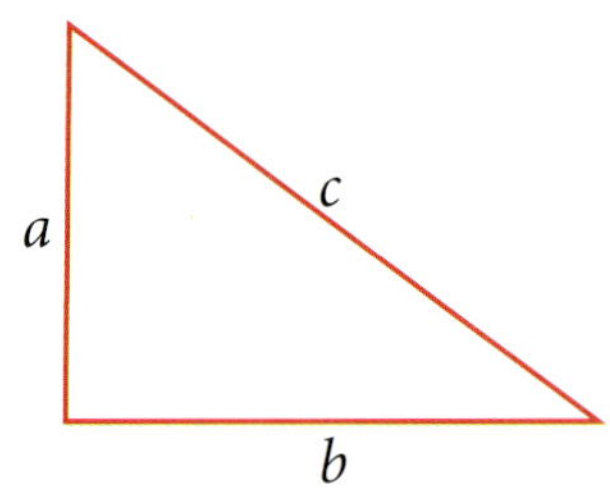

a. If $a = 12$ and $c = 13$, find b.
b. If $b = 15$ and $c = 25$, find a.
c. If $a = 7$ and $b = 24$, find c.

Solve each equation. Round each answer to the nearest hundredth.

16. $x^3 = 88$

17. $7x^2 = 77$

18. $2x^3 - 4 = 20$

19. $5x^3 - 20 = -130$

20. $\frac{x^2}{4} = 260$

21. $\frac{1}{2}x^2 + 16 = 400$

22. The volume of a cube is found using the formula $V = s^3$ where s is the length of each side. A cube has a volume of 250 cubic feet. What is the approximate length of each side? Round to the nearest hundredth.

REVIEW

Find each product or quotient. Write each answer in scientific and standard notation.

23. $(2 \times 10^4)(4 \times 10^2)$

24. $(3.5 \times 10^3)(6 \times 10^{-5})$

25. $(1.2 \times 10^3)(2.8 \times 10^5)$

26. $\frac{8 \times 10^5}{4 \times 10^2}$

27. $\frac{2.1 \times 10^3}{7 \times 10^{-6}}$

28. $\frac{2.84 \times 10^{-2}}{4 \times 10^3}$

29. An investment company's profits were $\$1.4 \times 10^5$ last year. This year, they predict they will have a profit of $\$8.4 \times 10^7$. If this happens, how many times greater will their profits be this year compared to last year?

Tic-Tac-Toe ~ Comparing Country Size

Choose 12 different countries around the world on at least 5 continents. Create a display that shows the capital city, approximate size (in square miles) and one other interesting fact about each country. Write the square mileage of the country in both standard and scientific notation. At the bottom of your display compare the sizes of 8 pairs of countries from your list. Make sure to use every country at least once. Show your calculations using numbers in scientific notation.

Tic-Tac-Toe ~ Simplifying Solutions

Engineers and scientists often work with exact numbers rather than decimal approximations. Other professionals also use exact numbers in their work so they do not have rounding errors. In higher-level math classes you will be asked to give exact answers with square roots. This is done by finding the square root of any perfect squares under the radical sign and moving those values outside the radical sign. Look at the examples below to see how square roots can be simplified.

Example 1: Simplify $\sqrt{20}$.

Write the prime factorization of 20.	$20 = 2 \cdot 2 \cdot 5$
Circle pairs of equivalent numbers.	$20 = (2 \cdot 2) \cdot 5$
Bring each pair of numbers outside the radical sign as a single number since $\sqrt{2 \cdot 2} = \sqrt{4} = 2$.	$\sqrt{20} = 2\sqrt{5}$

Example 2: Solve $x^2 = 90$.

Square root both sides.	$\sqrt{x^2} = \pm\sqrt{90}$
Write the prime factorization of 90.	$x = \pm\sqrt{2 \cdot 3 \cdot 3 \cdot 5}$
Simplify.	$x = \pm 3\sqrt{10}$

Solve each equation. Write each answer as a square root in simplest form. Include all answers by using the "±" symbol.

1. $x^2 - 7 = 17$

2. $2x^2 + 2 = 402$

3. $\frac{x^2}{2} - 1 = 41$

4. $\frac{6x^2}{5} = 30$

5. $3x^2 - 20 = 652$

6. $161\frac{1}{2} = 8x^2 + 1\frac{1}{2}$

7. $\frac{x^2}{-3} - 11 = -53$

8. $3 - 6x^2 = -2646$

9. $20 = -5 + \frac{x^2}{7}$

VOLUME OF CYLINDERS

LESSON 4.6

Find the volume of cylinders and solve real-world problems involving cylinders.

Volume is the number of cubic units needed to fill a three-dimensional figure. A **cylinder** is a solid with two congruent and parallel bases that are circles. Just as with prisms, the volume of a cylinder is found by multiplying the area of its base by its height. Volume is written using cubic units (e.g., in^3, cm^3, ft^3, m^3).

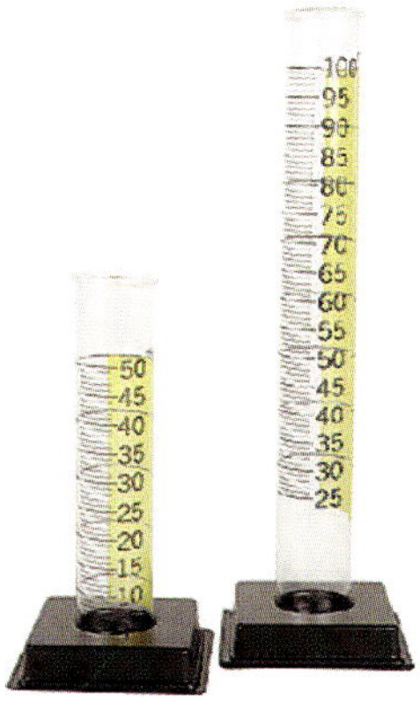

VOLUME OF A CYLINDER

The volume of a cylinder is equal to the product of the area of the base (B) and the height (h).

$$V = Bh$$

$$V = \pi r^2 h$$

EXAMPLE 1 **Find the volume of the cylinder. Use 3.14 for π.**

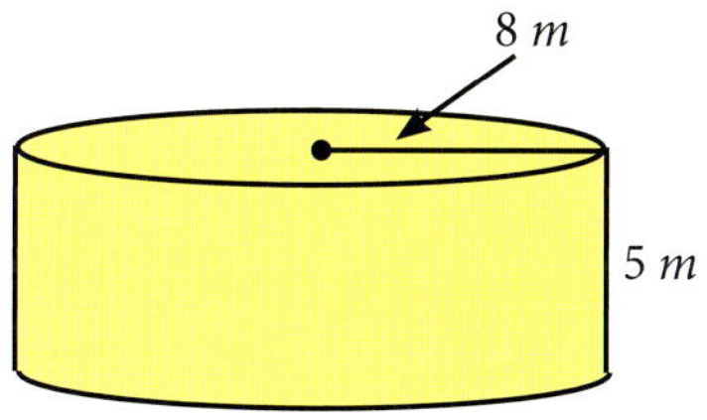

Since 3.14 is an approximation of π, the answer will be an approximation.

SOLUTION

Write the volume formula for a cylinder.	$V = \pi r^2 h$
Substitute all known values for the variables.	$V \approx (3.14)(8)^2(5)$
Find the value of the power.	$V \approx (3.14)(64)(5)$
Multiply.	$V \approx 1004.8$

The volume of the cylinder is about 1,004.8 cubic meters.

The missing dimensions of a cylinder can be found if the volume and one other measurement is provided. Substitute all known values into the volume formula. Use inverse operations to find the missing value.

EXAMPLE 2

The volume of a cylindrical water cooler is 1,695.6 cubic inches. The cooler has a radius of 6 inches. Find the height of the cooler. Use 3.14 for π.

SOLUTION

Write the volume formula for a cylinder. $V = \pi r^2 h$

Substitute all known values for the variables. $1695.6 \approx (3.14)(6^2)h$

Find the value of the power. $1695.6 \approx 3.14(36)h$

Multiply. $1695.6 \approx 113.04h$

Divide both sides of the equation by 113.04. $\frac{1695.6}{113.04} \approx \frac{\cancel{113.04}h}{\cancel{113.04}}$

$15 \approx h$

The height of the water cooler is about 15 inches.

EXAMPLE 3

Find the radius of a cylinder with an approximate volume of 3,014.4 and a height of 15 inches. Use 3.14 for π.

SOLUTION

Write the volume formula for a cylinder. $V = \pi r^2 h$

Substitute all known values for the variables. $3014.4 \approx 3.14r^2(15)$

Multiply. $3014.4 \approx 47.1r^2$

Divide both sides of the equation by 47.1. $\frac{3014.4}{47.1} \approx \frac{\cancel{47.1}r^2}{\cancel{47.1}}$

$64 \approx r^2$

Square root both sides of the equation. $\pm\sqrt{64} \approx \sqrt{r^2}$

Simplify. $8 \approx r$

Only the positive root is appropriate in this situation.

The radius of the cylinder is about 8 inches.

EXERCISES

Find the volume of each cylinder. Use 3.14 for π. Round to the nearest hundredth.

1.

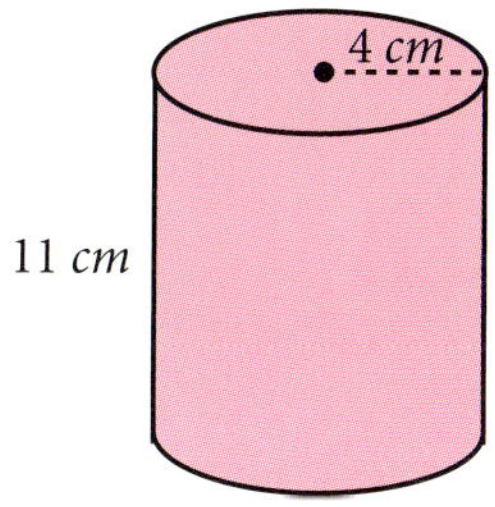

2.

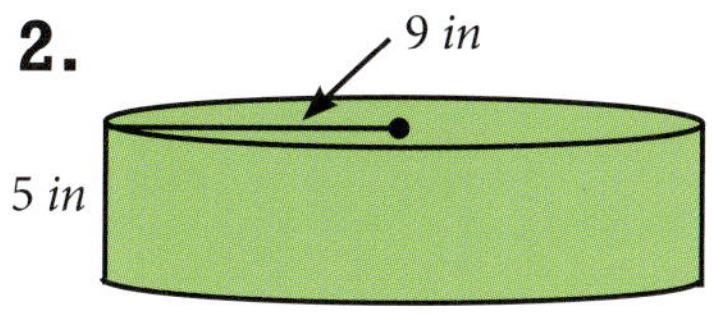

3.

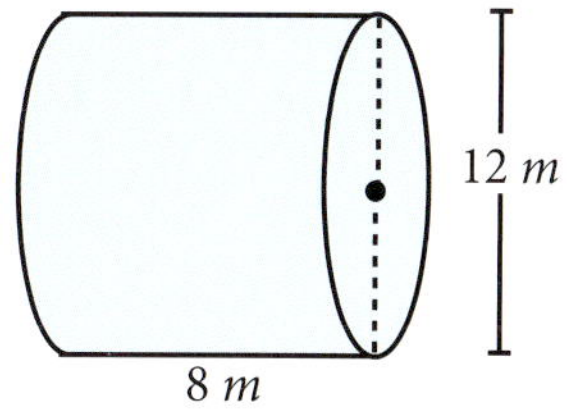

Find the volume of each solid. Use 3.14 for π.

4.

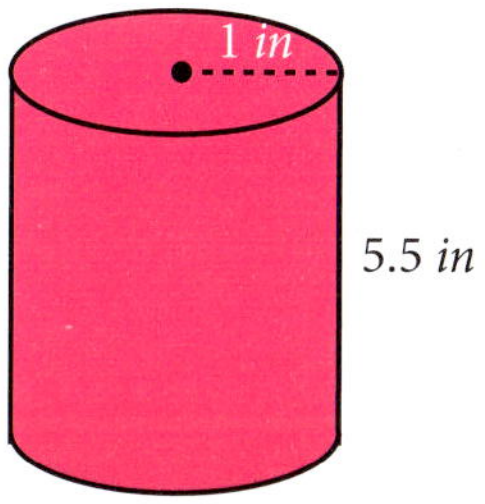

5.

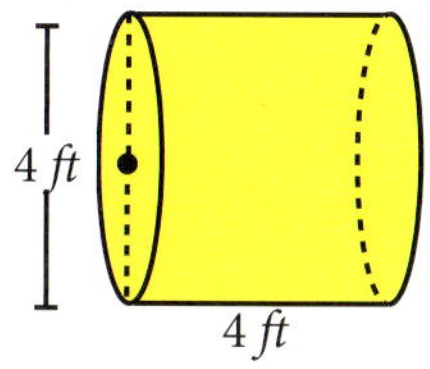

6.

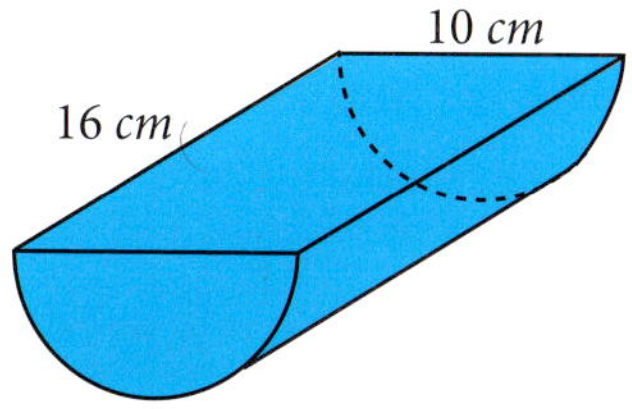

7. Kambree filled her circular swimming pool. The radius of the pool is 10 feet. The height is 4 feet. The instructions recommend filling the pool to the 3 foot mark.

a. Calculate the volume of water the pool contained when Kambree followed the recommendations.

b. Why would it be recommended not to fill the pool to the very top?

8. Which solid can hold more: a cylinder with a diameter and height of 6 inches or a square prism with side lengths of 6 inches? Verify the answer with calculations.

9. A round cement stepping stone has a radius of 1 foot and is 2 inches thick.

a. Convert the radius to inches.

b. How many cubic inches of cement does it take to make one stepping stone?

c. How many cubic inches of cement would it take to make 45 stepping stones?

10. A grain silo is made of cylindrical sections that have a 7.2 *m* diameter and are 2.4 *m* tall. The cylindrical sections are stacked on top of one another to form a silo. Orville wants to be able to store 500 cubic meters of grain. How many sections will Orville need to buy?

11. Recycling is important for the environment. A cylindrical rain barrel used to recycle water can hold 15,197.6 cubic inches of water. It has a radius of 11 inches. How tall is the barrel?

12. A cylindrical juice pitcher has a radius of 3 inches and is 10 inches tall.

a. Find the volume of the pitcher.

b. If the pitcher is half full, how much juice is in it?

13. A round sugar container has a volume of 3,140 cubic centimeters. The radius of the container is 10 *cm*.

a. Find the height of the container.

b. Could 3,140 sugar cubes fit in the container if each cube measures 1 *cm* on each side? Why or why not?

14. A cylindrical vase for flowers holds about 125.6 cubic inches of water. The vase is 10 inches tall. Find the radius of the vase.

15. A cylindrical diesel tank holds approximately 226.08 cubic feet of diesel fuel. The tank is 8 feet long. What is the radius of the tank?

16. Becky baked cookies. She wants to ship some to her brother who is overseas with the military. She has a cylindrical can with a volume of 1130.4 cm^3. The can is 22.5 *cm* tall. Assuming each cookie is circular and lays parallel to the base of the can, what is the largest diameter a cookie can have?

Find the volume of each solid. Use 3.14 for π. Round to the nearest hundredth.

17.

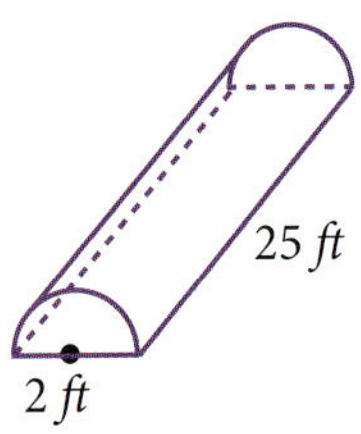

18.

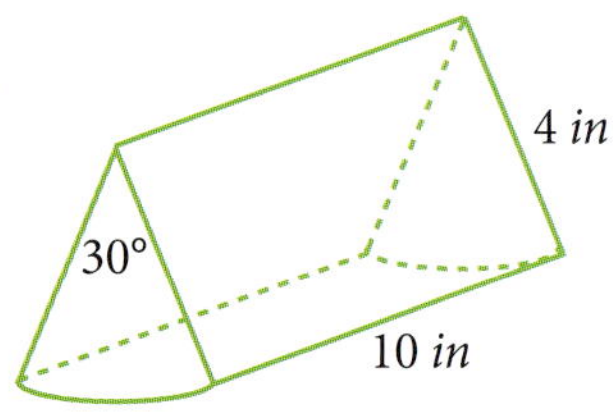

19. Brenden has a cylinder with a radius that is 18 *cm*. It is 25 *cm* tall.

a. Find the volume of Brenden's cylinder.

b. What is the radius of a cylinder that is half the volume of Brenden's cylinder and is 50 *cm* tall?

REVIEW

Use a protractor to measure each angle

20.

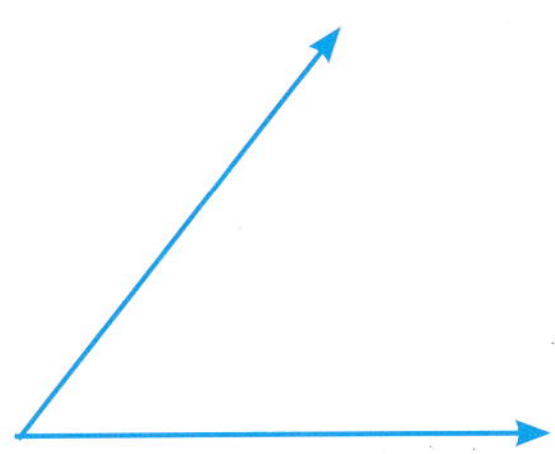

21.

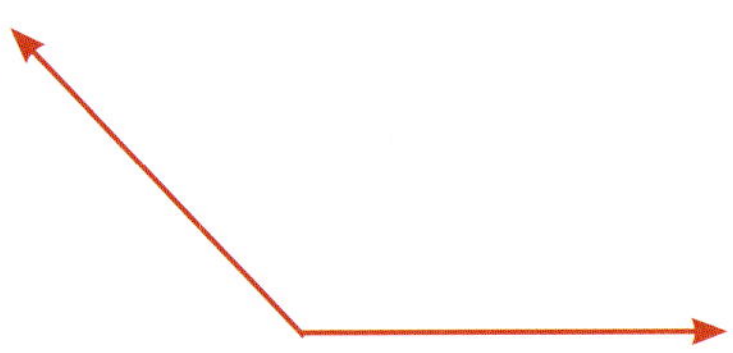

22.

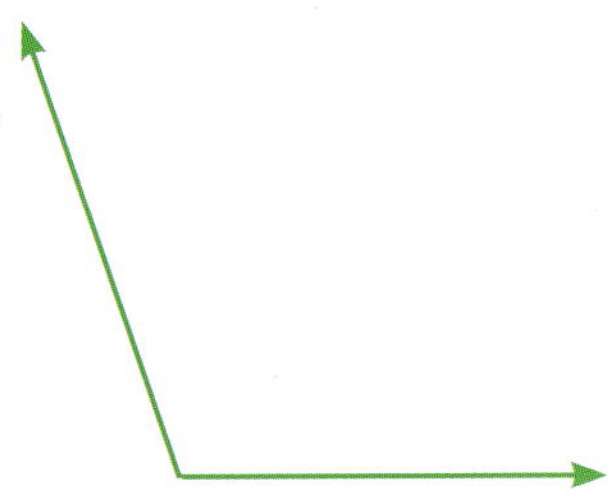

Solve for *x*.

23. $x^3 = 27$

24. $x^3 = -27$

25. $5x^3 - 12 = 28$

26. $\frac{x^2}{6} = 36$

27. $x^3 - 2 = -345$

28. $3x^3 + 114 = 222$

Tic-Tac-Toe ~ Measure of Money

The types of coins used in the United States vary in size. Locate five of each type of coin (penny, nickel, dime and quarter) to use for this activity.

1. Copy the table below onto a piece of notebook paper.

Coin	Radius	Height	Area of base	Surface area	Volume
Penny					
Nickel					
Dime					
Quarter					

2. Find the diameter of each coin in millimeters. Divide by two to find the radius. Record in the table.

3. Stack five of each coin on top of one another. This makes a cylinder. Measure the height of the stack. Divide this measurement by 5 to get the height of one coin. Repeat this process for each type of coin. Record the heights in the table.

4. Calculate the area of the base, surface area and volume of each coin.

5. Research why coins are different sizes. Write two or three paragraphs summarizing your research.

Tic-Tac-Toe ~ Field Test

A kitchen gadget company is designing cylindrical measuring cups that will hold one pint of liquid. They want to create 3 different cylinders that have the same volume. They will need to conduct a field test to find which container is most popular, which means the most marketable. One liquid pint is approximately 29 in^3.

1. Find three different sets of measures that will give a cylinder a volume of 29 in^3.

2. Draw a sketch of each cylinder. Label the measures.

3. Make a model of each cylinder by creating a net. Tape the net together to form a solid.

4. Show twenty-five adults the models and ask each adult which container they would most likely purchase.

5. Display your findings in a bar graph.

VOLUME OF CONES

LESSON 4.7

Find the volume of cones and solve real-world problems involving cones.

A **cone** is a solid formed by one circular base and a vertex.

EXPLORE! **CONES IN A CYLINDER**

Rita is making snow cones for her friends. She has four cylindrical glasses that she is using to take the crushed ice outside to her friends that are waiting to fill their snow cone cups. Each snow cone cup is as tall as one of the glasses and has the same size circular base. Rita wants to determine if there is a relationship between a cone and a cylinder. She needs to make sure she has enough ice for 12 snow cones. Complete the **Explore!** to find if Rita will have enough ice for the snow cones.

Step 1: Make or find a cone and a cylinder that have congruent bases and are the same height.

Step 2: Estimate how many times larger the volume of the cylinder is compared to the volume of the cone.

Step 3: Fill the cone with rice, beans or popcorn kernels. Pour the contents of the cone into the cylinder. Repeat until the cylinder is full.

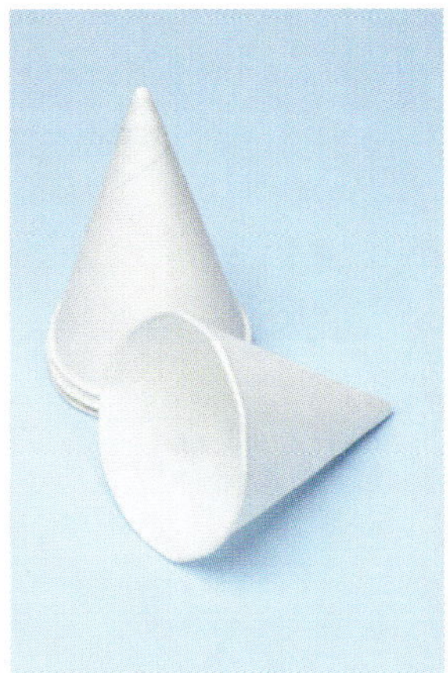

Step 4: How many times did you need to empty the cone in order to fill the cylinder?

Step 5: What fraction is the volume of the cone compared to the volume of the cylinder?

Step 6: What is the formula for the volume of a cylinder?

Step 7: Combine the answers from **Step 5** and **Step 6** to write a formula that can be used to find the volume of a cone.

Step 8: Use your formula to find the volume of each cone. Use 3.14 for π. Round to the nearest hundredth.

a.

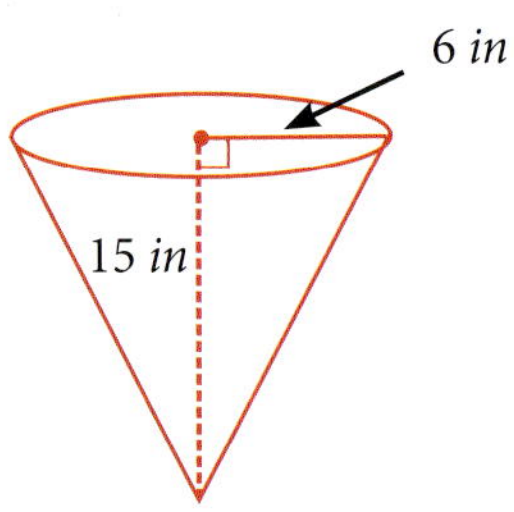

b.

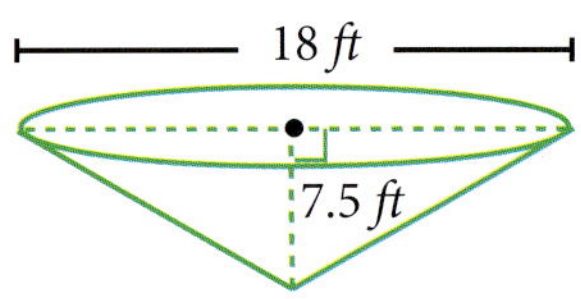

c. $r = 2.1$ *mm*

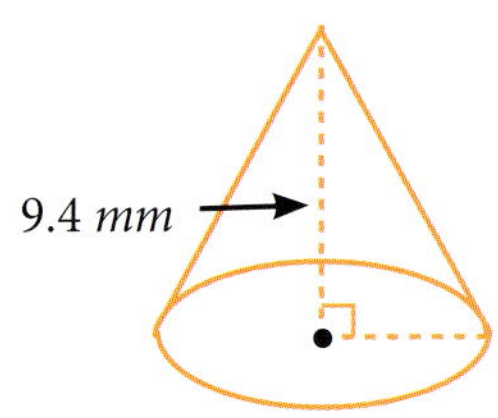

Step 9: Will Rita have enough crushed ice for 12 snow cones if she fills the four cylinders?

Volume of a Cone

The volume of a cone is equal to one-third the product of the area of the base (B) and height (h).

$$V = \frac{1}{3}Bh \text{ or } V = \frac{Bh}{3}$$

$$V = \frac{1}{3}\pi r^2 h \text{ or } \frac{\pi r^2 h}{3}$$

It takes the volume of three cones to equal the volume of one cylinder when they have congruent bases and heights. The height, h, of a cone is how tall a cone is from its vertex to the center of its base. A pyramid and a prism with congruent bases and heights have the same relationship.

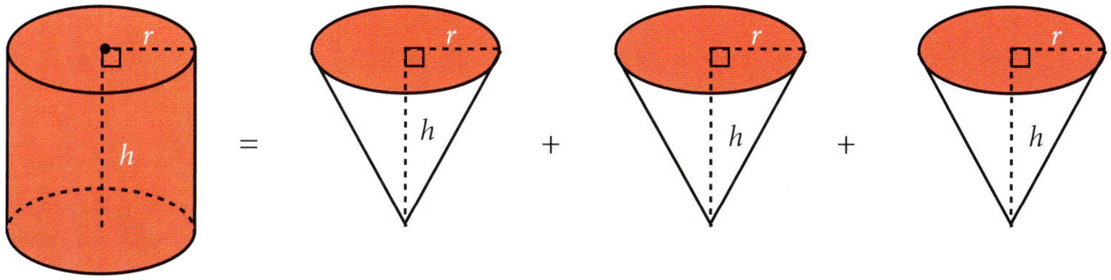

EXAMPLE 1 **Find the volume of the cone. Use 3.14 for π.**

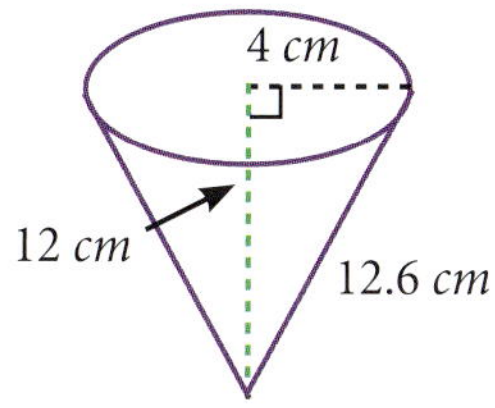

SOLUTION

Write the volume formula for a cone.	$V = \frac{1}{3}\pi r^2 h$
Substitute all known values for the variables.	$V \approx \frac{1}{3}(3.14)(4)^2(12)$
Find the value of the power.	$V \approx \frac{1}{3}(3.14)(16)(12)$
Multiply.	$V \approx 200.96$

The volume of the cone is about 200.96 cm^3.

EXAMPLE 2 **Chantel helped with her sister's party. Each child received a party hat full of treats. Each hat had a volume of 65.94 cubic inches. The radius of each hat was 3 inches. How tall was each party hat?**

SOLUTION

Write the volume formula for a cone.	$V = \frac{1}{3}\pi r^2 h$
Substitute all known values for the variables.	$65.94 \approx \frac{1}{3}(3.14)(3)^2 h$
Find the value of the power.	$65.94 \approx \frac{1}{3}(3.14)(9)(h)$
Multiply.	$65.94 \approx 9.42h$
Divide both sides of the equation by 9.42.	$\frac{65.94}{9.42} \approx \frac{9.42h}{9.42}$
	$7 \approx h$

Each party hat was about 7 inches tall.

EXAMPLE 3 **A cone-shaped popcorn container holds 314 cubic inches of popped corn. The container is 12 inches tall. Find the radius of the conical container. Use 3.14 for π.**

SOLUTION

Write the volume formula for a cone.	$V = \frac{1}{3}\pi r^2 h$
Substitute all known values for the variables.	$314 \approx \frac{1}{3}(3.14)r^2(12)$
Multiply.	$314 \approx 12.56r^2$
Divide by 12.56 on both sides of the equation.	$\frac{314}{12.56} \approx \frac{\cancel{12.56}r^2}{\cancel{12.56}}$
Square root both sides of the equation.	$\pm\sqrt{25} \approx \sqrt{r^2}$
	$5 = r$

Only the positive root is used in this situation.

The radius of the popcorn container is 5 inches.

EXERCISES

Find the volume of each cone. Use 3.14 for π. Round answers to the nearest whole number.

1.

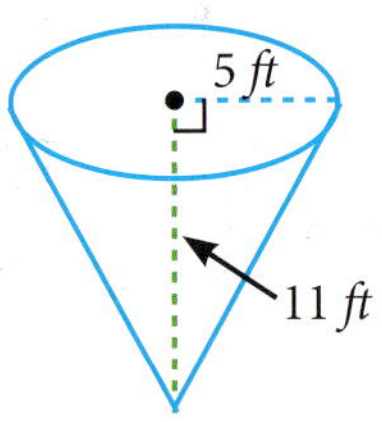

2.

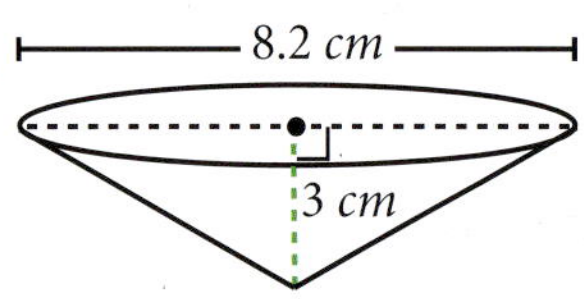

3.

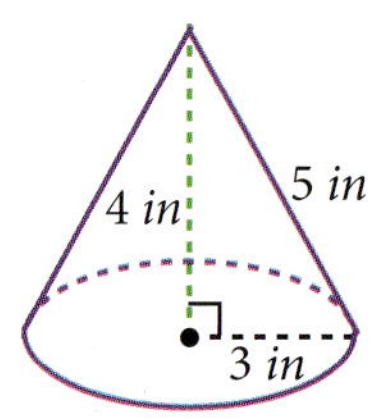

4. A cone has the same height as a cylinder. The areas of the bases of the two solids are the same. How many cones does it take to fill the cylinder?

5. The volume of a cylinder is 18 cubic meters. What is the volume of a cone with a congruent base and height?

6. A cylinder holds 87 gallons of water. How many gallons of water does a cone with the same radius and height hold?

7. Sketch a diagram of a cone with a radius of 9 yards and a height of 12 yards. Find the volume of the cone.

Find the volume of each cone with the given radius and height. Use 3.14 for π. Round to the nearest hundredth.

8. $r = 3\ m$
$h = 5\ m$

9. $r = 6\ cm$
$h = 2\ cm$

10. $d = 20\ in$
$h = 7\ in$

Find the length of the radius in each cone. Use 3.14 for π. Round to the nearest whole number.

11. V ≈ 339.12 in^3

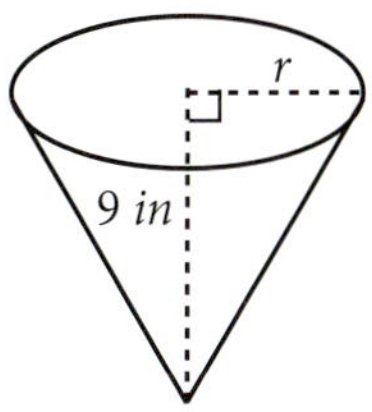

12. V ≈ 169.56 ft^3

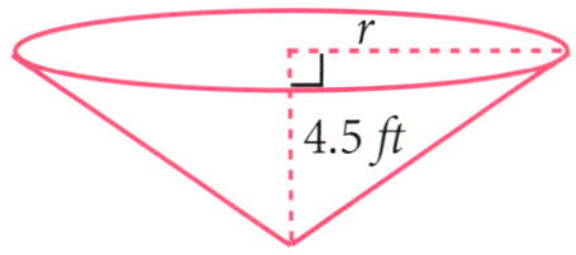

13. V ≈ 30.77 in^3

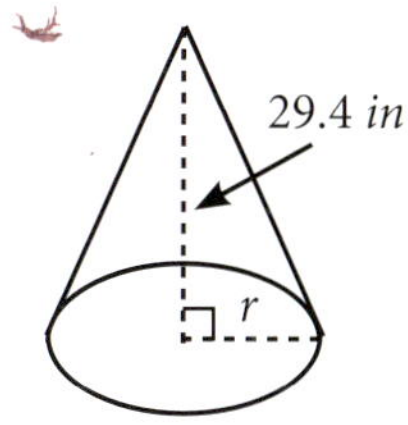

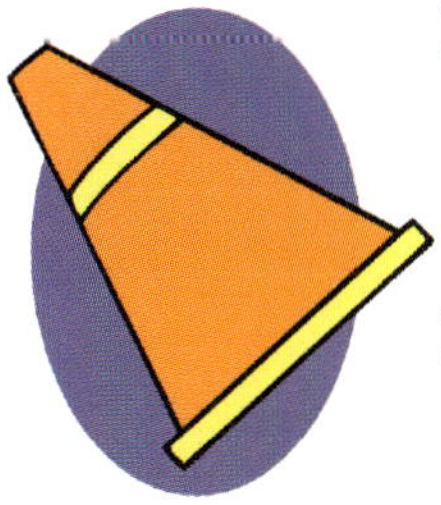

14. An orange cone is often used to alert drivers in a construction zone to warn them to slow down. A construction cone is 24 inches tall. The circular base has a diameter of 18 inches. Find the volume of the cone. Use 3.14 for π.

15. An icicle is shaped like a cone. It has a diameter of 1 inch. It is 2 feet long.

a. Convert all measurements to inches.

b. Draw a diagram and label it.

c. Find the volume of the icicle. Use 3.14 for π.

16. A candle maker makes cone-shaped candles. The candle maker uses 9.42 cubic inches of wax to make each candle. The radius of each candle is 1.5 inches. Find the height of the candles. Use 3.14 for π.

17. The conical paper cups next to a water cooler each hold 3.14 cubic inches of water. The radius of each paper cup is 1 inch. Find the height of the cup. Use 3.14 for π.

18. Frank is setting up a conical teepee. The box says it is 10 feet tall and has a volume of approximately 150.72 cubic feet. He needs to find the diameter of the teepee to find a space large enough to set it up.

a. Find the radius of the teepee.

b. What is the diameter of the teepee?

c. If Frank has a square patch of grass in his yard that has a perimeter of 28 feet, will the teepee fit in it?

Find the volume of each composite solid. Use 3.14 for π. Round to the nearest hundredth.

19.

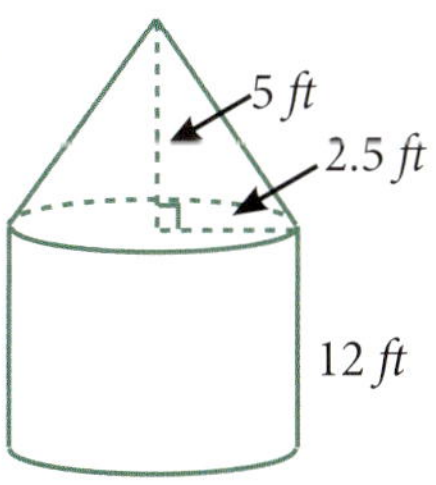

20.

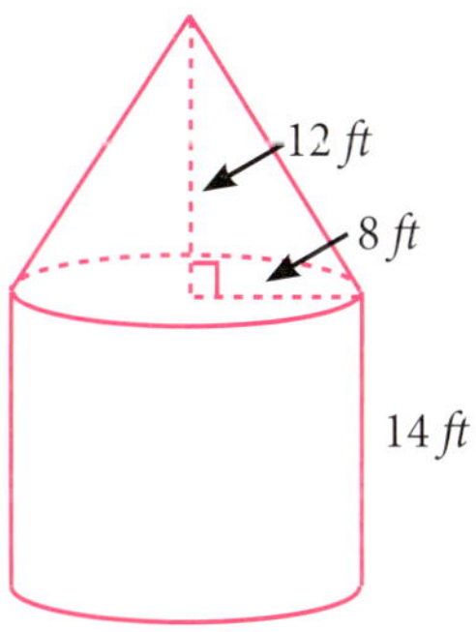

REVIEW

Find the distance between each pair of points. If necessary, round to the nearest tenth.

21. (2, 7) and (5, 11) **22.** (6, 0 and (−2, 15) **23.** (−3, 2) and (−10, −5)

Find the volume of each cylinder. Use 3.14 for π.

24.

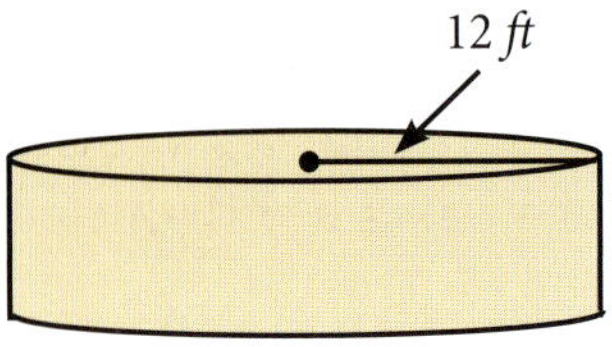

25.

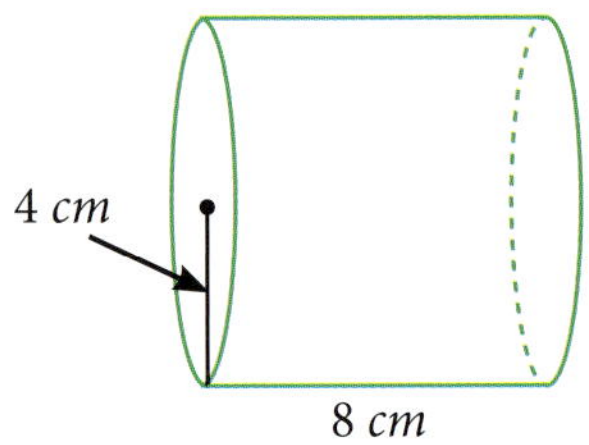

Simplify.

26. $\frac{28x^3y^{10}}{16xy^8}$ **27.** $\left(\frac{5m^3}{2a^5}\right)^2$ **28.** $\frac{k^{-5}w^7}{k^9w^{-3}}$

Tic-Tac-Toe ~ Teaching Volume

Design an activity kit to teach third graders about volume. The activity kit could do any combination of the following:

- Teach the general concept of volume
- Show volume using different materials
- Show and teach that different sizes of containers have different volumes
- Show and teach that sometimes containers that are different shapes have the same volume

The kit will include:

- A written explanation describing what students will learn from the lesson.
- A step-by-step lesson plan to follow that will take approximately 30 minutes to complete.
- Materials to use for the activity.
- Packaging – all parts of the kit are together in some sort of package.

If possible, teach your lesson on volume to a group of third graders.

Tic-Tac-Toe ~ Displacement

A water glass is half full. When ice cubes are added the water level rises. This is called displacement. It is possible to calculate the volume of objects using displacement.

Example: A cylinder has a radius of 1.5 inches. The 8-inch tall cylinder is half full of water. When ice is added the water level rises 3 inches. Find the volume of the ice.

	$V = \pi r^2 h$
Substitute known values.	$= (\pi)(1.5)^2(3)$
Find the value of the power.	$\approx (3.14)(2.25)(3)$
Multiply.	≈ 21.195

Notice the amount of displacement is used for height.

The ice has a volume of about 21.195 cubic inches.

Find the volume of each object using displacement. Use 3.14 for π.

1. A 12-inch tall cylinder has a radius of 3 inches. A rock is dropped in the cylinder and the water rises 5 inches. Find the volume of the rock.

2. Keely filled her glass half full of tea. She then put in 5 equal-sized ice cubes. The base of her glass has a 10 *cm* diameter. The tea reached 16 *cm* high in the glass. The tea rose to 20 *cm* when the ice was added.

a. What was the height change when the ice was added?
b. Find the volume of the ice.
c. Find the volume of one ice cube.

3. A fish tank is a 13" × 24" × 16" rectangular prism. The water rises 5 inches when the fish are added. Find the volume of the fish. (Note: 16" is the height of the fish tank.)

4. Do your own displacement tests. Find a cylinder or prism to use.

a. Calculate the area of the base.
b. Partially fill with water.
c. Measure the height of the water.
d. Find four objects and record them in the table.
e. Find the height change of the water when each object is dropped in.
f. Calculate the volume of each object. Show all work in your table.

Name of Object	Height Change	Work	Volume of Object

VOLUME OF SPHERES

LESSON 4.8

Find the volume of spheres and solve real-world problems involving spheres.

Many sports involve spheres. Baseballs, softballs, volleyballs, dodge balls, soccer balls, tennis balls and basketballs are all spheres. A **sphere** is a round, curved, closed three-dimensional solid. A sphere has no edges, sides or vertices. Every point on the surface of the sphere is an equal distance from the center of the sphere. This distance is the radius of the sphere.

VOLUME OF A SPHERE

The volume of a sphere is equal to four-thirds the product of pi (π) and the cube of its radius (r^3).

$$V = \frac{4}{3}\pi r^3 \textbf{ or } V = \frac{4\pi r^3}{3}$$

EXAMPLE 1 **Find the volume of the sphere. Use 3.14 for π.**

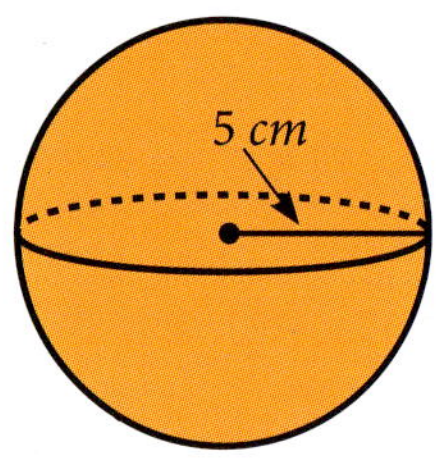

SOLUTION

Write the volume formula for a sphere.	$V = \frac{4}{3}\pi r^3$
Substitute known values for the variables.	$V \approx \frac{4}{3}(3.14)(5)^3$
Find the value of the power.	$V \approx \frac{4}{3}(3.14)(125)$
Multiply.	$V \approx 523.33$

The volume of the sphere is about 523.33 cm^3.

EXAMPLE 2 **A water tower has a spherical tank. The diameter of the tank is 30 meters. How much water can the tank hold? Use 3.14 for π.**

SOLUTION

Find the radius of the tank	Diameter ÷ 2 = 30 ÷ 2 = 15
Write the volume formula for a sphere.	$V = \frac{4}{3}\pi r^3$
Substitute known values for the variables.	$V \approx \frac{4}{3}(3.14)(15)^3$
Find the value of the power.	$V \approx \frac{4}{3}(3.14)(3375)$
Multiply.	$V \approx 14{,}130$

The tank can hold approximately 14,130 cubic meters of water.

EXAMPLE 3 **A bouncy ball has a volume of 113.04 cubic centimeters. Find the radius of the ball. Use 3.14 for π.**

SOLUTION

Write the volume formula for a sphere.	$V = \frac{4}{3}\pi r^3$
Substitute known values for the variables.	$113.04 \approx \frac{4}{3}(3.14)r^3$
Multiply.	$113.04 \approx 4.19r^3$
Divide both sides of the equation by 4.19.	$\frac{113.04}{4.19} \approx \frac{\cancel{4.19}r^3}{\cancel{4.19}}$
	$27 \approx r^3$
Cube root both sides of the equation.	$\sqrt[3]{27} \approx \sqrt[3]{r^3}$
Simplify.	$3 \approx r$

This is a rounded answer. Rounding can make cubic roots easier to calculate.

The radius of the bouncy ball is close to 3 *cm*.

EXERCISES

Find the volume of each sphere. Use 3.14 for π. Round to the nearest hundredth.

1.

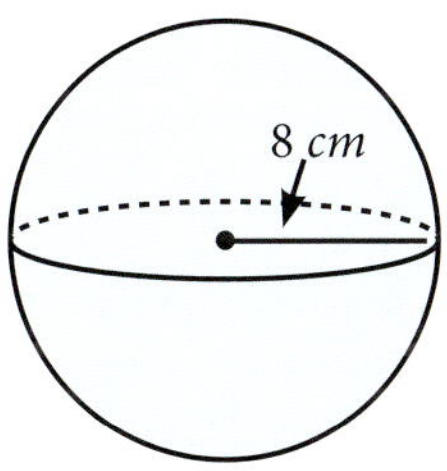

2. $d = 15$ *in*

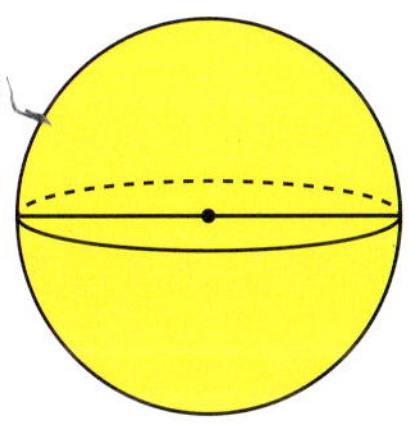

3. $r = 4$ *m*

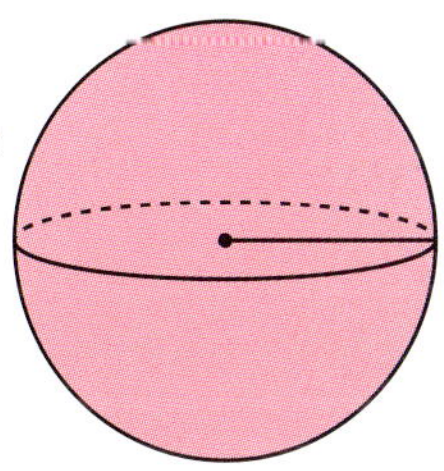

4. A standard basketball used by professionals has a radius close to 12 *cm*. Find the approximate volume of a standard basketball.

5. A spherical liquid soap container has a diameter of 5 *in*. How much soap can the container hold?

6. A bowl is shaped like a hemisphere, which is half of a sphere. The diameter of the bowl is 8 inches. How much water will the bowl hold?

7. Karissa is having a party. She has a dozen balloons that are spheres when they are inflated. She wants to fill them with helium, but she does not know how much helium she needs to buy. The radius of each balloon is 6 inches when it is inflated properly.

a. Calculate the volume of one balloon. Use 3.14 for π. Round to the nearest hundredth, as needed.

b. How much total helium will it take to fill all of the balloons?

c. If helium can only be purchased in whole number units, how much helium will Karissa need to buy?

8. Tennis balls are sold in sets of three inside a cylindrical can. Each tennis ball has a diameter of 2.5 inches. Assume the balls touch the can on the sides, top and bottom.

a. Calculate the volume of one tennis ball.

b. Calculate the volume of the cylindrical can.

c. How many cubic inches are not used by the tennis balls inside the cylinder?

Find each missing measure. Use 3.14 for π.

9. Volume $\approx$ 904.32 cm^3

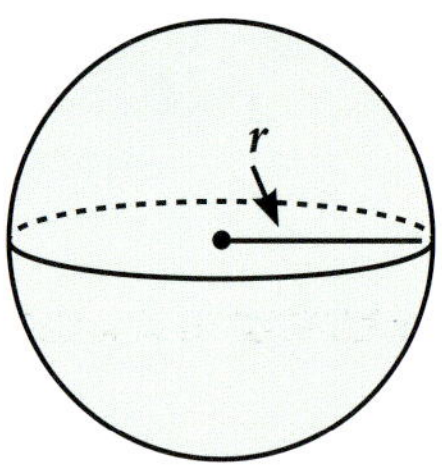

10. Volume $\approx$ 33.49 ft^3

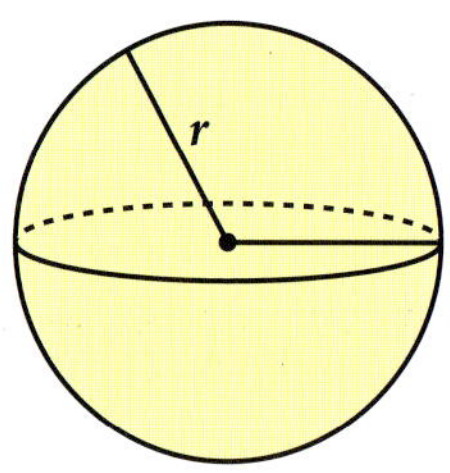

11. Volume $\approx$ 267.95 in^3

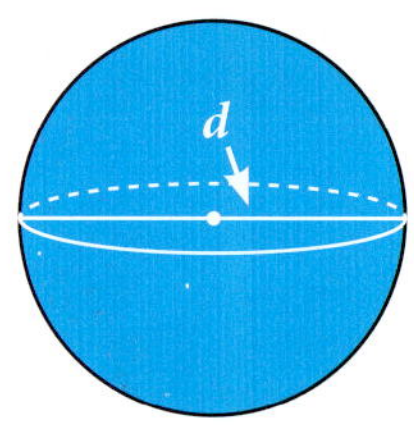

12. Volume $\approx$ 7234.56 m^3

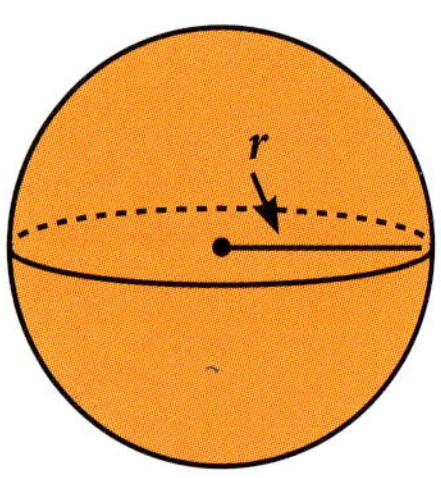

13. Volume $\approx$ 523.33 yd^3

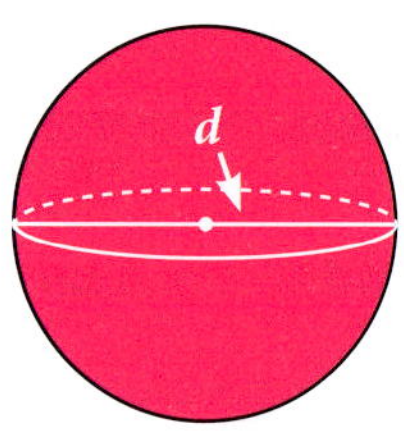

14. Volume $\approx$ 1436.03 in^3

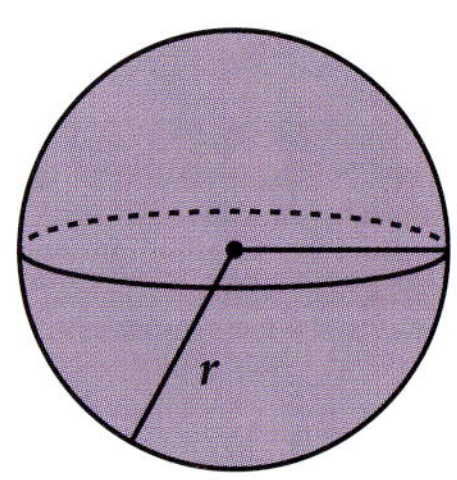

15. A bouncy ball has a volume of 4.187 cubic inches. What is the diameter of the ball?

16. Graysen owns a world globe that has a volume of 3,052.08 cubic inches. What is the diameter of the globe?

17. A beach ball holds 800 cubic inches of air. What is the radius of the ball? Round to the nearest hundredth.

18. A spherical piece of candy has a chocolate outside with caramel in the middle. The diameter of the whole piece of candy is 3 centimeters. The diameter of the caramel filling is 2 centimeters.

a. Find the volume of one whole piece of candy.

b. Find the volume of the caramel center.

c. What is the volume of chocolate used in each piece of candy?

d. The candy comes in a package that states the volume of the candy is approximately 255 cubic centimeters. About how many pieces of candy are in the package?

19. A small beach ball has a radius of 10 inches. A larger beach ball has a volume that is twice the volume of the smaller ball. What is the radius of the large beach ball? Round to the nearest tenth.

20. The equator is an imaginary line on the Earth's surface which divides the Earth into two equal hemispheres. It is approximately 24,901.55 miles long. Assume the earth is perfectly round and use the length of the equator to find the volume of the earth.

21. Abbigail shipped a globe to her sister. The globe fit snuggly in a box, touching the sides, bottom and top. The globe has a volume of 1,436 cubic inches. Find the volume of the box.

REVIEW

Solve for x.

22.

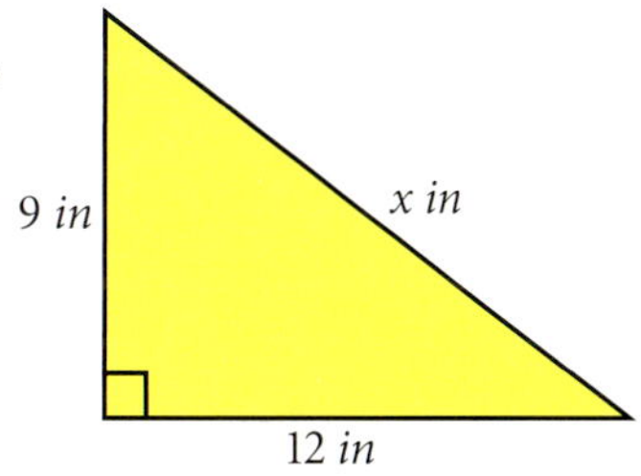

23.

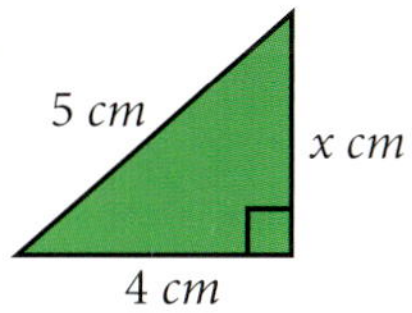

24.

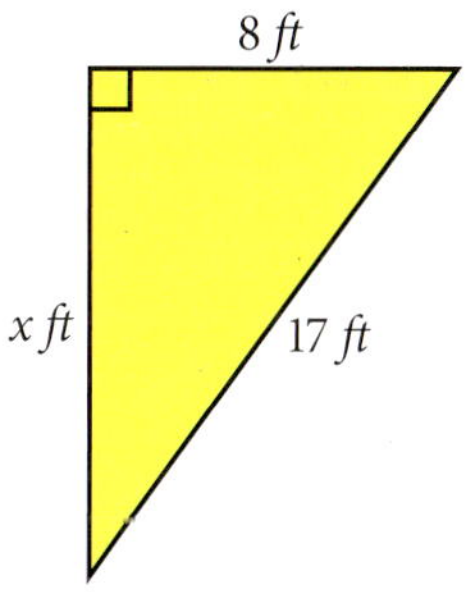

25. Samantha lives 40 blocks east of her school. Her best friend, Hayley, lives 25 blocks south of the school. What is the shortest distance between Samantha and Hayley's house?

Use the distance formula to find the distance between each pair of coordinates. Round to the nearest hundredth, if necessary.

26. (4, 6) and (8, 3)

27. (–2, 0) and (4, 1)

28. (8, –5) and (–3, –5)

REVIEW

BLOCK 4

Vocabulary

base	cubed	power
cone	exponent	scientific notation
cylinder	perfect cube	sphere
cube root		volume

Simplify expressions involving multiplication using properties of exponents.
Use properties of exponents to simplify expressions involving division.
Express numbers in scientific notation and standard notation.
Compute with numbers in scientific notation.
Use roots to solve equations with exponents.
Find the volume of cylinders and solve real-world problems involving cylinders.
Find the volume of cones and solve real-world problems involving cones.
Find the volume of spheres and solve real-world problems involving spheres.

Lesson 4.1 ~ Multiplication Properties of Exponents

Simplify.

1. x^9x^3

2. $(p^2)^4$

3. $(x^2y)^6$

4. $(3m^3)^2$

5. $(mn^3w^2)(w^7mn)$

6. $(7k^2x)(2k^3x^9)$

7. $(4y^4)^2(2y)^3$

8. $(-5x^2w^3)^2(3w^5)^3$

9. $(5x^2)^2(x^4)^2(-3x^2)^3$

10. A candy company is creating a new type of chocolate bar that is packaged in a cube. The length of one side of the cube is represented by c^2. What term represents the volume of the cube?

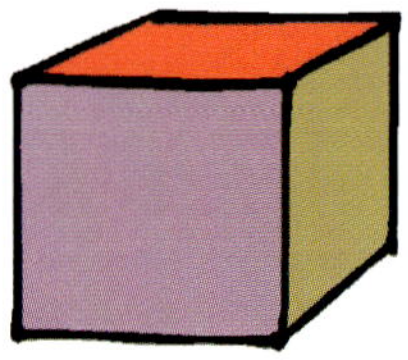

Write the volume of each rectangular prism as a single term.

11.

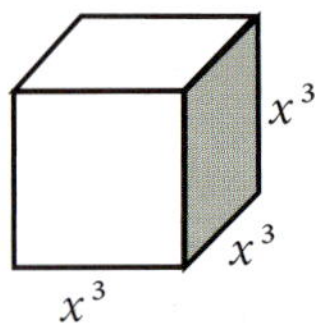

12.

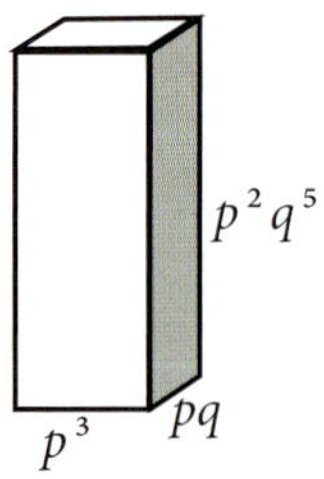

13.

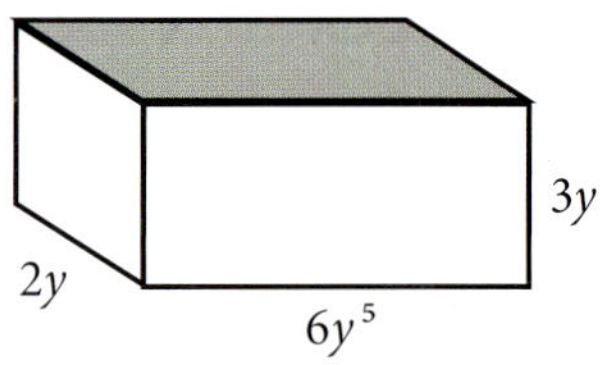

Lesson 4.2 ~ Division Properties of Exponents

Simplify.

14. $\frac{x^{10}}{x^4}$

15. $\frac{18y^4}{6y^3}$

16. 3^{-2}

17. $\frac{5b^3c^7}{15bc^2}$

18. $(7p^5m^4)^0$

19. $\left(\frac{5w^3}{2}\right)^2$

20. $\frac{30x^5y^2}{20x^{-4}y^{-3}}$

21. $\left(\frac{9d^{-5}}{d}\right)^2$

22. $\left(\frac{h^{-3}z^{-6}}{n^0}\right)^2$

23. A packaging company makes boxes with a base area represented by the expression $4a^2b^3$. The length of the box is represented by ab^2. What expression represents the width?

24. Find the missing dimension given that the volume of the prism is $24w^4y^2z$. Write the missing dimension in simplest form.

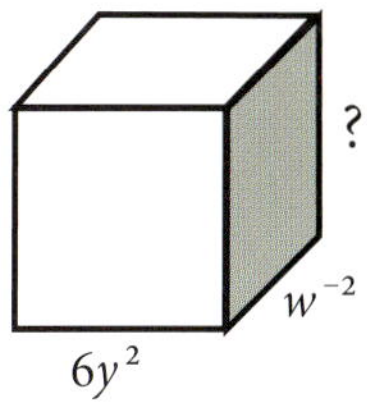

Lesson 4.3 ~ Scientific Notation

Write each number in scientific notation.

25. 0.00043

26. 40,000,000

27. –850,000

28. –0.00003

29. 798

30. 0.000000953

Write each number in standard notation.

31. 8×10^4

32. -5.3×10^{-5}

33. 1.95×10^{-3}

34. 6.2×10^6

35. -9.9×10^7

36. 7×10^{-8}

37. According to www.most-expensive.net, the most expensive college to attend was Sarah Lawrence University in Bronxville, New York. In 2011, it cost about \$57,600 per year to attend, including room and board. Write this number in scientific notation.

38. Texas' population in 2011 could be approximated as 2.48×10^7 people. Write this number in standard notation.

39. Nelly drank 9×10^4 milliliters of water. Peter drank 8.49×10^5 milliliters of water. Who drank more water?

40. Frankie weighed four different diamonds. List the diamonds in order from lightest to heaviest.

Diamond	Weight in Ounces
A	2.8×10^{-2}
B	3×10^{-3}
C	9×10^{-4}
D	5.7×10^{-2}

Lesson 4.4 – Applications of Scientific Notation

Find each product or quotient. Write each answer in scientific and standard notation.

41. $(2 \times 10^5)(3 \times 10^4)$

42. $\dfrac{2.1 \times 10^2}{7 \times 10^{-3}}$

43. $\dfrac{1.08 \times 10^{-4}}{6 \times 10^2}$

44. $(-9 \times 10^3)(6 \times 10^{-1})$

45. $\dfrac{7.2 \times 10^{10}}{8 \times 10^4}$

46. $(4.5 \times 10^{-6})(6 \times 10^{-4})$

47. The population of China in 2005 was approximately 1.3×10^9 people. The population of Uzbekistan was approximately 2.6×10^7 people. How many times more people lived in China than Uzbekistan? Write your answer in scientific and standard notation.

48. There were approximately 3.1×10^8 people in the United States of America in 2010. The average person ate 1.2×10^2 pounds of potatoes that year.

a. Approximately how many pounds of potatoes were eaten in the USA in 2010?

b. The average potato weighs 0.3 pounds. Approximately how many potatoes did an average American eat in 2010?

Lesson 4.5 ~ Exponents and Roots

Solve each equation.

49. $x^3 = 1331$

50. $5x^2 = 245$

51. $\dfrac{x^2}{2} - 7 = 11$

52. $3x^3 + 1 = -23$

53. $x^4 + 17 = 273$

54. $\dfrac{x^2}{2} - 15 = 0.125$

55. The area of a circle is found using the formula $A = \pi r^2$ where r is the radius. A circular placemat has an area of 301.5656 square inches. What is the approximate radius of the placemat? Use 3.14 for π. Round to the nearest tenth.

56. The volume of a sphere is found using the formula $V = \frac{4}{3}\pi r^3$ where r is the radius. A large beach ball has an approximate volume of 275 cubic inches. What is the radius of the beach ball? Use 3.14 for π. Round to the nearest tenth.

Solve each equation. Include all answers. Round each answer to the nearest hundredth.

57. $x^2 = 54$

58. $\frac{x^3}{2} = 19$

59. $12x^3 - 28 = 20$

60. The volume of a cube is found using the formula $V = s^3$ where s is the length of a side. A cube has a volume of 9,060 square feet. What is the approximate length of a side? Round to the nearest hundredth.

Lesson 4.6 ~ Volume of Cylinders

Find the volume of each cylinder. Use 3.14 for π.

61.

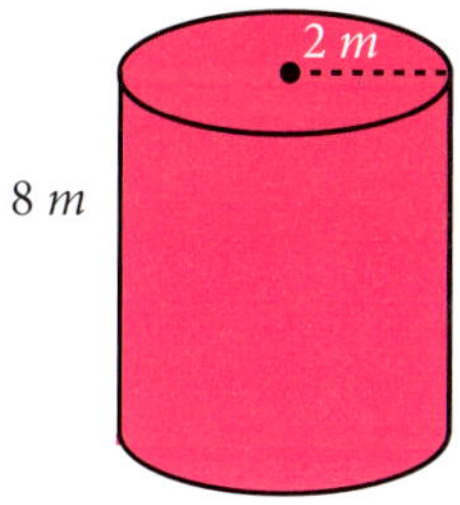

62.

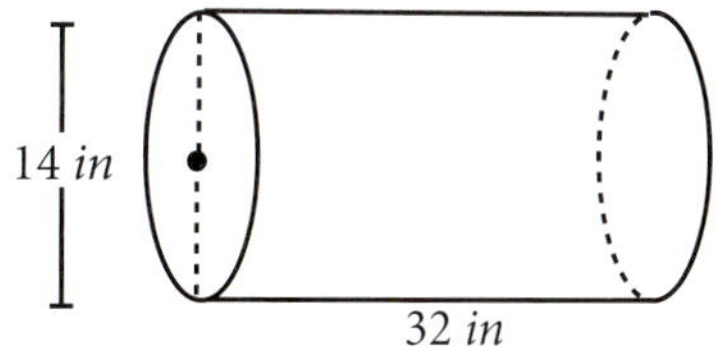

63. A 20-inch tall cylindrical fish tank has a diameter of 12 inches.

a. Find the volume of the fish tank.

b. One fish requires 88 cubic inches of water. How many fish can safely be put in the tank?

64. A bird feeder is the shape of a cylinder. It is 30 centimeters tall and has a radius of 4 *cm*. How many cubic centimeters of bird seed can be put in the feeder?

Find each missing measure. Use 3.14 for π.

65. $V \approx 602.88\ cm^3$

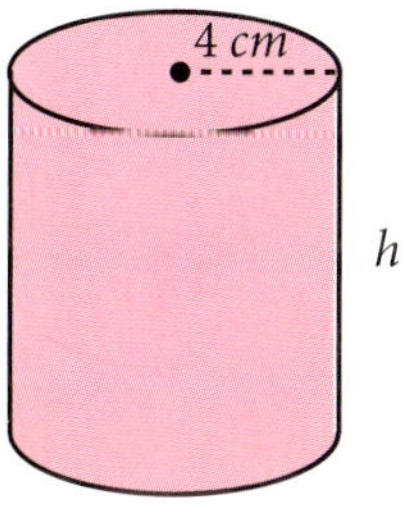

66. $V \approx 1{,}570\ in^3$

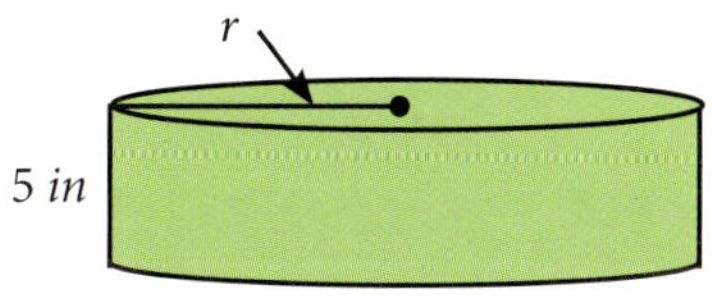

67. A cylindrical drinking glass can hold close to 49.5 cubic inches of water. The glass is 7 inches tall. What is the diameter of the glass? Round to the nearest whole number.

Lesson 4.7 ~ Volume of Cones

68. The volume of a cylinder is 75 cm^3. What is the volume of a cone with a congruent base and equal height?

69. Draw a cone with a height of 3 *cm* and a radius of 4 *cm*. Find its volume. Use 3.14 for π.

70. A sand pile is in the shape of cone. The pile is $8\frac{1}{2}$ feet tall and has a diameter of 6 feet. How many cubic feet of sand is in the pile? Use 3.14 for π.

71. Jesse had a party. He filled party hats full of treats. Each hat is 9 inches tall and has a diameter of 4 inches.
  a. Find the volume of one hat. Use 3.14 for π.
  b. Seven friends came to his party. Jesse filled one hat for each friend and one for himself. What was the volume of treats he used?

Find the volume of each cone. Use 3.14 for π.

72.

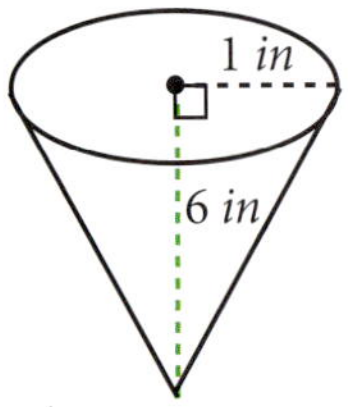

73.

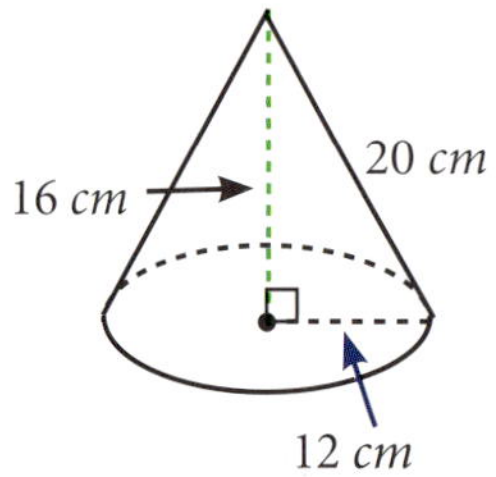

74.

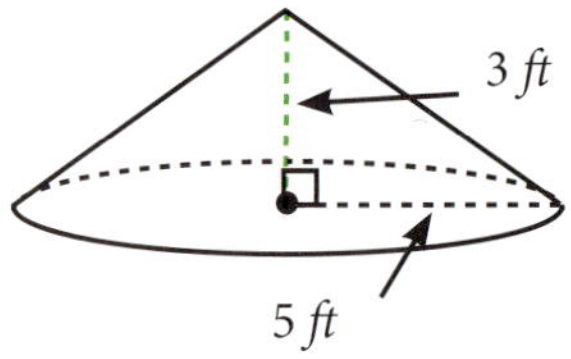

75. A cone has a volume close to 339.12 cm^3. It is 4 *cm* tall. Find the radius of the cone.

76. A conical candle is made of 39.25 cm^3 of wax. The diameter of the candle is 5 *cm*. How tall is the candle?

Lesson 4.8 ~ Volume of Spheres

Find the volume of each sphere. Use 3.14 for π. Round to the nearest hundredth.

77.

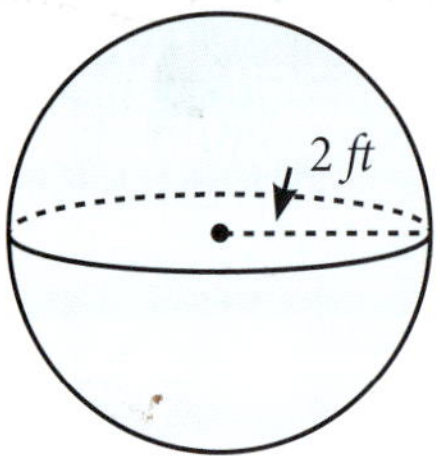

78. $d = 24$ *in*

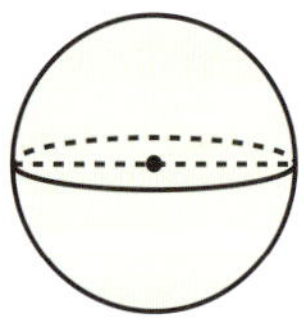

79. $r = 9$ *m*

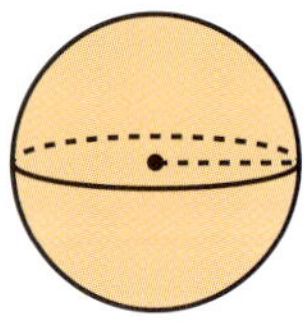

80. Russell made a giant spherical popcorn ball. Its diameter was 12 inches. Find the volume of the popcorn ball.

81. Jacklyn made a large spherical popcorn ball that had a volume close to 5,572.45 cubic inches. Find the diameter of Jacklyn's popcorn ball.

82. Jorge was playing mush ball. The ball he was using had a volume of nearly 9,198.11 cubic centimeters. What was the radius of the mush ball?

CAREER FOCUS

JUDY
SOFTWARE PROJECT MANAGER

I am a project manager for a software company. I work with people at large companies who conduct business in many different countries. Part of my job is to help companies look at their current business practices and make them better. This includes using the software that my company provides and giving training and support to businesses who use it.

Math is an important part of my job. Companies who do business all over the world have to pay different tax rates or percentages depending on where things were made and where they are coming from. For instance, with the NAFTA agreement, a product manufactured in Mexico out of parts made in the US may have a reduced tax rate or be tax free when it is imported back into the US. A strong understanding of math allows me to help businesses make decisions that will maximize the profit they make.

The profession of designing and supporting computer software requires at least some training after high school. Sometimes training can be done at a community college or through a program but generally people in this profession get a Bachelor's degree in computer science. In my particular position I needed further training in understanding both international trade and all of the tax laws that are involved in doing business in other countries.

The average salary of someone starting out in my profession is around \$60,000 per year. After 5 to7 years of experience salaries can range from \$75,000 to \$120,000 per year or more.

There are many things I enjoy about my career; my favorite part is that I get to use math and logic in every aspect of it. It is really fun to test the results of the data that our software provides and then to use that data to help companies make good business decisions. I also enjoy working with people and getting the opportunity to teach people how to use our software.

ACKNOWLEDGEMENTS

All Photos and Clipart ©2008 Jupiterimages Corporation and Clipart.com with the exception of the cover photos and the following photos:

Core Focus on Geometry Page 109
©iStockphoto.com/Leah-Anne Thompson

Core Focus on Geometry Page 145
©iStockphoto.com/Ariel White

Core Focus on Geometry Page 147
©iStockphoto.com/René Mansi

Core Focus on Geometry Page 154
©iStockphoto.com/design56

Core Focus on Geometry Page 157
©iStockphoto.com/nfng

Core Focus on Geometry Page 163
©iStockphoto.com/malerapaso

Layout and Design by Judy St. Lawrence

Cover Design by Schuyler St. Lawrence

Glossary Translation by Heather Contreras

CORE FOCUS ON MATH

GLOSSARY ~ GLOSARIO

A

Term	Definition	Término	Definición
Absolute Value	The distance a number is from 0 on a number line.	Valor Absoluto	La distancia de un número desde el 0 en una recta numérica.
Acute Angle	An angle that measures more than 0° but less than 90°. 	Ángulo Agudo	Un ángulo que mide mas 0° pero menos de 90°. 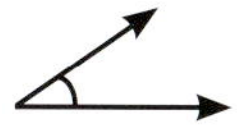
Adjacent Angles	Two angles that share a ray. 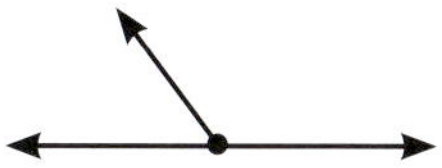	Ángulos Adyacentes	Dos ángulos que comparten un rayo. 
Algebraic Expression	An expression that contains numbers, operations and variables.	Expresiones Algebraicas	Una expresión que contiene números, operaciones y variables.
Alternate Exterior Angles	Two angles that are on the outside of two lines and are on opposite sides of a transversal. 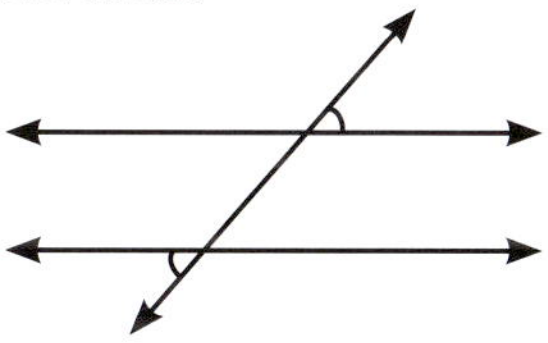	Ángulos Exteriores Alternos	Dos ángulos que están afuera de dos rectas y están a lados opuestos de una transversal. 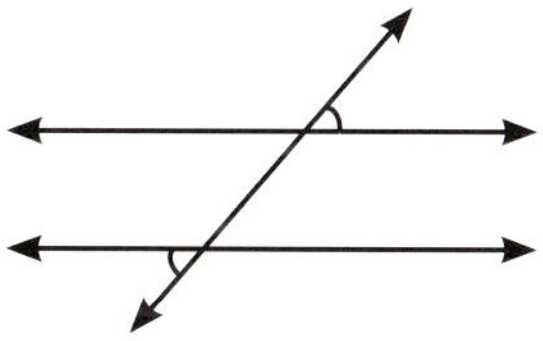
Alternate Interior Angles	Two angles that are on the inside of two lines and are on opposites sides of a transversal. 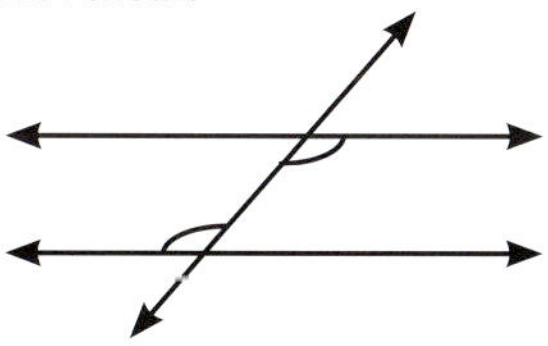	Ángulos Interiores Alternos	Dos ángulos que están adentro de dos rectas y están a lados opuestos de una transversal. 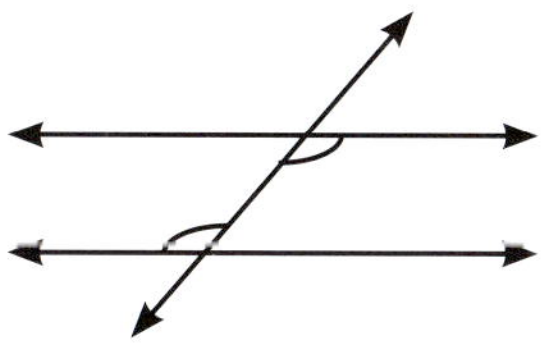
Angle	A figure formed by two rays with a common endpoint. 	Ángulo	Una figura formada por dos rayos con un punto final en común. 

Area The number of square units needed to cover a surface.

Área El número de unidades cuadradas necesitadas para cubrir una superficie.

Ascending Order Numbers arranged from least to greatest.

Progresión Ascendente Los números ordenados de menor a mayor.

Associative Property A property that states that numbers in addition or multiplication expressions can be grouped without affecting the value of the expression.

Propiedad Asociativa Una propiedad que establece que los números en expresiones de suma o de multiplicación pueden ser agrupados sin afectar el valor de la expresión.

Axes A horizontal and vertical number line on a coordinate plane.

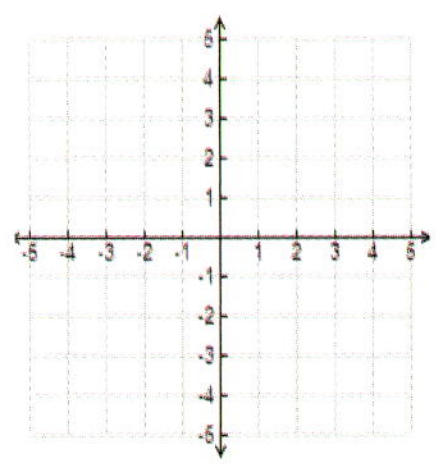

Ejes Una recta numérica horizontal y vertical en un plano de coordenadas.

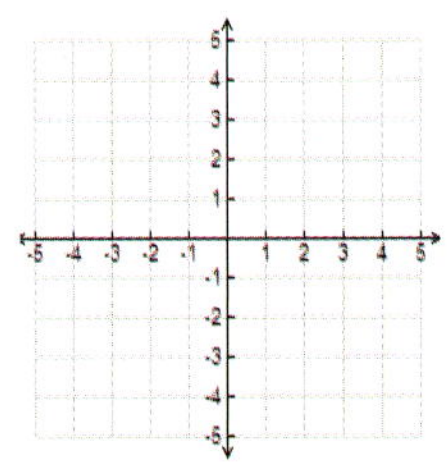

Axis of Symmetry The line of symmetry on a parabola that goes through the vertex.

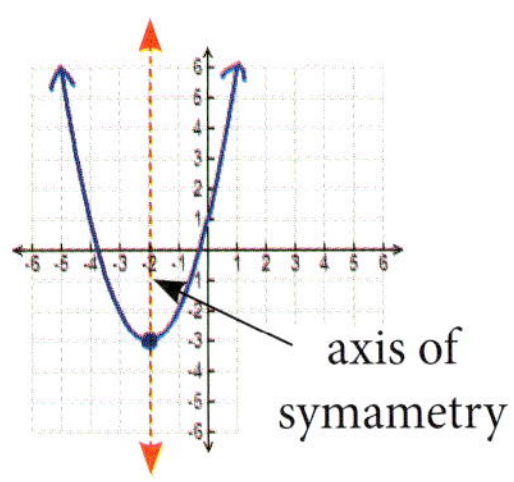

El Eje De Las Simetría La linia de simetría de una parábola que pasa por el vértice.

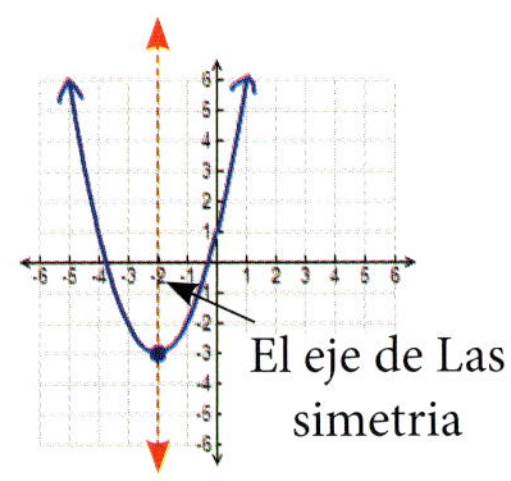

B

Bar Graph A graph that uses bars to compare the quantities in a categorical data set.

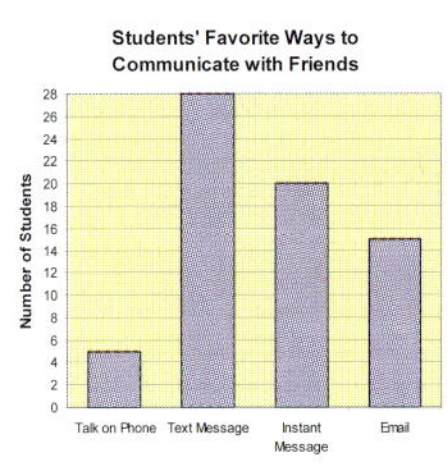

Gráfico de Barras Una gráfica que utiliza barras para comparar las cantidades en un conjunto de datos categórico.

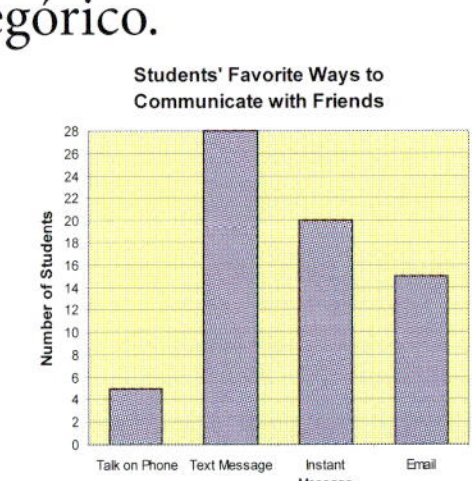

Base (of a power) The base of the power is the repeated factor. In x^a, x is the base.

Base (de la potencia) La base de la potenciación es el factor repatidio. En x^a, x es la base.

Base (of a solid) See Prism, Cylinder, Pyramid and Cone.

Base (de un sólido) Ver Prisma, Cilindro, Pirámide y Cono.

Base (of a triangle) Any side of a triangle.

Base (de un triángulo) Cualquier lado de un triángulo.

Bias A problem when gathering data that affects the results of the data.

Sesgo Un problema que ocurre cuando se recogen datos que afectan los resultados de los datos.

Biased Sample A group from a population that does not accurately represent the entire population.

Muestra Sesgada Un grupo de una población que no representa con exactitud la población entera.

Binomials Expressions involving two terms (i.e. $x - 2$).

Binomiales Expresiones que impliquen dos terminos. (es decir: $x - 2$).

Bivariate Data Data that describes two variables and looks at the relationship between the two variables.

Datos de dos Variables Los datos que describen dos variables y analiza la relación entre estas dos variables.

Box-and-Whisker Plot A diagram used to display the five-number summary of a data set.

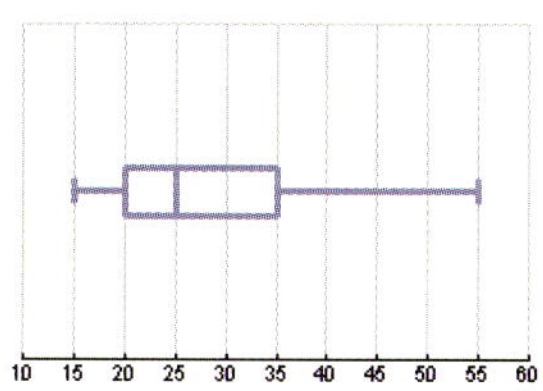

Diagrama de Líneas y Bloques Un diagrama utilizado para mostrar el resumen de cinco números de un conjunto de datos.

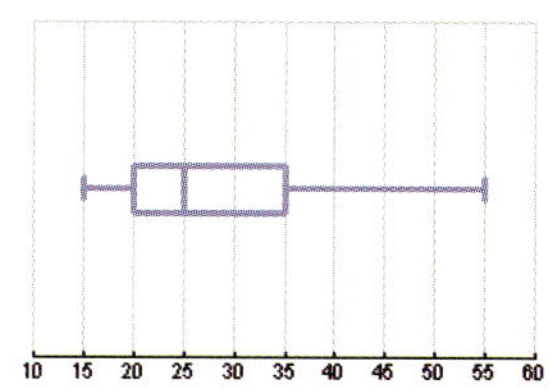

C

Categorical Data Data collected in the form of words.

Datos Categóricos Datos recopilados en la forma de palabras.

Center of a Circle The point inside a circle that is the same distance from all points on the circle.

Centro de un Círculo Un ángulo dentro de un círculo que está a la misma distancia de todos los puntos en el círculo.

Central Angle	An angle in a circle with its vertex at the center of a circle. 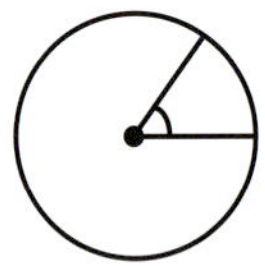	Ángulo Central	Un ángulo en un círculo con su vértice en el centro del círculo. 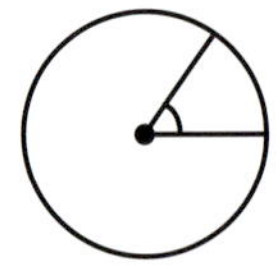
Chord	A line segment with endpoints on a circle. 	Cuerda	Un segmento de la recta con puntos finales en el círculo. 
Circle	The set of all points that are the same distance from a center point.	Círculo	El conjunto de todos los puntos que están a la misma distancia de un punto central.
Circumference	The distance around a circle. 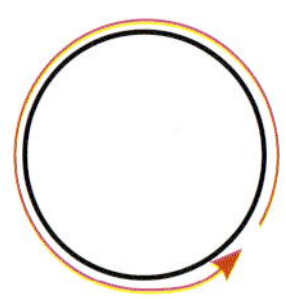	Circunferencia	La distancia alrededor de un círculo. 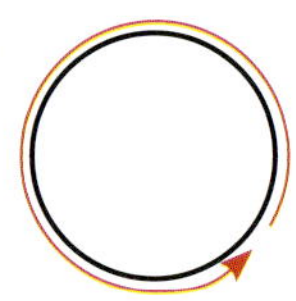
Coefficient	The number multiplied by a variable in a term.	Coeficiente	El número multiplicado por una variable en un término.
Commutative Property	A property that states numbers can be added or multiplied in any order.	Propiedad Conmutativa	Una propiedad que establece que los números pueden ser sumados o multiplicados en cualquier orden.
Compatible Numbers	Numbers that are easy to mentally compute; used when estimating products and quotients.	Números Compatibles	Números que son fáciles de calcular mentalmente; utilizado cuando se estiman productos y cocientes.
Complementary Angles	Two angles whose sum is 90°.	Ángulos Complementarios	Dos ángulos cuya suma es de 90°.
Complements	Two probabilities whose sum is 1. Together they make up all the possible outcomes without repeating any outcomes.	Complementos	Dos probabilidades cuya suma es de 1. Juntos crean todos los posibles resultados sin repetir alguno.

Completing the Square	The creation of a perfect square trinomial by adding a constant to an expression in the form $x^2 + bx$.	Terminado el Cuadrado	La creación de un trinomio cuadrado perfecto por adición de una constante a una expresión en la forma $x^2 + bx$.
Complex Fraction	A fraction that contains a fractional expression in its numerator, denominator or both. $\dfrac{\frac{3}{4}}{\frac{3}{8}}$	Fracción Compleja	Una fracción que contiene una expresión fraccionaria en su numerador, el denominador o ambos. $\dfrac{\frac{3}{4}}{\frac{3}{8}}$
Composite Figure	A geometric figure made of two or more geometric shapes.	Figura Compuesta	Una figura geométrica formada por dos o más formas geométricas.
Composite Number	A whole number larger than 1 that has more than two factors.	Número Compuesto	Un número entero mayor que el 1 con más de dos factores.
Composite Solid	A solid made of two or more three-dimensional geometric figures.	Sólido Compuesto	Un sólido formado por dos o más figuras geométricas tridimensionales.
Composition of Transformations	A series of transformations on a point.	Composición de Transformaciones	Una serie de transformaciones en un punto.
Compound Probability	The probability of two or more events occurring.	Compuesto de Probabilidad	La probabilidad de dos o más eventos ocurriendo.
Conditional Frequency	The ratio of the observed frequency to the total number of frequencies in a given category from an experiment or survey.	Frecuencia Condicional	La relación de una frecuencia observada para el número total de frecuencias en una categoría dada del experimento o encuesta.
Cone	A solid formed by one circular base and a vertex. 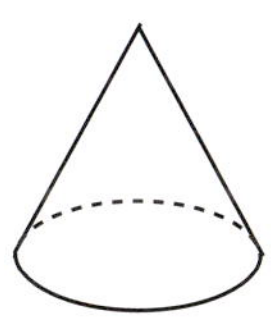	Cono	Un sólido formado por una base circular y una vértice. 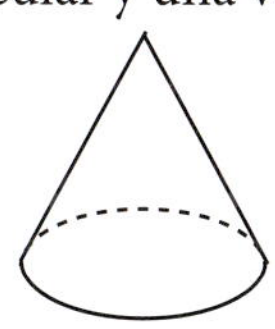
Congruent	Equal in measure.	Congruente	Igual en medida.

Congruent Figures

Two shapes that have the exact same shape and the exact same size.

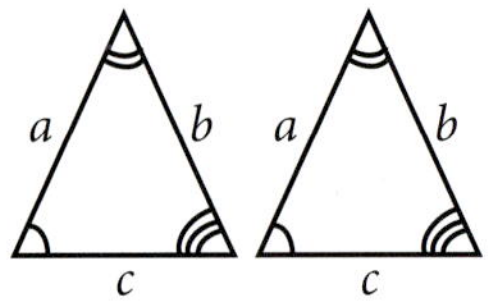

Figuras Congruentes

Dos figuras que tienen exactamente la misma forma y el mismo tamaño.

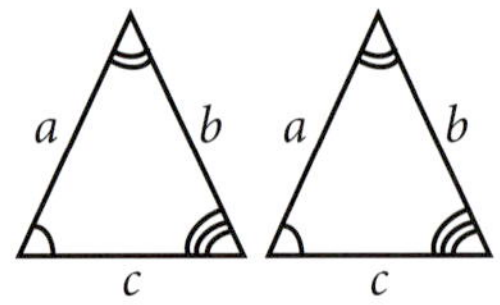

Constant

A term that has no variable.

Constante

Un término que no tiene variable.

Continuous

When a graph can be drawn from beginning to end without any breaks.

Continuo

Cuando una gráfica puede ser dibujada desde principio a fin sin ninguna interrupción.

Conversion

The process of renaming a measurement using different units.

Conversión

El proceso de renombrar una medida utilizando diferentes unidades.

Coordinate Plane

A plane created by two number lines intersecting at a 90° angle.

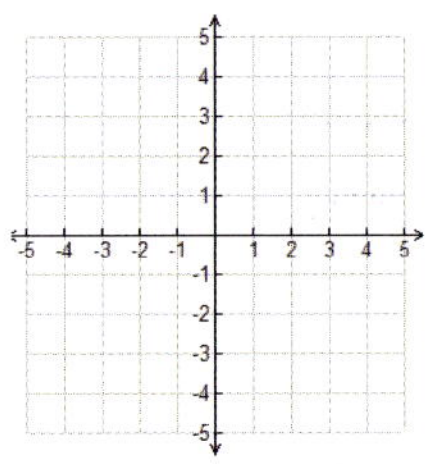

Plano de Coordenadas

Un plano creado por dos rectas numéricas que se intersecan a un ángulo de 90°.

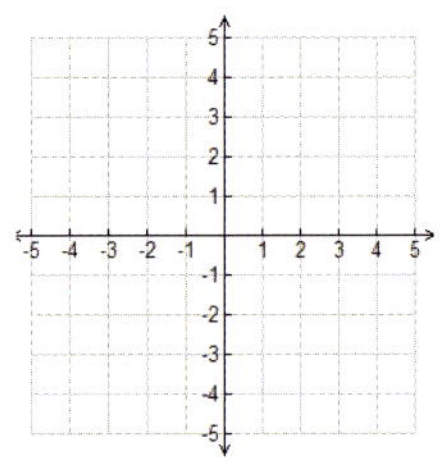

Correlation

The relationship between two variables in a scatter plot.

Correlación

La relación entre dos variables en un gráfico de dispersión.

Corresponding Angles

Two non-adjacent angles that are on the same side of a transversal with one angle inside the two lines and the other on the outside of the two lines.

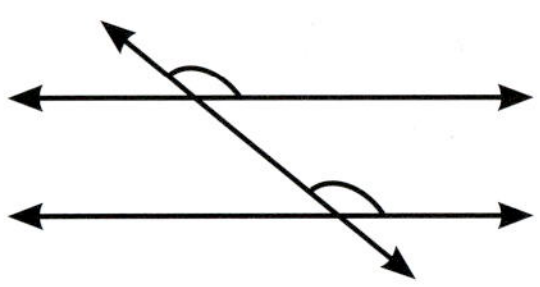

Ángulos Correspondientes

Dos ángulos no adyacentes que están en el mismo lado de una transversal con un ángulo adentro de las dos rectas y el otro afuera de las dos rectas.

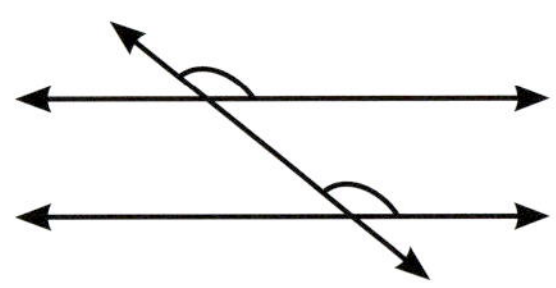

Corresponding Parts

The angles and sides in similar or congruent figures that match.

Partes Correspondientes

Los ángulos y lados en figuras similares o congruentes que concuerdan.

Cube Root	One of the three equal factors of a number. $3 \cdot 3 \cdot 3 = 27 \quad \sqrt[3]{27} = 3$	Raíz Cúbica	Uno de los tres factores iguales de un número. $3 \cdot 3 \cdot 3 = 27 \quad \sqrt[3]{27} = 3$
Cubed	A term raised to the power of 3.	Cubicado	Un término elevado a la potencia de 3.
Cylinder	A solid formed by two congruent and parallel circular bases. 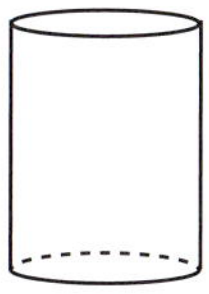	Cilindro	Un sólido formado por dos bases circulares congruentes y paralelas. 

D

Decimal	A number with a digit in the tenths place, hundredths place, etc.	Decimal	Un número con un dígito en las décimas, las centenas, etc.
Degrees	A unit used to measure angles.	Grados	Una unidad utilizada para medir ángulos.
Dependent Events	Two (or more) events such that the outcome of one event affects the outcome of the other event(s).	Eventos Dependiente	Dos (o más) eventos de tal manera que el resultado de un evento afecta el resultado del otro evento (s).
Dependent Variable	The variable in a relationship that depends on the value of the independent variable. 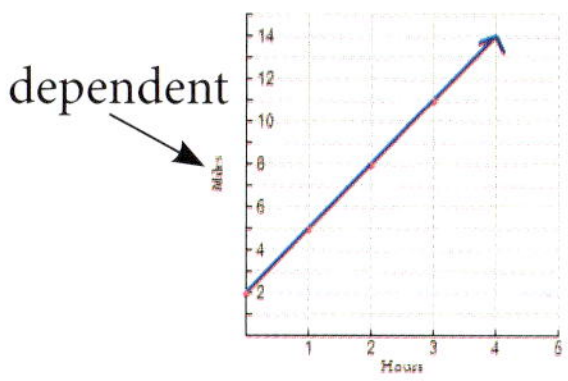	Variable Dependiente	La variable en una relación que depende del valor de la variable independiente. 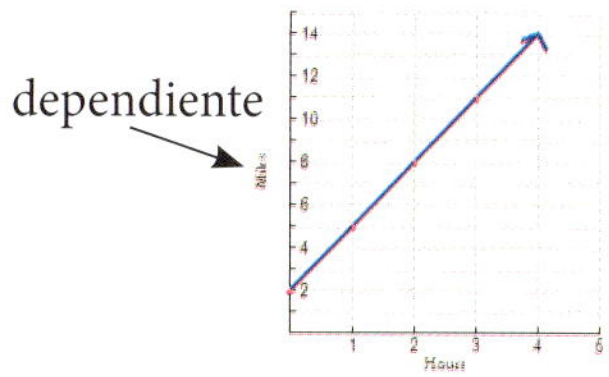
Descending Order	Numbers arranged from greatest to least.	Progresión Descendente	Los números ordenados de mayor a menor.

Diameter

The distance across a circle through the center.

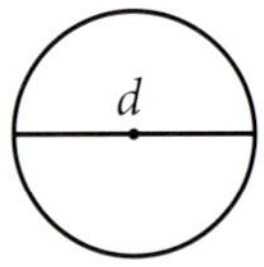

Diámetro

La distancia a través de un círculo por el centro.

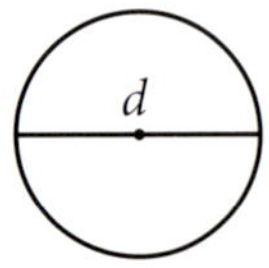

Dilation

A transformation which changes the size of the figure but not the shape.

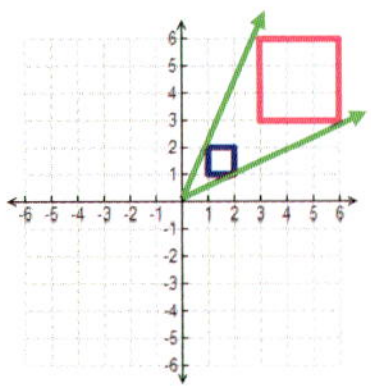

Dilatación

Una transformación que cambia el tamaño de la figura, pero no la forma.

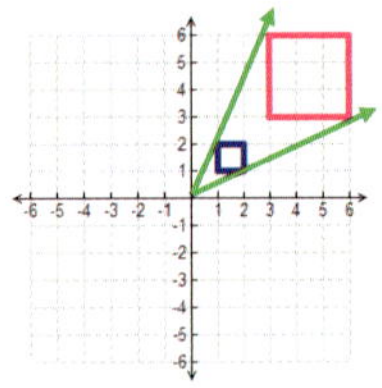

Direct Variation

A linear function that passes through the origin and has the equation $y = mx$.

Variación Directa

Una función lineal que pasa a través del origen y tiene la ecuación $y = mx$.

Discount

The decrease in the price of an item.

Descuento

La disminución de precio en un artículo.

Discrete

When a graph can be represented by a unique set of points rather than a continuous line.

Discreto

Cuando una gráfica puede ser representada por un conjunto de puntos único en vez de una recta continua.

Discriminant

In the quadratic formula, the expression under the radical sign. The discriminant provides information about the number of real roots or solutions of a quadratic equation.

$$\frac{-b \pm \sqrt{b^2 - 4ac}}{2a}$$

Discriminante

En la fórmula cuadrática, la expresión bajo el signo radical. El discriminante proporciona información sobre el número o las verdaderas raíces o soluciones de una ecuación cuadrática.

$$\frac{-b \pm \sqrt{b^2 - 4ac}}{2a}$$

Distance Formula

A formula used to find the distance between two points on the coordinate plane.

$$d = \sqrt{(x_2 - x_1)^2 + (y_2 - y_1)^2}$$

Fórmula de Distancia

Una fórmula utilizada para encontrar la distancia entre dos puntos en un plano de coordenadas.

$$d = \sqrt{(x_2 - x_1)^2 + (y_2 - y_1)^2}$$

Distributive Property

A property that can be used to rewrite an expression without parentheses.

$$a(b + c) = ab + ac$$

Propiedad Distributiva

Una propiedad que puede ser utilizada para reescribir una expresión sin paréntesis:

$$a(b + c) = ab + ac$$

Dividend

The number being divided.

$$100 \div 4 = 25$$

Dividendo

El número que es dividido.

$$100 \div 4 = 25$$

Divisor

The number used to divide.

$$100 \div 4 = 25$$

Divisor

El número utilizado para dividir.

$$100 \div 4 = 25$$

Domain

The set of input values of a function.

Dominio

El conjunto de valores entrados de la función.

Dot Plot

A data display that consists of a number line with dots equally spaced above data values.

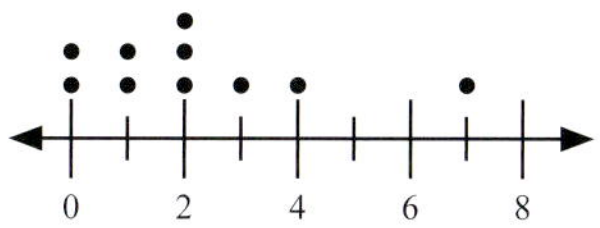

Punto de Gráfico

Una visualización de datos que consiste de una línea numérica con puntos igualmente espaciados sobre valores de datos.

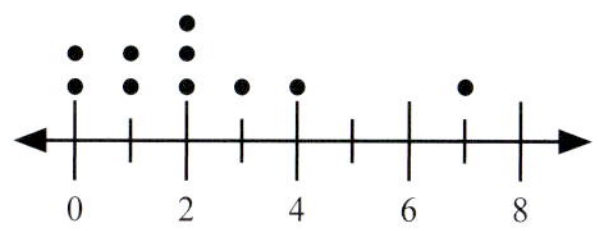

Double Stem-and-Leaf Plot

A stem-and-leaf plot where one set of data is placed on the right side of the stem and another is placed on the left of the stem.

Jon's Movies		Scott's Movies
7	195	
7 7 5 4 1 0	196	7 8
2	197	
4 4	198	4 5 8 8
4	199	1 3 4 7 9
1	200	2

Key: 195 | 7 = 1957 copyright date

Doble Gráfica de Tallo y Hoja

Una gráfica de tallo y hoja donde un conjunto de datos es colocado al lado derecho del tallo y el otro es colocado a la izquierda del tallo.

Jon's Movies		Scott's Movies
7	195	
7 7 5 4 1 0	196	7 8
2	197	
4 4	198	4 5 8 8
4	199	1 3 4 7 9
1	200	2

Key: 195 | 7 = 1957 copyright date

E

Edge

The segment where two faces of a solid meet.

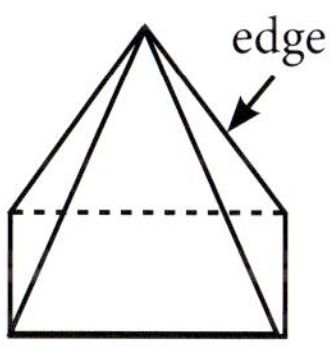

Arista (Borde)

El segmento donde dos caras de un sólido se encuentran.

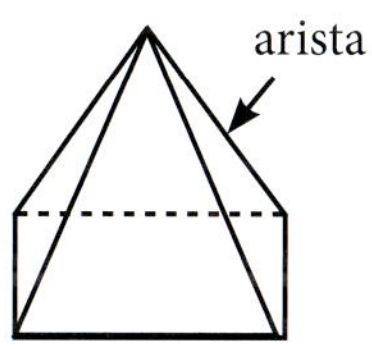

English	Definition	Español	Definición
Elimination Method	A method for solving a system of linear equations.	Método de Eliminación	Un método para resolver un sistema de ecuaciones lineales.
Enlargement	A dilation that creates an image larger than its pre-image.	Agrandamiento	Una dilatación que crea una imagen más grande que su pre-imagen.
Equally Likely	Two or more possible outcomes of a given situation that have the same probability.	Igualmente Probables	Dos o más posibles resultados de una situación dada que tienen la misma probabilidad.
Equation	A mathematical sentence that contains an equals sign between 2 expressions.	Ecuación	Una oración matemática que contiene un símbolo de igualdad entre dos expresiones.
Equiangular	A polygon in which all angles are congruent.	Equiángulo	Un polígono en el cual todos los ángulos son congruentes.
Equilateral	A polygon in which all sides are congruent. 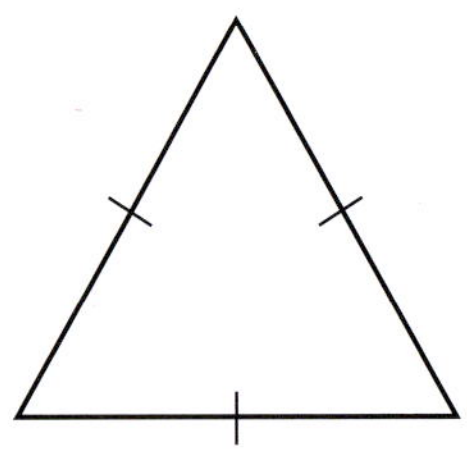	Equilátero	Un polígono en el cual todos los lados son congruentes. 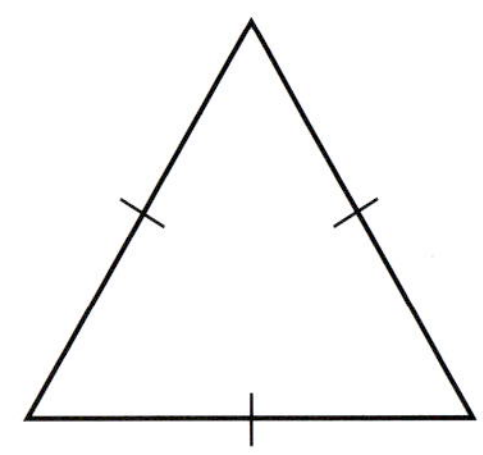
Equivalent Decimals	Two or more decimals that represent the same number.	Decimales Equivalentes	Dos o más decimales que representan el mismo número.
Equivalent Expressions	Two or more expressions that represent the same algebraic expression.	Expresiones Equivalentes	Dos o más expresiones que representan la misma expresión algebraica.
Equivalent Fractions	Two or more fractions that represent the same number.	Fracciones Equivalentes	Dos o más fracciones que representan el mismo número.
Evaluate	To find the value of an expression.	Evaluar	Encontrar el valor de una expresión.
Even Distribution	A set of data values that is evenly spread across the range of the data.	Distribución Igualada	Un conjunto de valores de datos que es esparcido de modo uniforme a través del rango de los datos.

Event	A desired outcome or group of outcomes.	Evento	Un resultado o grupo de resultados deseados.
Experimental Probability	The ratio of the number of times an event occurs to the total number of trials.	Probabilidad Experimental	La razón de la cantidad de veces que un suceso ocurre a la cantidad total de intentos.
Exponent	In x^a, a is the exponent. The exponent shows the number of times the factor (x) is repeated.	Exponente	En x^a, a es el exponente. El exponente indica el número de veces que se repite el factor (x).
Exponential Function	A function that can be described by an equation in the form $f(x) = bm^x$.	Función Exponencial	Una función que puede ser descrito por una ecuasión en la forma $f(x) = bm^x$.

F

Face	A polygon that is a side or base of a solid. 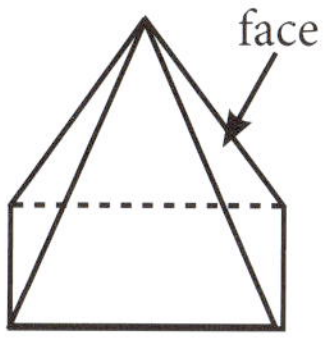	Cara	Un polígono que es una base de lado de un sólido. 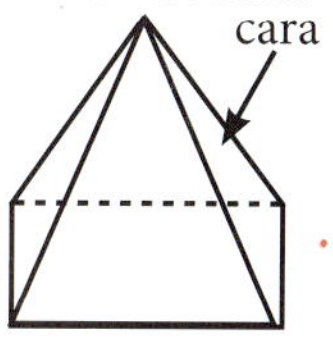
Factors	Whole numbers that can be multiplied together to find a product.	Factores	Números enteros que pueden ser multiplicados entre si para encontrar un producto.
First Quartile (Q1)	The median of the lower half of a data set.	Primer Cuartil (Q1)	Mediana de la parte inferior de un conjunto de datos.
Five-Number Summary	Describes the spread of a data set using the minimum, 1st quartile, median, 3rd quartile and maximum.	Sumario de Cinco Números	Describe la extensión de un conjunto de datos utilizando el mínimo, el primer cuartil, la mediana el tercer cuartil y el máximo.
Formula	An algebraic equation that shows the relationship among specific quantities.	Fórmula	Una ecuación algebraica que enseña la relación entre cantidades específicas.
Fraction	A number that represents a part of a whole number, written as $\frac{\text{numerator}}{\text{denominator}}$.	Fracción	Un número que representa una parte de un número entero, escrito como $\frac{\text{numerador}}{\text{denominador}}$.

Frequency	The number of times an item occurs in a data set.	Frecuencia	La cantidad de veces que un artículo ocurre en un conjunto de datos.
Frequency Table	A table which shows how many times a value occurs in a given interval.	Tabla de Frecuencia	Una tabla que enseña cuantas veces un valor ocurre en un intervalo dado.

Weight of Newborn (in Pounds)	Tally
4 – 5.5	|
5.5 – 7	|||
7 – 8.5	𝍸
8.5 – 10	||
10 – 11.5	|

Weight of Newborn (in Pounds)	Tally
4 – 5.5	|
5.5 – 7	|||
7 – 8.5	𝍸
8.5 – 10	||
10 – 11.5	|

Function	A relationship between two variables that has one output value for each input value.	Función	Una relación entre dos variables que tiene un valor de salida para cada valor de entrada.

G

General Form	A quadratic function is in general form when written $f(x) = ax^2 + bx + c$ where $a \neq 0$.	Forma General	Una función cuadrática es en forma general cuándo escrito $f(x) = ax^2 + bx + c$ donde $a \neq 0$.
Geometric Probability	Ratios of lengths or areas used to find the likelihood of an event.	Probabilidad Geométrica	Razones de longitudes o áreas utilizadas para encontrar la probabilidad de un suceso.
Geometric Sequence	A list of numbers that begins with a starting value. Each term in the sequence is generated by multiplying the previous term in the sequence by a constant multiplier.	Secuenciación Geométrica	Una lista de números que comienza con un valor inicial. Cada término de la secuencia se genera al multiplicar el término anterior de la secuencia por un multiplicar constante.
Greatest Common Factor (GCF)	The greatest factor that is common to two or more numbers.	Máximo Común Divisor (MCD)	El máximo divisor que le es común a dos o más números.
Grouping Symbols	Symbols such as parentheses or fraction bars that group parts of an expression.	Símbolos de Agrupación	Símbolos como el paréntesis o barras de fracción que agrupan las partes de una expresión.

H

Height of a Triangle — A perpendicular line drawn from the side whose length is the base to the opposite vertex.

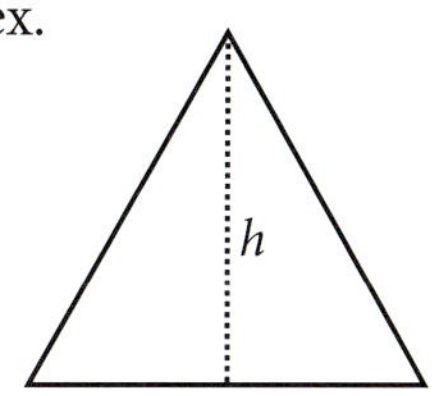

Altura de un Triángulo — Una recta perpendicular dibujada desde el lado cuya longitud es la base al vértice opuesto.

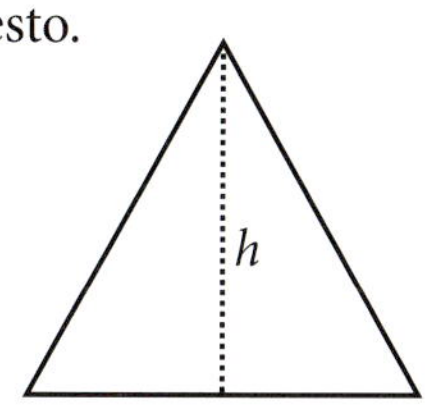

Histogram — A bar graph that displays the frequency of numerical data in equal-sized intervals.

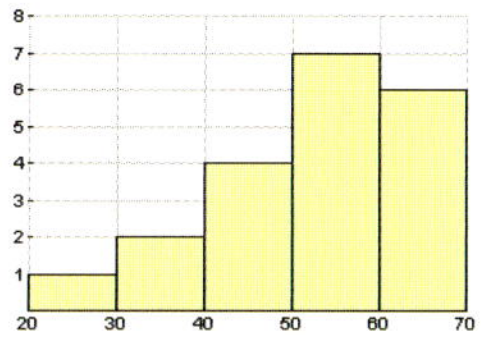

Histograma — Un gráfico de barras que muestra la frecuencia de datos numéricos en intervalos de tamaños iguales.

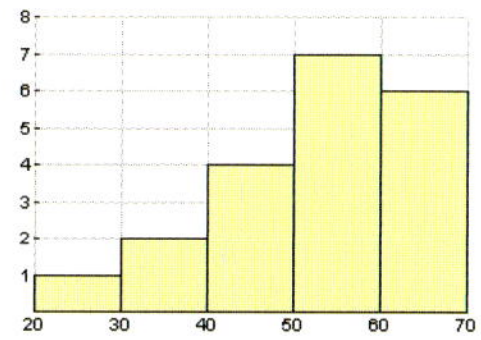

Hypotenuse — The side opposite the right angle in a right triangle.

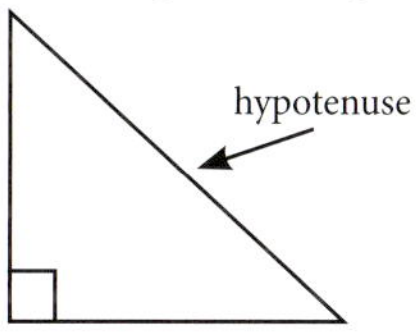

Hipotenusa — El lado opuesto el ángulo recto en un triángulo rectángulo.

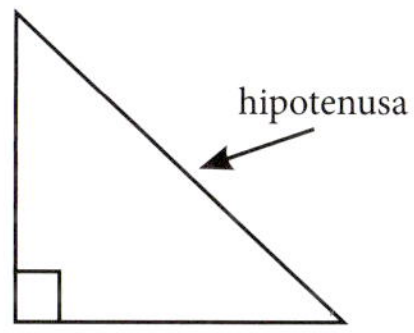

I-J-K

Image — A point or figure which is the result of a transformation or series of transformations.

Imagen — Un punto o figura que es el resultado de una transformación o una serie de transformaciones.

Improper Fraction — A fraction whose numerator is greater than or equal to its denominator.

Fracción Impropia — Una fracción cuyo numerador es mayor o igual a su denominador.

Independent Events — Two (or more) events such that the outcome of one event does not affect the outcome of the other event(s).

Eventos Independientes — Dos (o más) eventos de tal manera que el resultado de un evento no afecta el resultado del otro evento (s).

Independent Variable

The variable representing the input values.

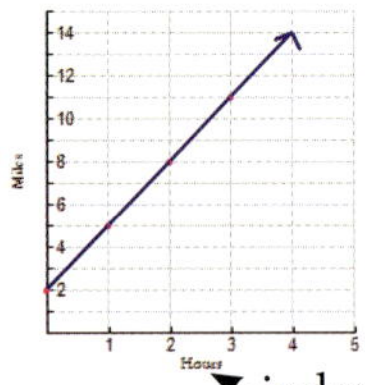

independent

Variable Independiente

La variable que representa los valores entratos.

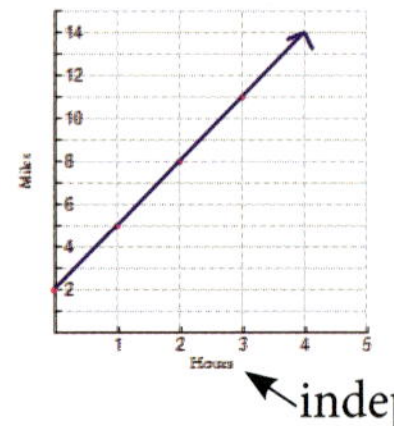

independiente

Inequality

A mathematical sentence using <, >, ≤ or ≥ to compare two quantities.

Desigualdad

Un enunciado matemático usando <, >, ≤ ó ≥ para comparar dos cantidades.

Inference

A logical conclusion based on known information.

Inferencia

Una conclusión lógica basada en la información conocida.

Input-Output Table

A table used to describe a function by listing input values with their output values.

Input, x	Output, y

Tabla de Entrada y Salida

Una tabla utilizada para describir una función al enumerar valores de entrada con sus valores de salidas.

Input, x	Output, y

Integers

The set of all whole numbers, their opposites, and 0.

Enteros

El conjunto de todos los números enteros, sus opuestos y 0.

Interquartile Range (IQR)

The difference between the 3rd quartile and the 1st quartile in a set of data.

Rango Intercuartil (IQR)

La diferencia entre el tercer cuartil y el primer cuartil en un conjunto de datos.

Inverse Operations

Operations that undo each other.

Operaciones Inversas

Operaciones que se cancelan la una a la otra.

IQR Method

A method for determining outliers using interquartile ranges.

Método IQR

Un método para determinar los datos aberrantes.

Irrational Numbers

A number that cannot be expressed as a fraction of two integers.

Números Irracionales

Un número que no puede ser expresado como una fracción de dos enteros.

Isosceles Trapezoid — A trapezoid that has congruent legs.

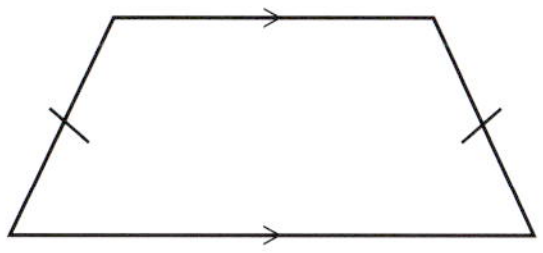

Trapezoide Isósceles — Un trapezoide con catetos congruentes.

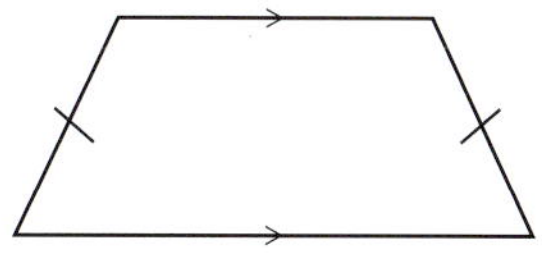

Isosceles Triangle — A triangle that has two or more congruent sides.

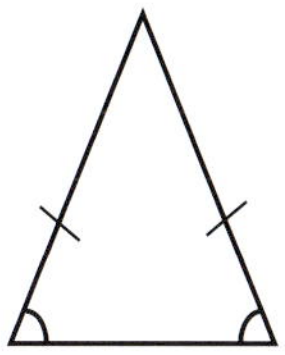

Triángulo Isósceles — Un triángulo que tiene dos o más lados congruentes.

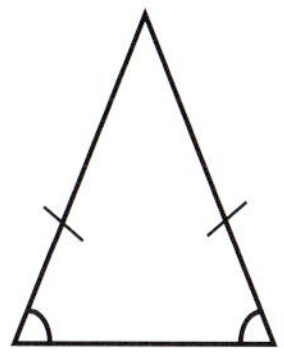

L

Lateral Face — A side of a solid that is not a base.

Cara Lateral — Un lado de un sólido que no sea una base.

Least Common Denominator (LCD) — The least common multiple of two or more denominators.

Mínimo Común Denominador (MCD) — El mínimo común múltiplo de dos o más denominadores.

Least Common Multiple (LCM) — The smallest nonzero multiple that is common to two or more numbers.

Mínimo Común Múltiplo (MCM) — El múltiplo más pequeño que no sea cero que le es común a dos o más números.

Leg — The two sides of a right triangle that form a right angle.

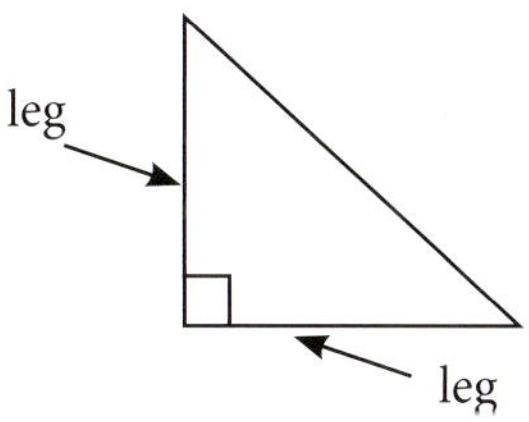

Cateto — Los dos lados de un triángulo rectángulo que forman un ángulo recto.

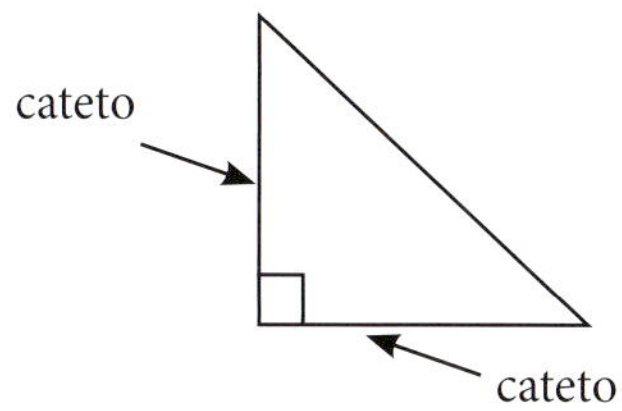

Like Terms — Terms that have the same variable(s).

Términos Semejantes — Términos que tienen el mismo variable(s).

Line of Best Fit
A line which best represents the pattern of a two-variable data set.

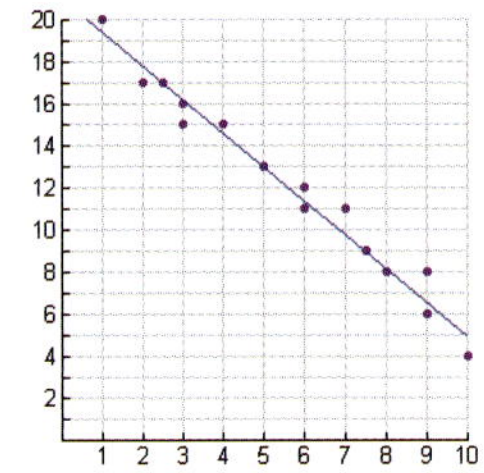

Recta de Mejor Ajuste
Una recta que mejor representa el patrón de un conjunto de datos de dos variables.

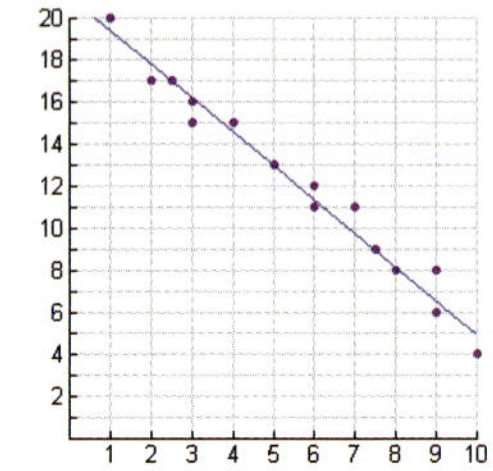

Linear Equation
An equation whose graph is a line.

Ecuación Lineal
Una ecuación cuya gráfica es una recta.

Linear Function
A function whose graph is a line.

Función Lineal
Una función cuya gráfica es una recta.

Linear Pair
Two adjacent angles whose non-common sides are opposite rays.

Par Lineal
Dos ángulos adyacentes cuyos lados no comunes son rayos opuestos.

M

Mark-up
The increase in the price of an item.

Margen de Beneficio
El aumento de precio en un artículo.

Maximum
The highest point on a curve.

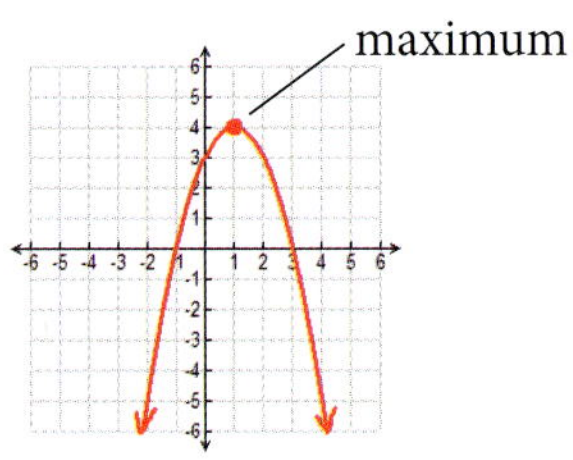

Máximo
El punto más alto en la curva.

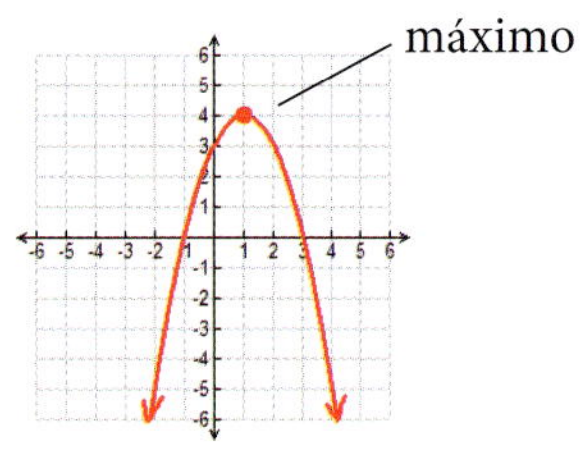

Mean
The sum of all values in a data set divided by the number of values.

Media
La suma de todos los valores en un conjunto de datos dividido entre la cantidad de valores.

Mean Absolute Deviation
A statistic that shows the average distance from the mean for all numbers in a data set.

Desviación Media Absoluta
Una estadística que muestra la distancia promedio entre la media de todos los números en una serie de datos.

Measures of Center
Numbers that are used to represent a data set with a single value; the mean, median, and mode are the measures of center.

Medidas de Centro
Números que son utilizados para representar un conjunto de datos con un solo valor; la media, la mediana, y la moda son las medidas de centro.

Measures of Variability
Statistics that help determine the spread of numbers in a data set.

Medidas de Variabilidad
Las estadísticas que ayudan a determinar la extensión de los números en una serie de datos.

Median
The middle number or the average of the two middle numbers in an ordered data set.

Mediana
El número medio o el promedio de los dos números medios en un conjunto de datos ordenados.

Minimum
The lowest point on a curve.

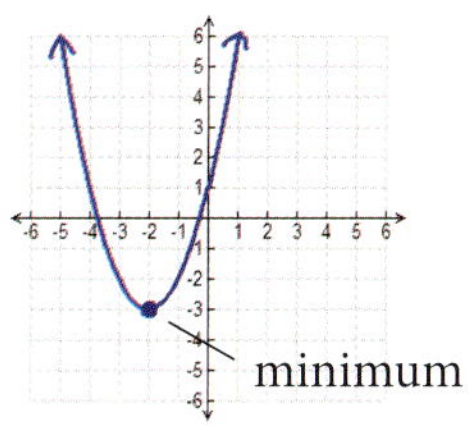

Mínimo
El punto más bajo en la curva.

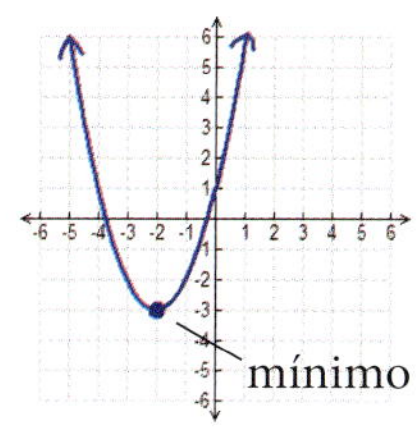

Mixed Number
The sum of a whole number and a fraction less than 1.

Números Mixtos
La suma de un número entero y una fracción menor que 1.

Mode
The number(s) or item(s) that occur most often in a data set.

Moda
El número(s) o artículo(s) que ocurre con más frecuencia en un conjunto de datos.

Motion Rate
A rate that compares distance to time.

Índice de Movimiento
Un índice que compara distancia por tiempo.

Multiple
The product of a number and nonzero whole number.

Múltiplo
El producto de un número y un número entero que no sca ccro.

N

Negative Number
A number less than 0.

Número Negativo
Un número menor que 0.

Net	A two-dimensional pattern that folds to form a solid. 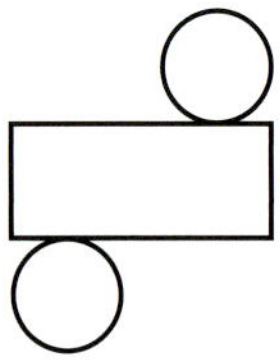	Red	Un patrón bidimensional que se dobla para formar un sólido. 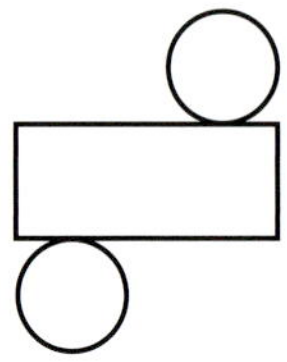
Non-Linear Function	A function whose graph does not form a line.	Ecuación No Lineal	Una ecuación cuya gráfica no forma una recta.
Normal Distribution	A set of data values where the majority of the values are located in the middle of the data set and can be displayed by a bell-shaped curve.	Distribución Normal	Un conjunto de valores de datos donde la mayoría de los valores están localizados en el medio del conjunto de datos y pueden ser mostrados por una curva de forma de campana.
Numerical Data	Data collected in the form of numbers.	Datos Numéricos	Datos recopilados en la forma de números.
Numerical Expressions	An expression consisting of numbers and operations that represents a specific value.	Expresiones Numéricas	Una expresión que consta de números y operaciones que representa un valor específico.

O

Obtuse Angle	An angle that measures more than 90° but less than 180°. 	Ángulo Obtuso	Un ángulo que mide más de 90° pero menos de 180°. 
Opposites	Numbers that are the same distance from 0 on a number line but are on opposite sides of 0.	Opuestos	Números a la misma distancia del 0 en un recta numérica pero en lados opuestos del 0.
Order of Operations	The rules to follow when evaluating an expression with more than one operation.	Orden de Operaciones	Las reglas a seguir cuando se evalúa una expresión con más de una operación.
Ordered Pair	A pair of numbers used to locate a point on a coordinate plane (x, y).	Par Ordenado	Un par de números utilizados para localizar un punto en un plano de coordinadas (x, y).

Origin
The point where the x- and y-axis intersect on a coordinate plane (0, 0).

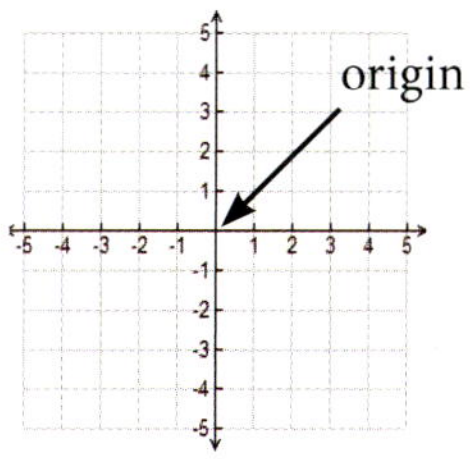

Origen
El punto donde el eje de la x- y el de la y- se cruzan en un plano de coordinadas (0,0).

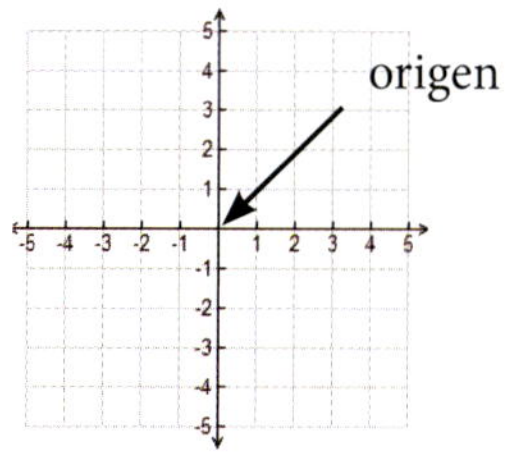

Outcome
One possible result from an experiment or a probability sample space.

Resultado
Un resultado posible de un experimento o un espacio de probabilidad de la muestra.

Outlier
An extreme value that varies greatly from the other values in a data set.

Dato Aberrante
Un valor extremo que varía mucho de los otros valores en un conjunto de datos.

P

Parabola
The graph of a quadratic function.

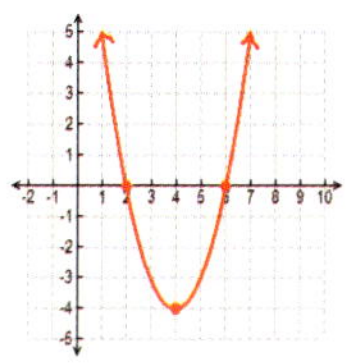

Parábola
La gráfica de una función cuadratica.

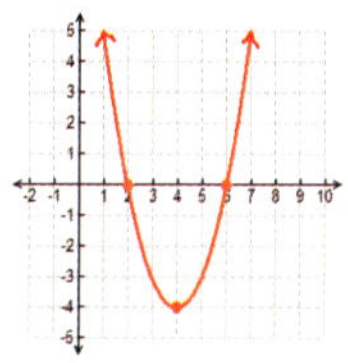

Parallel
Lines in the same plane that never intersect.

Paralela
Rectas en el mismo plano que nunca se intersecan.

Parallel Box-and-Whisker Plot
One box-and-whisker plot placed above another; often used to compare data sets.

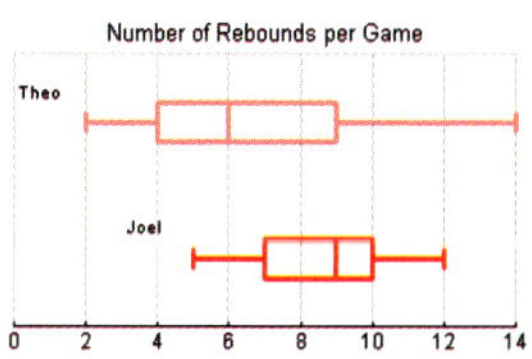

Diagrama Paralelo de Líneas y Bloques
Un diagrama de líneas y bloques ubicado sobre otro para comparar conjuntos de datos.

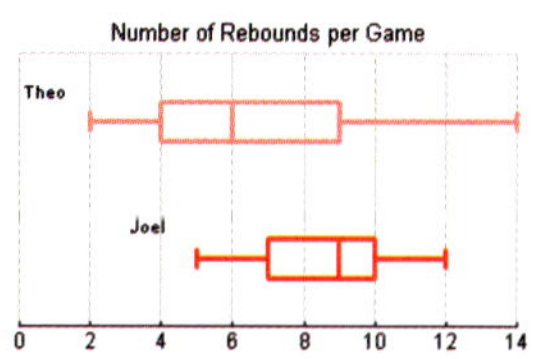

Parallelogram	A quadrilateral with both pairs of opposite sides parallel. 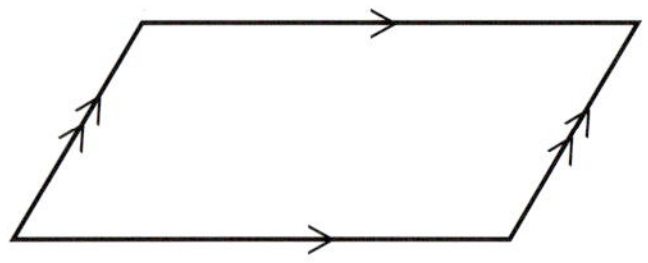	Paralelogramo	Un cuadrilateral con ambos pares de lados opuestos paralelos. 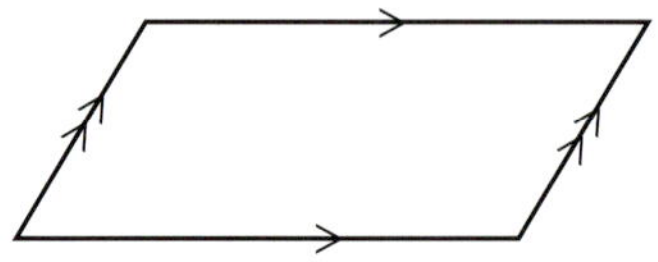
Parent Function	The simplest form of a particular type of function.	Función Principal	La forma más simple de un tipo particular de la función.
Parent Graph	The most basic graph of a function.	Gráfico Matriz	La gráfica más básica de una función.
Percent	A ratio that compares a number to 100.	Por Ciento	Una razón que compara un número con 100.
Percent of Change	The percent a quantity increases or decreases compared to the original amount.	Por Ciento de Cambio	El por ciento que una cantidad aumenta o disminuye comparado a la cantidad original.
Percent of Decrease	The percent of change when the new amount is less than the original amount.	Por Ciento de Disminución	El por ciento de cambio cuando la nueva cantidad es menos que la cantidad original.
Percent of Increase	The percent of change when the new amount is more than the original amount.	Por Ciento de Incremento	El por ciento de cambio cuando la nueva cantidad es más que la cantidad original.
Perfect Cube	A number whose cube root is an integer.	Cubo Perfecto	Un número cuyo raíz cúbica es un número entero.
Perfect Square	A number whose square root is an integer.	Cuadrado Perfecto	Un número cuyo raíz cuadrado es un número entero.
Perfect Square Trinomial	A trinomial that is the square of a binomial.	Trinomio Cuadrado Perfecto	Un trinomio que es el cuadrado de un binomio.
Perimeter	The distance around a figure.	Perímetro	La distancia alrededor de una figura.

Perpendicular

Two lines or segments that form a right angle.

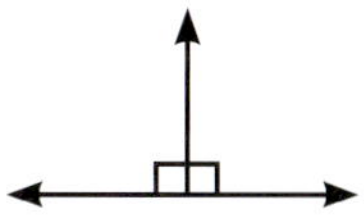

Perpendicular

Dos rectas o segmentos que forman un ángulo recto.

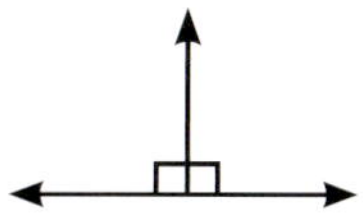

Pi (π)

The ratio of the circumference of a circle to its diameter.

Pi (π)

La razón de la circunferencia de un círculo a su diámetro.

Pictograph

A graph that uses pictures to compare the amounts represented in a categorical data set.

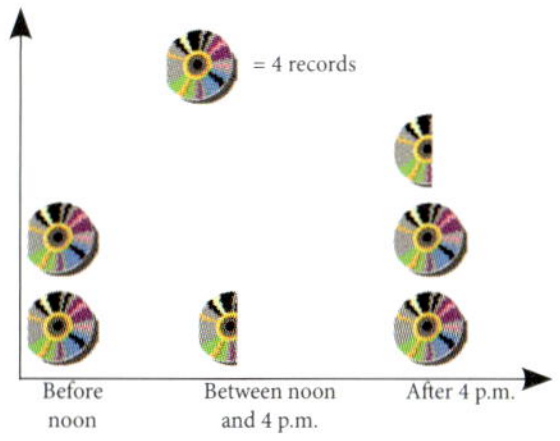

Gráfica Pictórica

Una gráfica que utiliza dibujos para comparar las cantidades representadas en un conjunto de datos categóricos.

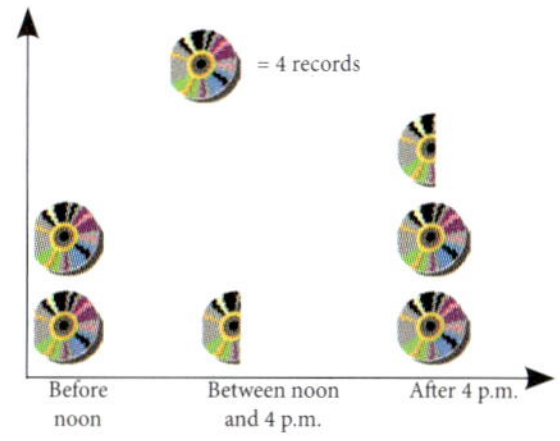

Pie Chart

A circle graph that shows information as sectors of a circle.

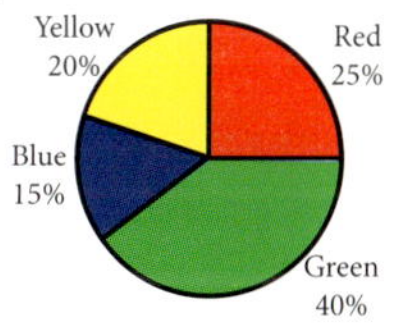

Gráfico Circular

Enseña la información como sectores de un círculo.

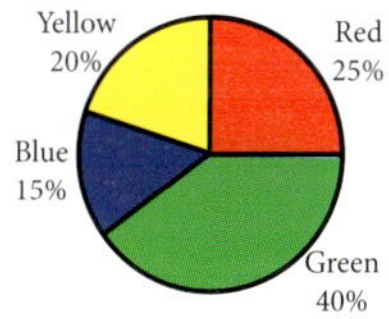

Polygon

A closed figure formed by three or more line segments.

Polígono

Una figura cerrada formada por tres o más segmentos de rectas.

Population

The entire group of people or objects one wants to gather information about.

Población

Todo el grupo de personas o los objetos a los que se quiere obtener información sobre.

Positive Number

A number greater than 0.

Número Positivo

Un número mayor que 0.

Power

An expression such as x^a which consists of two parts, the base (x) and the exponent (a).

Potencia

Una expresión como x^a que consiste de dos partes, la base (x) y el exponente (a).

Pre-image

The original figure prior to a transformation.

Pre-imagen

La figura original antes de una transformación.

Term	Definition	Término	Definición
Prime Factorization	When any composite number is written as the product of all its prime factors.	Factorización Prima	Cuando cualquier número compuesto es escrito como el producto de todos los factores primos.
Prime Number	A whole number larger than 1 that has only two possible factors, 1 and itself.	Número Primo	Un número entero mayor que 1 que tiene solo dos factores posibles, 1 y el mismo.
Prism	A solid formed by polygons with two congruent, parallel bases. 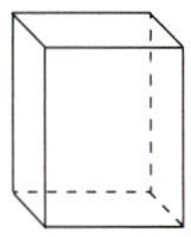	Prisma	Un sólido formado por polígonos con dos bases congruentes y paralelas. 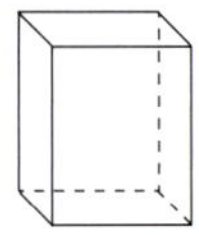
Probability	The measure of how likely it is an event will occur.	Probabilidad	La medida de cuán probable un suceso puede ocurrir.
Product	The answer to a multiplication problem.	Producto	La respuesta a un problema de multiplicación.
Proper Fraction	A fraction with a numerator that is less than the denominator.	Fracción Propia	Una fracción con un numerador que es menos que el denominador.
Proportion	An equation stating two ratios are equivalent.	Proporción	Una ecuación que establece que dos razones son equivalentes.
Protractor	A tool used to measure angles. 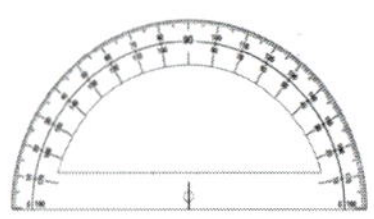	Transportador	Una herramienta para medir ángulos. 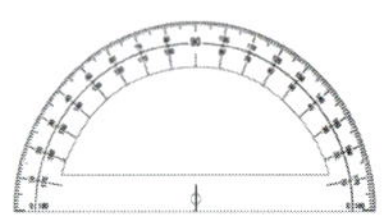
Pyramid	A solid with a polygonal base and triangular sides that meet at a vertex. 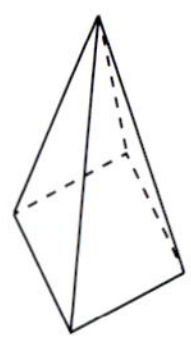	Pirámide	Un sólido con una base poligonal y lados triangulares que se encuentran en un vértice. 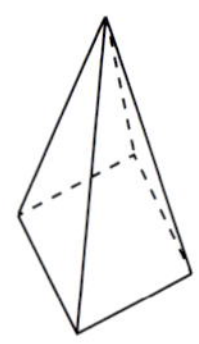

Pythagorean Triple

A set of three positive integers (a, b, c) such that $a^2 + b^2 = c^2$.

Triple de Pitágoras

Un conjunto de tres enteros positivos (a, b, c) tal que $a^2 + b^2 = c^2$.

Q

Q-Points

Points that are created by the intersection of the quartiles for the x- and y-values of a two-variable data set.

Puntos Q

Puntos que son creados por la intersección de los cuartiles para los valores de la x- y la y- de un conjunto de datos de dos variables.

Quadrants

Four regions formed by the x and y axes on a coordinate plane.

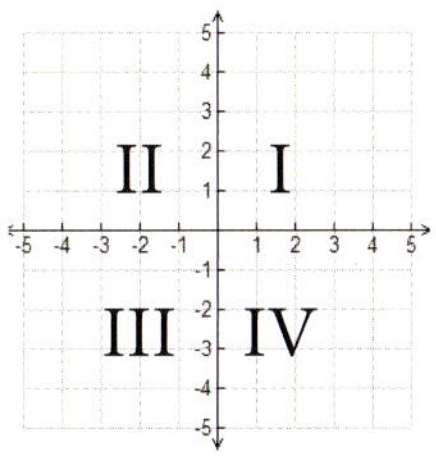

Cuadrantes

Cuatro regiones formadas por el eje-x y el eje-y en un plano de coordinadas.

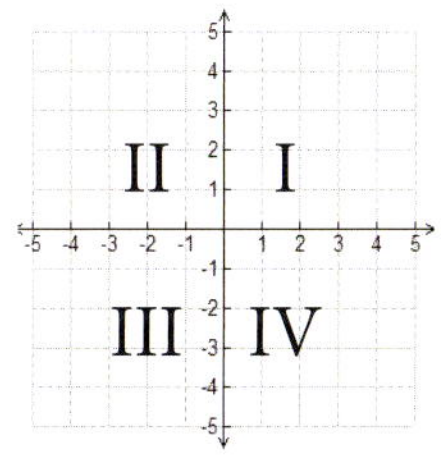

Quadratic Formula

A method which can be used to solve quadratic equations in the form $0 = ax^2 + bx + c$, where $a \neq 0$.

$$x = \frac{-b \pm \sqrt{b^2 - 4ac}}{2a}$$

Fórmula Cuadrática

Un métado que puede usarse para resolver ecuaciones cuadraticas en la forma $0 = ax^2 + bx + c$, donde $a \neq 0$.

$$x = \frac{-b \pm \sqrt{b^2 - 4ac}}{2a}$$

Quadratic Function

Any function in the family with the parent function of $f(x) = x^2$.

Función Cuadrática

Cualquier otra función en la familia con la función principal de $f(x) = x^2$.

Quadrilateral

A polygon with four sides.

Cuadrilateral

Un polígono con cuatro lados.

Quotient

The answer to a division problem.

Cociente

La solución a un problema de división.

R

Radius

The distance from the center of a circle to any point on the circle.

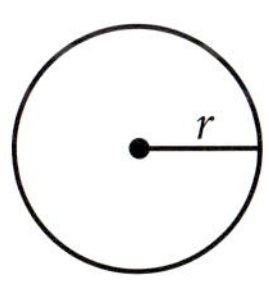

Radio

La distancia desde el centro de un círculo a cualquier punto en el círculo.

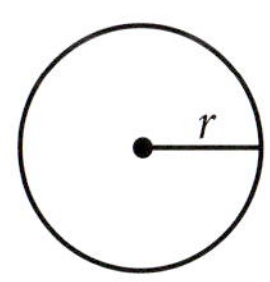

Term	Definition	Término	Definición
Random Sample	A group from a population created when each member of the population is equally likely to be chosen.	Muestra Aleatoria	Un grupo de una población creada cuando cada miembro de la población tiene la misma probabilidad de ser elegido.
Range (of a data set)	The difference between the maximum and minimum values in a data set.	Rango	La diferencia entre los valores máximo y mínimo en un conjunto de datos.
Range (of a function)	The set of output values of a function.	Rango (de una función)	El conjunto de valores salidos de la función.
Rate	A ratio of two numbers that have different units.	Índice	Una proporción de dos números con diferentes unidades.
Rate Conversion	A process of changing at least one unit of measurement in a rate to a different unit of measurement.	Conversión de Índice	Un proceso de cambiar por lo menos una unidad de medición en un índice a una diferente unidad de medición.
Rate of Change	The change in y-values over the change in x-values on a linear graph.	Índice de Cambio	El cambio en los valores de y sobre el cambio en los valores de x en una gráfica lineal.
Ratio	A comparison of two numbers using division. $a : b$ $\quad \frac{a}{b}$ $\quad a$ to b	Razón	Una comparación de dos números utilizando división. $a : b$ $\quad \frac{a}{b}$ $\quad a$ a b
Rational Number	A number that can be expressed as a fraction of two integers.	Número Racional	Un número que puede ser expresado como una fracción de dos enteros.
Ray	A part of a line that has one endpoint and extends forever in one direction. 	Rayo	Una parte de una recta que tiene un punto final y se extiende eternamente en una dirección. 
Real Numbers	The set of numbers that includes all rational and irrational numbers.	Números Racionales	El conjunto de números que incluye todos los números racionales e irracionales.

Reciprocals

Two numbers whose product is 1.

Recíprocos

Dos números cuyo producto es 1.

Rectangle

A quadrilateral with four right angles.

Rectángulo

Un cuadrilátero con cuatro ángulos rectos.

Recursive Routine

A routine described by stating the start value and the operation performed to get the following terms.

Rutina Recursiva

Una rutina descrita al exponer el valor del comienzo y la operación realizada para conseguir los términos siguientes.

Recursive Sequence

An ordered list of numbers created by a first term and a repeated operation.

Secuencia Recursiva

Una lista de números ordenados creada por un primer término y una operación repetida.

Reduction

A dilation that creates an image smaller than its pre-image.

Reducción

Una dilatación que crea una imagen más pequeña que su pre-imagen.

Reflection

A transformation in which a mirror image is produced by flipping a figure over a line.

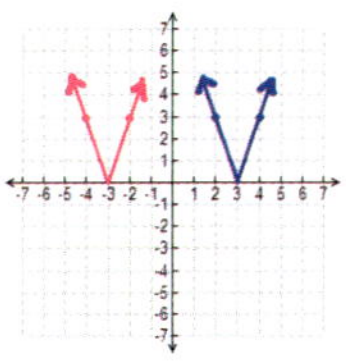

Reflexión

Una transformación en el que se produce una imagen reflejada volteando una figura sobre una línea.

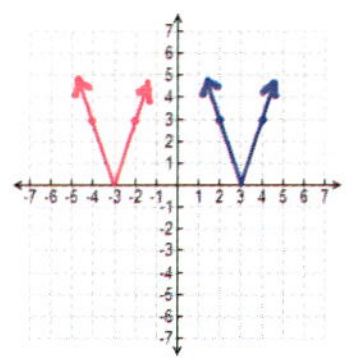

Relative Frequency

The ratio of the observed frequency to the total number of frequencies in an experiment or survey.

Frecuencia Relativa

La proporción de la frecuencia observada para el número total de frecuencias en un experimento o estudio.

Remainder

A number that is left over when a division problem is completed.

Remanente

Un número que queda cuando un problema de división se ha completado.

Repeating Decimal

A decimal that has one or more digits that repeat forever.

Decimal Repetitivo

Un decimal que tiene uno o más dígitos que se repiten eternamente.

Representative Sample — A group from a population that accurately represents the entire population.

Muestra Representativa — Un grupo de una población que representa con precisión toda la población.

Rhombus — A quadrilateral with four sides equal in measure.

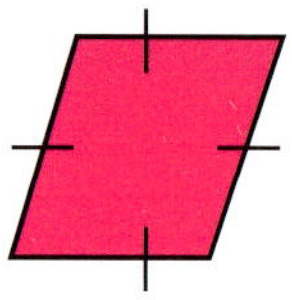

Rombo — Un cuadrilátero con cuatro lados iguales en la medida.

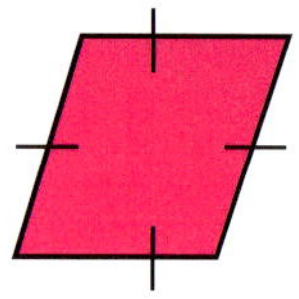

Right Angle — An angle that measures 90°.

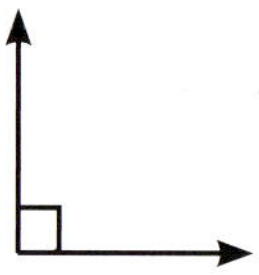

Ángulo Recto — Un ángulo que mide 90°.

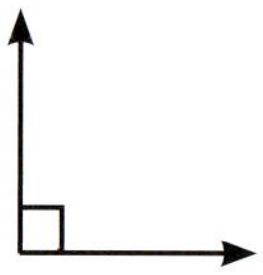

Roots — The x-intercepts of a quadratic function.

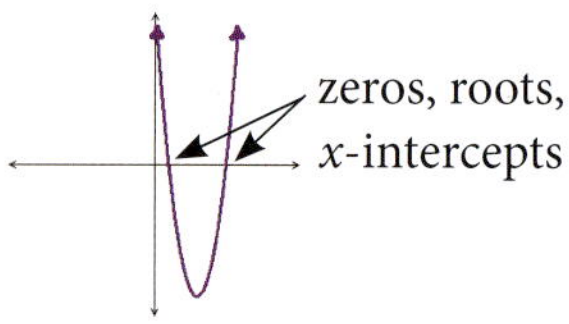

Raíces — Las intersecciones-x de una función cuadratica.

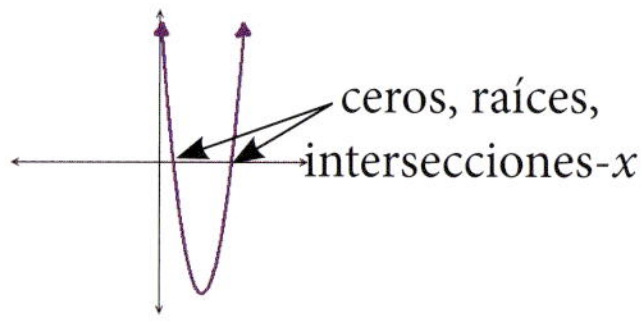

Rotation — A transformation which turns a point or figure about a fixed point, often the origin.

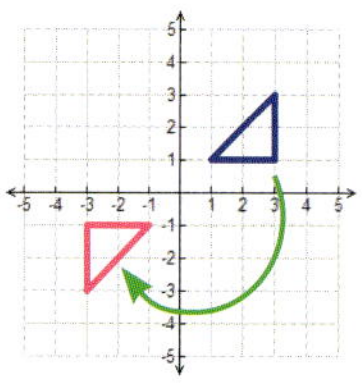

Rotación — Una transformación que convierte un punto a una figura sobre un punto fijo.

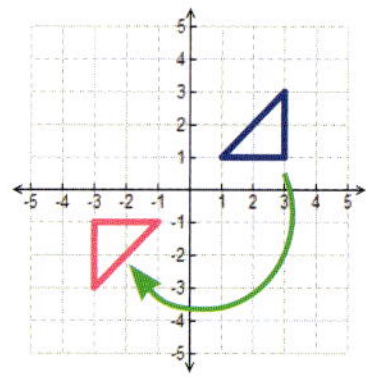

S

Sales Tax — An amount added to the cost of an item. The amount added is a percent of the original amount as determined by a state, county or city.

Impuesto sobre las Ventas — Una cantidad añadida al costo de un artículo. La cantidad añadida es un por ciento de la cantidad original determinado por el estado, condado o ciudad.

Same-Side Interior Angles

Two angles that are on the inside of two lines and are on the same side of a transversal.

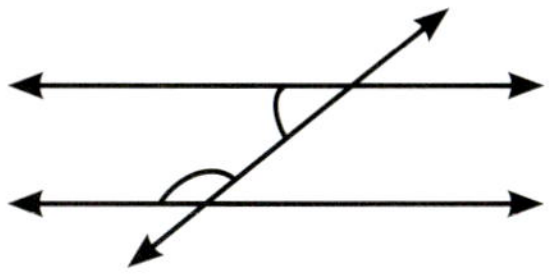

Ángulos Interiores del Mismo Lado

Dos ángulos que están en el interior de dos rectas y están en el mismo lado de una transversal.

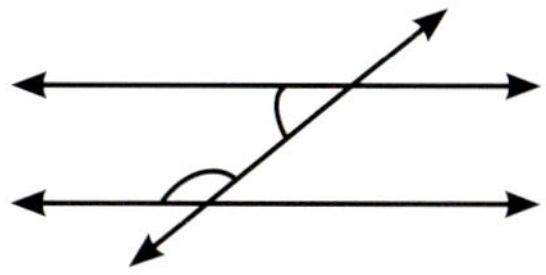

Sample

A group from a population that is used to make conclusions about the entire population.

Muestra

Un grupo de una población que se utiliza para sacar conclusiones sobre toda la población.

Sample Space

The set of all possible outcomes.

Muestra de Espacio

El conjunto de todos los resultados posibles.

Scale

The ratio of a length on a map or model to the actual object.

Escala

La razón de una longitud en un mapa o modelo al objeto verdadero.

Scale Factor

The ratio of corresponding sides in two similar figures.

Factor de Escala

La razón de los lados correspondientes en dos figuras similares.

Scalene Triangle

A triangle that has no congruent sides.

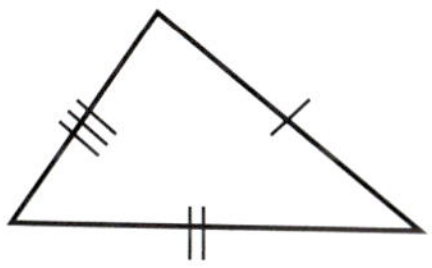

Triángulo Escaleno

Un triángulo sin lados congruentes.

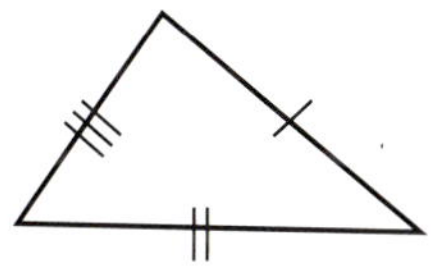

Scatter Plot

A set of ordered pairs graphed on a coordinate plane.

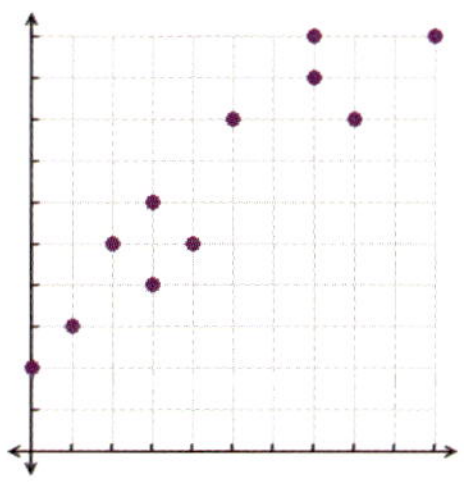

Diagrama de Dispersión

Un conjunto de pares ordenados graficados en un plano de coordenadas.

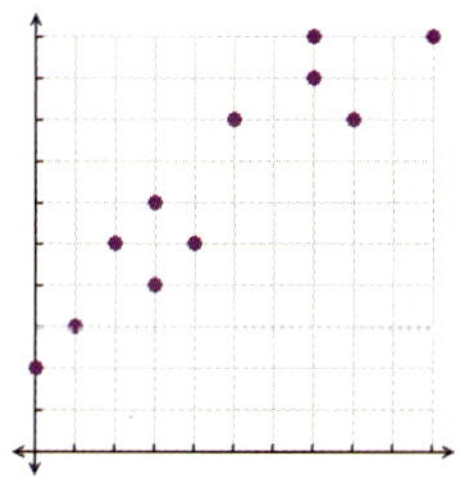

Scientific Notation

Scientific notation is an exponential expression using a power of 10 where $1 \le N < 10$ and P is an integer. $N \times 10^P$

La Notación Científica

Notación científica es una expresión exponencial con una potencia de 10, donde $1 \le N < 10$ y P es un número entero. $N \times 10^P$

Sector — A portion of a circle enclosed by two radii.

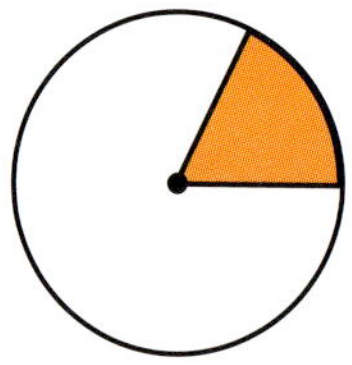

Sector — Una porción de un circulo encerado por dos radios.

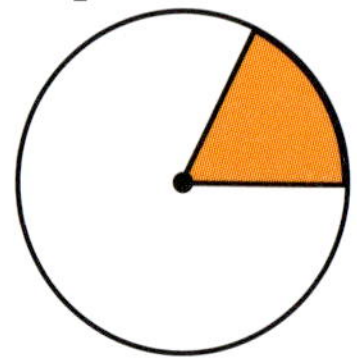

Sequence — An ordered list of numbers.

Sucesión — Una lista de números ordenados.

Similar Figures — Two figures that have the exact same shape, but not necessarily the exact same size.

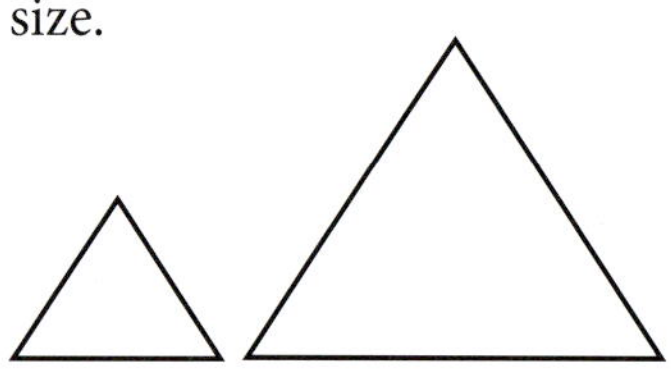

Figuras Similares — Dos figuras que tienen exactamente la misma forma, pero no necesariamente el mismo tamaño exacto.

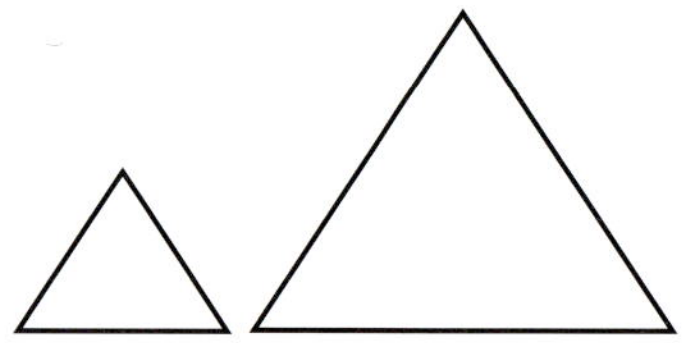

Similar Solids — Solids that have the same shape and all corresponding dimensions are proportional.

Sólidos Similares — Sólidos con la misma forma y todas sus dimensiones correspondientes son proporcionales.

Simplest Form — A fraction whose numerator and denominator's only common factor is 1.

Expresión Simple — Una fracción cuyo único factor común del numerador y del denominador es 1.

Simplify an Expression — To rewrite an expression without parentheses and combine all like terms.

Simplificar una Expresión — Reescribir una expresión sin paréntesis y combinar todos los términos iguales.

Simulation — An experiment used to model a situation.

Simulación — Un experimento utilizado para modelar una situación.

Single-Variable Data — A data set with only one type of data.

Datos de una Variable — Un conjunto de datos con tan solo un tipo de datos.

Sketch — To make a figure free hand without the use of measurement tools.

Esbozo — Hacer una figura a mano libre sin utilizar herramientas de medidas.

Skewed Left — A plot or graph with a longer tail on the left-hand side.

Torcido a la Izquierda — Un gráfico con una cola al lado izquierdo.

English	Definition	Español	Definición
Skewed Right	A plot or graph with a longer tail on the right-hand side.	Torcido a la Derecha	Un gráfico con una cola al lado derecho.
Slant Height	The height of a lateral face of a pyramid or cone. 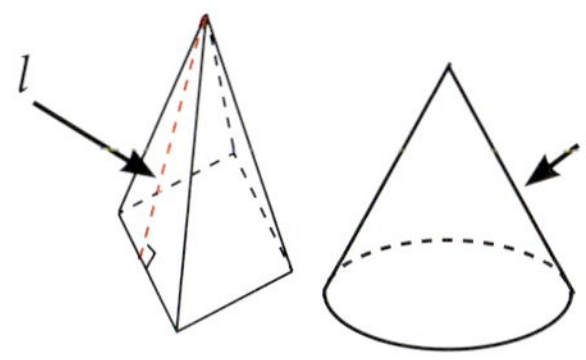	Altura Sesgada	La altura de un cara lateral de una pirámide o cono. 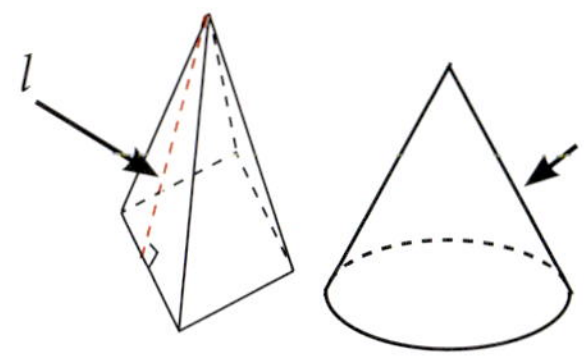
Slope	The ratio of the vertical change to the horizontal change in a linear graph.	Pendiente	La razón del cambio vertical al cambio horizontal en una gráfica lineal.
Slope Triangle	A right triangle formed where one leg represents the vertical rise and the other leg is the horizontal run in a linear graph. 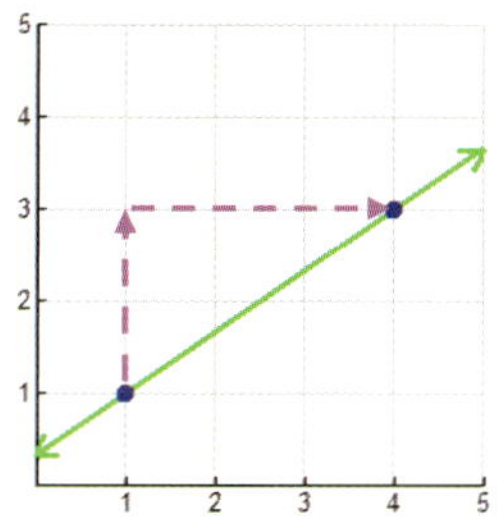	Triángulo de Pendiente	Un triángulo rectángulo formado donde una cateto representa el ascenso y la otra es una carrera horizontal en una gráfica lineal. 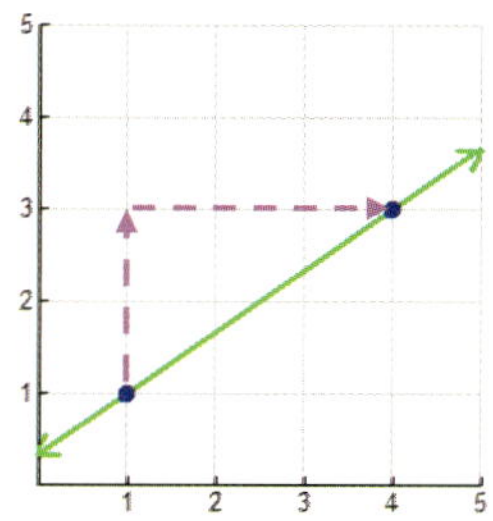
Slope-Intercept Form	A linear equation written in the form $y = mx + b$.	Forma de las Intersecciones con la Pendiente	Una ecuación lineal escrita en la forma $y = mx + b$.
Solid	A three-dimensional figure that encloses a part of space.	Sólido	Una figura tridimensional que encierra una parte del espacio.
Solution	Any value or values that makes an equation true.	Solución	Cualquier valor o valores que hacen una ecuación verdadera.
Solution of a System of Linear Equations	The ordered pair that satisfies both linear equations in the system.	Solución de un Sistema de Ecuaciones Lineales	El par ordenado que satisface ambas ecuaciones lineales en el sistema.

Sphere

A solid formed by a set of points in space that are the same distance from a center point.

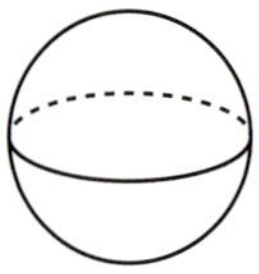

Esfera

Un sólido formado por un conjunto de puntos en el espacio que están a la misma distancia de un punto central.

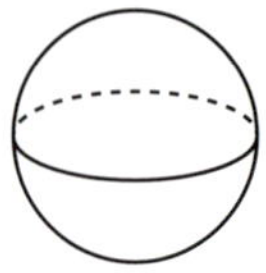

Square

A quadrilateral with four right angles and four congruent sides.

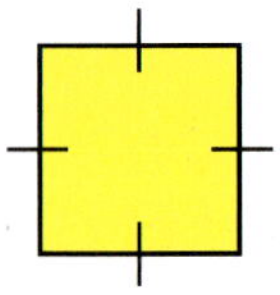

Cuadrado

Un cuadrilátero con cuatro ángulos rectos y cuatro lados congruente.

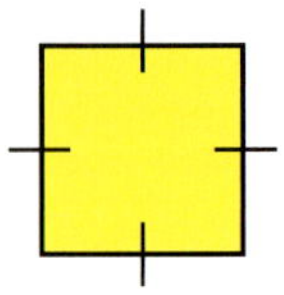

Square Root

One of the two equal factors of a number.

$3 \cdot 3 = 9 \qquad 3 = \sqrt{9}$

Raíz Cuadrada

Uno de los dos factores iguales de un número.

$3 \cdot 3 = 9 \qquad 3 = \sqrt{9}$

Squared

A term raised to the power of 2.

Cuadrado

Un término elevado a la potencia de 2.

Start Value

The output value that is paired with an input value of 0 in an input-output table.

Valor de Comienzo

El valor de salida que es aparejado con un valor de entrada de 0 en una tabla de entradas y salidas.

Statistics

The process of collecting, displaying and analyzing a set of data.

Estadísticas

El proceso de recopilar, exponer y analizar un conjunto de datos.

Stem-and-Leaf Plot

A plot which uses the digits of the data values to show the shape and distribution of the data set.

Stem	Leaf
5	6
6	
7	2 5 9 9
8	0 0 6 8 9
9	2 3 4 8
10	0 0

Key: 7 | 5 = 75

Gráfica de Tallo y Hoja

Un diagrama que utiliza los dígitos de los valores de datos para mostrar la forma y la distribución del conjunto de datos.

Stem	Leaf
5	6
6	
7	2 5 9 9
8	0 0 6 8 9
9	2 3 4 8
10	0 0

Key: 7 | 5 = 75

Straight Angle

An angle that measures 180°.

Ángulo Recto

Un ángulo que mide 180°.

English	Definition	Español	Definición
Straight Edge	A ruler-like tool with no markings.	Borde Recto	Un gobernante como herramienta sin marcas.
Substitution Method	A method for solving a system of linear equations.	Método de Substitución	Un método para resolver un sistema de ecuaciones lineales.
Supplementary Angles	Two angles whose sum is 180°.	Ángulos Suplementarios	Dos ángulos cuya suma es 180°.
Surface Area	The sum of the areas of all the surfaces on a solid.	Área de la Superficie	La suma de las áreas de todas las superficies en un sólido.
System of Linear Equations	Two or more linear equations.	Sistema de Ecuaciones Lineales	Dos o más ecuaciones lineales.

T

English	Definition	Español	Definición
Term	A number or the product of a number and a variable in an algebraic expression; a number in a sequence.	Término	Un número o el producto de un número y una variable en una expresión algebraica; un número en una sucesión.
Terminating Decimal	A decimal that stops.	Decimal Terminado	Un decimal que para.
Theorem	A relationship in mathematics that has been proven.	Teorema	Una relación en las matemáticas que ha sido probada.
Theoretical Probability	The ratio of favorable outcomes to the number of possible outcomes.	Probabilidad Teórica	La proporción de resultados favorables a la cantidad de resultados posibles.
Third Quartile (Q3)	The median of the upper half of a data set.	Tercer Cuartil (Q3)	Mediana de la parte superior de un conjunto de datos.
Tick Marks	Equally divided spaces marked with a small line between every inch or centimeter on a ruler. 	Marcas de Graduación	Espacios divididos igualmente marcados con una línea pequeña entre cada pulgada o centímetro en una regla. 
Transformation	The movement of a figure on a graph so that it changes size or position.	Transformación	El movimiento de una figura en un gráfico de modo que cambia el tamaño o posición

Translation	A transformation in which a figure is shifted up, down, left or right. 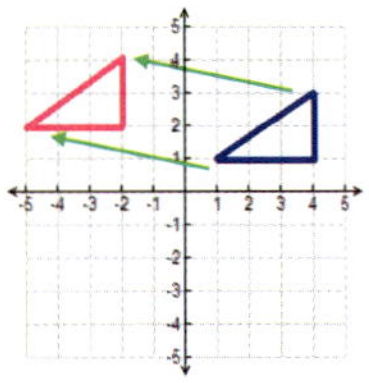	Traducción	Una transformación donde la figura se mudo arriba, abajo, a la izquierda o a la derecha. 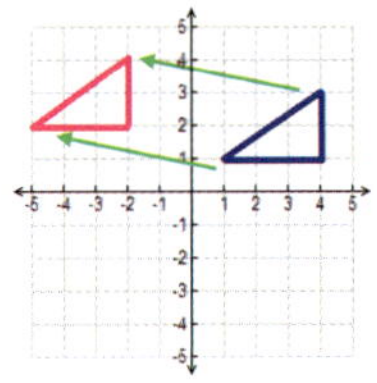
Transversal	A line that intersects two or more lines in the same plane.	Transversal	Una recta que interseca dos o más rectas en el mismo plano.
Trapezoid	A quadrilateral with exactly one pair of parallel sides. 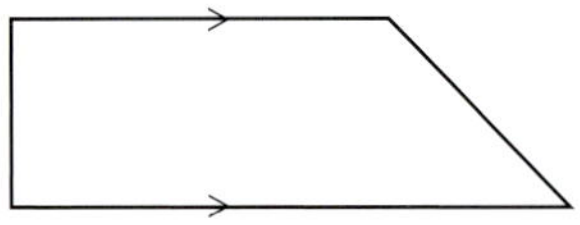	Trapezoide	Un cuadrilateral con exactamente un par de lados paralelos. 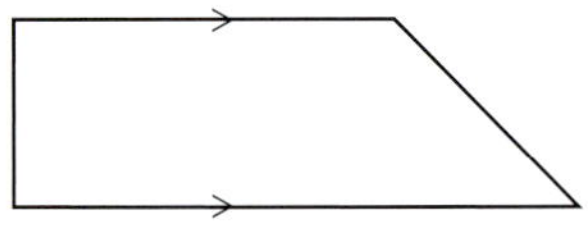
Tree Diagram	A display that organizes information to determine possible outcomes. 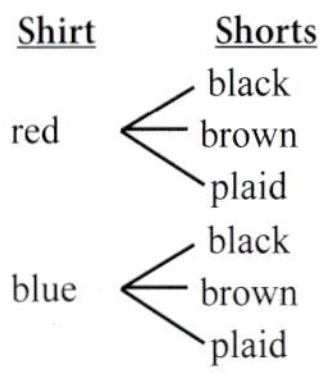	Diagrama de Árbol	Una pantalla que organiza la información para determinar los posibles resulatados. 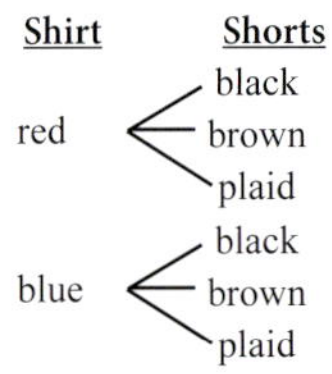
Trial	A single act of performing an experiment.	Prueba	Un solo intento de realizar un experimento.
Trinomial	An expression with three terms (i.e. $x^2 - 3x + 4$).	Trinomio	Una expreción que tiene tres terminos (es decir: $x^2 - 3x + 4$).
Two-Step Equation	An equation that has two different operations.	Ecuación de Dos Pasos	Una ecuación que tiene dos operaciones diferentes.
Two-Variable Data	A data set where two groups of numbers are looked at simultaneously.	Datos de dos Variables	Un conjunto de datos dónde dos grupos de números se observan simultáneamente.

Two-Way Frequency Table — A table that shows how many times a value occurs for a pair of categorical data.

Dog Owners \ Walk	Yes	No
Yes	15	20
No	25	20

Tabla de Frecuencia Bidireccional — Una tabla que muestra cuántas veces aparece un valor de un par de datos categóricos.

Perro Propietario \ Paseo	Si	No
Si	15	20
No	25	20

U-V-W

Unit Rate — A rate with a denominator of 1.

Índice de Unidad — Un índice con un denominador de 1.

Univariate Data — Data that describes one variable (i.e., scores on a test).

Data Univariados — Datos que describen una variable (es decir: puntajes en una prueba).

Variable — A symbol that represents one or more numbers.

Variable — Un símbolo que representa uno o más números.

Vertex — The minimum or maximum point on a parabola.

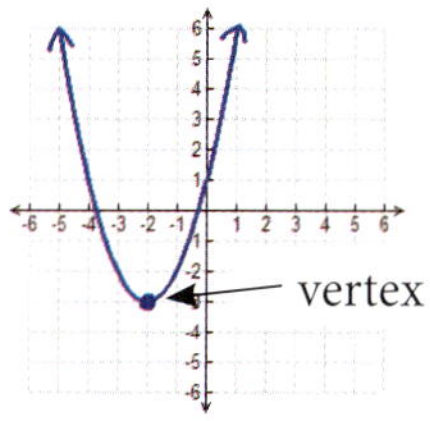

Vértice — El mínimo o máximo punto en una parábola.

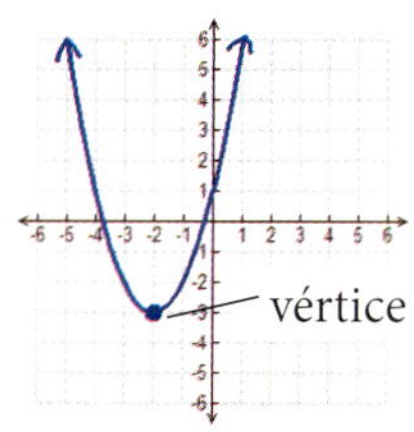

Vertex of a Solid — The point where three or more edges meet.

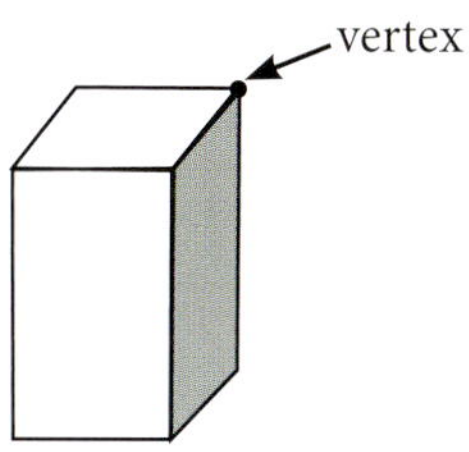

Vértice de un Sólido — El punto donde tres o más bordes se encuentran.

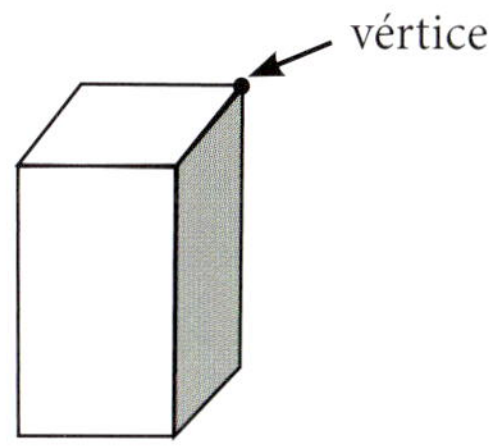

Vertex of a Triangle — A point where two sides of a triangle meet.

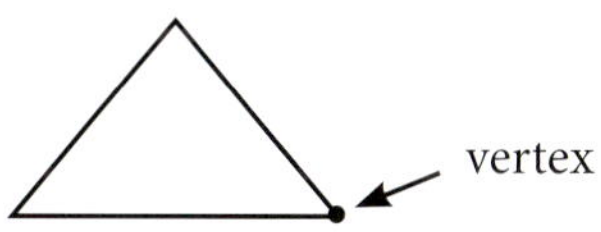

Vértice de un Triángulo — Un punto donde dos lados de un triángulo se encuentran.

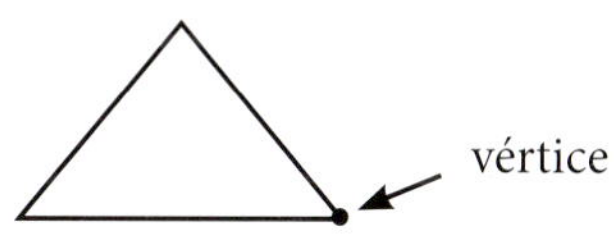

Vertex of an Angle — The common endpoint of the two rays that form an angle.

Vértice de un Ángulo — El punto final en común de los dos rayos que forma un ángulo.

Vertex Form — A quadratic function is in vertex form when written $f(x) = a(x - h)^2 + k$ where $a \neq 0$.

Forma De Vértice — Una función cuadrática es en forma general cuándo escrito $f(x) = a(x - h)^2 + k$ donde $a \neq 0$.

Vertical Angles — Non-adjacent angles with a common vertex formed by two intersecting lines.

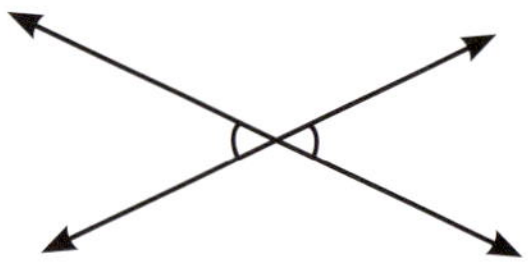

Ángulos Verticales — Ángulos no adyacentes con un vértice en común formado por dos rectas intersecantes.

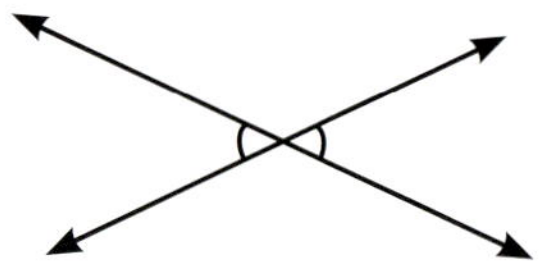

Vertical Line Test — A test used to determine if a graph represents a function by checking to see if a vertical line passes through no more than one point of the graph of a relation.

Examen Vertical De Línia — Un examen para determinar si una gráfica representa una función. Es utilizada para ver si una línia vertical que pasa a través de no más de un punto de la gráfica de una relación.

Volume — The number of cubic units needed to fill a three-dimensional figure.

Volumen — La cantidad de unidades cúbicas necesitadas para llenar un sólido.

X-Y-Z

x-Axis — The horizontal number line on a coordinate plane.

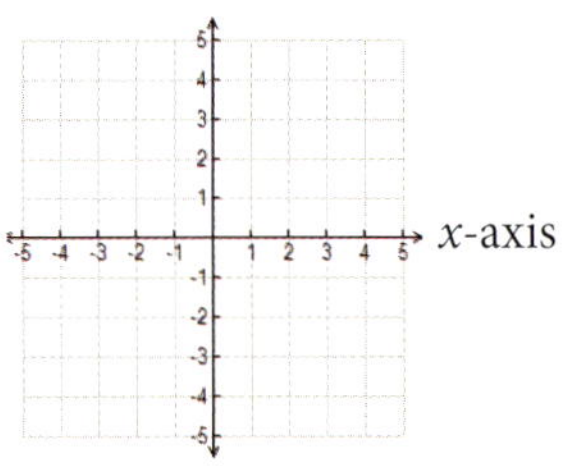

Eje-x, Eje de la x — La recta numérica horizontal en un plano de coordenadas.

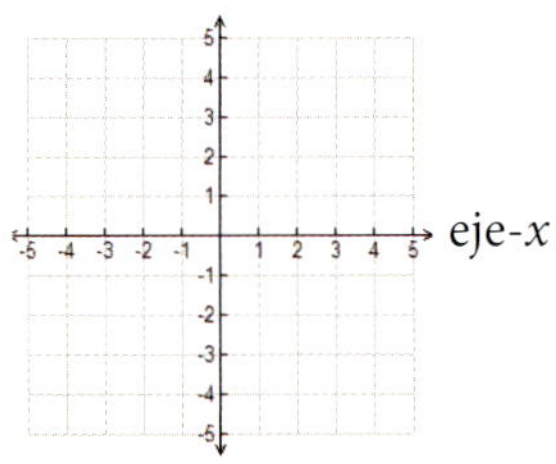

y-Axis

The vertical number line on a coordinate plane.

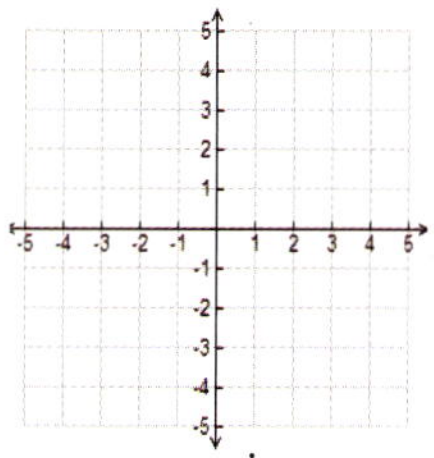

Eje-*y*, Eje de la *y*

La recta numérica vertical en un plano de coordenadas.

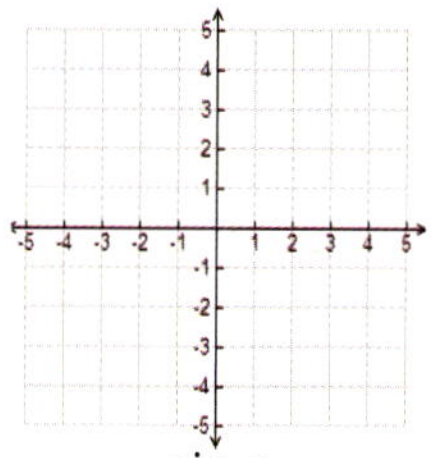

y-Intercept

The point where a graph intersects the *y*-axis.

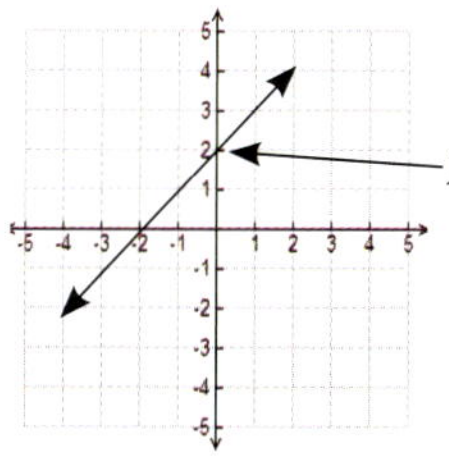

Intersección *y*

El punto donde una gráfica interseca el eje-*y*.

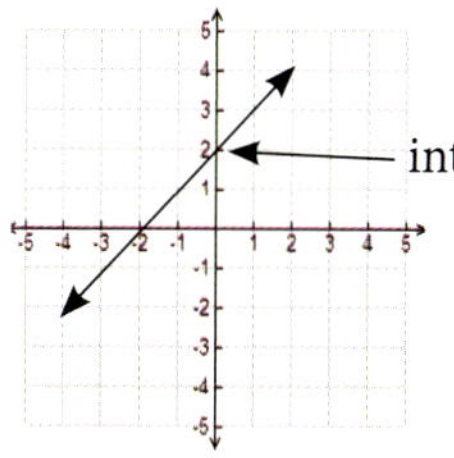

Zero Pair

One positive integer chip paired with one negative integer chip.

$$1 + (-1) = 0$$

Par Cero

Un chip entero positivo emparejado con un chip entero negativo.

$$1 + (-1) = 0$$

Zero Product Property

If a product of two factors is equal to zero, then one or both of the factors must be zero.

Propiedad De Producto Cero

Si un producto de dos factores es iqual a cero, uno o ambos de los factores debe ser cero.

Zeros

The *x*-intercepts of a quadratic function.

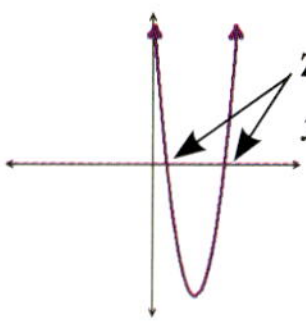

Ceros

Las intersecciones-*x* de una función cuadratica.

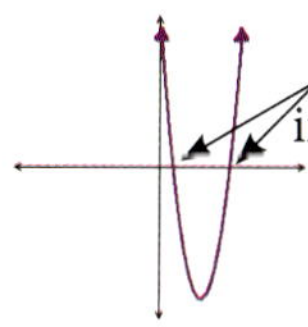

SELECTED ANSWERS

BLOCK 1

Lesson 1.1

1. Alternate interior **3.** Linear pair **5.** Vertical **7.** Vertical **9.** Alternate interior **11.** Alternate interior; not congruent because the lines are not parallel **13.** Alternate interior; $x = 20$ **15.** Alternate interior; $x = 15$ **17.** $x = 34$; each angle is 80° **19.** $x = 17$; angle measures are 34° and 56° **21.** $x = 34$; $m\angle 5 = 62°$ and $m\angle 8 = 118°$

Lesson 1.2

1. Alternate exterior **3.** Same-side interior **5.** Same-side interior **7.** Corresponding **9.** Alternate exterior **11.** $x = 11$ **13.** $x = 24$ **15.** $x = 23.5$; angle is 134° **17.** $m\angle 1 = 99°$, $m\angle 2 = 81°$, $m\angle 3 = 99°$, $m\angle 4 = 99°$, $m\angle 5 = 81°$, $m\angle 6 = 81°$, $m\angle 7 = 99°$ **19.** $m\angle 1 = 76°$, $m\angle 2 = 104°$, $m\angle 3 = 76°$, $m\angle 4 = 104°$, $m\angle 5 = 104°$, $m\angle 6 = 104°$ **21.** See student work.

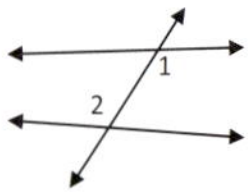

23. See student work.

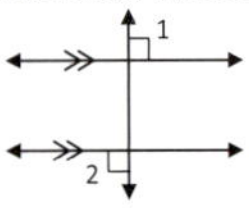

25. See student work.

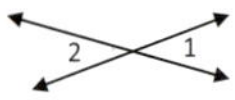

27. See student work.

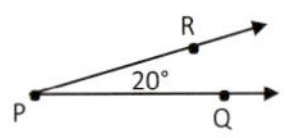

29. See student work.

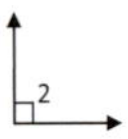

31. Complementary; $x = 18$

Lesson 1.3

1. Equilateral, scalene and isosceles **3.** Right **5.** Acute **7.** Obtuse **9.** Equilateral **11.** Isosceles **13.** Scalene **15.** Acute isosceles **17.** Acute equilateral **19.** Right scalene **21.** See student work.

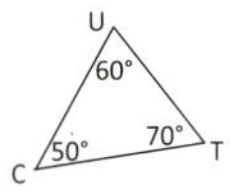

23. See student work.

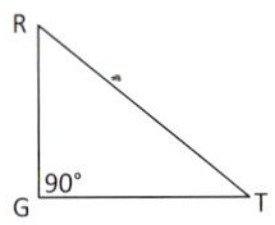

25. See student work.

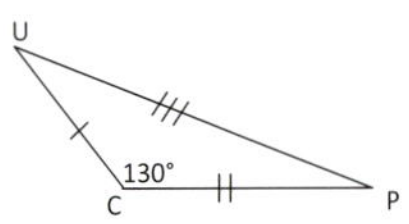

27. Equilateral
29. Scalene
31. Isosceles
33. No. See student work. **35.** $x = 5$ **37.** $x = 15$

Lesson 1.4

1. $x = 35$ **3.** $x = 53$ **5.** $x = 21$ **7.** $4x + 5x + 90 = 180°$; $x = 10$ **9.** $x + x + x = 180°$; $x = 60$ **11.** $3x + 5 + 2x + 90 = 180°$; $x = 17$ **13. a)** $2x + x - 6 + 90 = 180$ **b)** $x = 32$ **c)** $m\angle A = 64°$; $m\angle T = 26°$; $m\angle M = 90°$ **15. a)** $2x + 1 + 42 + 42 = 180$; $x = 47.5$ **b)** $m\angle P = 96°$ **17. a)** $m\angle S = 90°$; $m\angle K = 41°$; $m\angle L = 37°$ **b)** No. The angles do not add up to 180°. **c)** $x = -9$ **19.** Scalene acute **21.** 180 **23.** transversal

Lesson 1.5

1. 60° **3.** $x = 71$ **5.** $x = 112$ **7.** $x = 5$ **9. a)** See student work. **b)** 24° **c)** Isosceles acute **11.** See student work. The possible combinations of angles are: 35°, 35°, 110° or 35°, 72.5°, 72.5°. **13. a)** 2 miles **b)** $x = 2$ **15.** $x = 41$ **17.** $x = 13.5$ **19.** $m\angle L = 34°$ **21.** 95°

Lesson 1.6

1. a) $\overline{MN}$ corresponds to $\overline{TP}$; $\angle M \cong \angle T$;
b) $\overline{AN}$ corresponds to $\overline{OP}$; $\angle N \cong \angle P$;
c) $\overline{MA}$ corresponds to $\overline{TO}$; $\angle A \cong \angle O$
3. See student work. **5.** $x = 15$ **7.** $x = 15$ **9.** $x = 12$; $y = 15.4$ **11.** Exercise #9; sides and angles are congruent. **13.** No. All sides are not proportional. **15.** The slope of the triangle acts as a transversal that makes similar angles; See student work. **17.** See student work. All angles must be congruent and two possible sets of side lengths might be 10, 16, 20 and 2.5, 4, 5. **19.** Corresponding angles; $x = 66$
21. a) $5x - 4 + 9x + 16 = 180$; $x = 12$; **b)** 56° and 124°

Lesson 1.7

1. $x = 103$ **3.** $x = 44$ **5.** Answers may vary; the sum of the two angles must be 78°. **7. a)** $\angle ADN$ or $\angle NDA$ **b)** 76° **c)** 76° **d)** By the Angle-Angle Similarity Rule **9.** $x = 65$, $y = 75$, $z = 65$ **11.** $k = 109$, $m = 71$, $t = 133$ **13.** $m\angle 1 = 72°$, $m\angle 2 = 74°$, $m\angle 3 = 34°$ **15. a)** corresponding angles **b)** alternate interior angles **c)** 50° **d)** 70° **e)** 70° **f)** 50° **g)** 60° **h)** 180° **i)** 180°
j) $m\angle 1 \cong m\angle a$ because corresponding angles are congruent on parallel lines and $m\angle 2 \cong m\angle b$ because alternate interior angles are congruent when between parallel lines. Thus, by substitution, $m\angle 1 + m\angle 2 + m\angle c \cong m\angle a + m\angle b + m\angle c$
17.

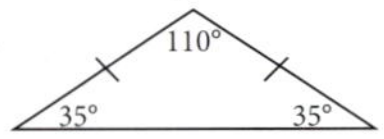

Block 1 Review

1. alternate exterior angles; not congruent **3.** vertical angles; congruent **5.** $x = -7$; measures are 74° **7.** corresponding angles **9.** same-side interior angles
11. $2x + 10 + x + 5 = 180°$; $x = 55$; measures are 120° and 60°
13. $8x - 12 = 48 - 2x$; $x = 6$; measures are 36°
15. acute equilateral **17.** scalene obtuse **19.** isosceles acute
21. See student work. **23.** $x = 67$ **25.** $x = 16$
27. a) See student work.

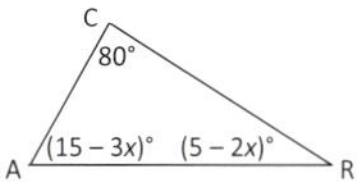

b) $80 + 15 - 3x + 5 - 2x = 180$; $x = -16$ **c)** $m\angle A = 63°$, $m\angle R = 37°$ **29. a)** $x = 67$ **31.** $x = 130$
33. $x = 8$ **35. a)** 1.5 *km* **b)** $x = 1$ **37.** $x = 10$ **39.** $x = 3$
41. Yes, by the Angle-Angle Similarity Theorem.
43. a) $\angle$EGM **b)** 58° **c)** 80° **d)** MEG **45.** $m\angle 1 = 45°$, $m\angle 2 = 54°$, $m\angle 3 = 81°$ **47.** $x = 5$; $y = 68$

BLOCK 2

Lesson 2.1

1. 16 **3.** 1 **5.** 3 **7.** 6 **9.** 5 **11.** 8 **13.** 225 ft^2 **15. a)** 8 *m* **b)** 32 *m* **17.** 7 units **19.** $x = \pm 9$ **21.** $m = \pm 8$ **23.** $d = \pm 3$ **25.** $x = \pm 1$ **27.** $\angle$NPR; $\angle$RPN; $\angle$P; $\angle 1$ **29.** isosceles right **31.** equilateral acute

Lesson 2.2

1. 7 and 8 **3.** 5 and 6 **5.** 11 and 12 **7. a)** 7 and 8 **b)** 7 **c)** Answers may vary (7.1). **d)** 7.071067812 ≈ 7.1; see student work. **9.** Answers may vary (2.2). **11.** Answers may vary (10.1). **13.** Answers may vary (9.4). **15.** 7.7 **17.** 19.2 **19.** 3.9 **21. a)** 114 yards **b)** 456 yards **23. a)** 29 inches **b)** 9 tiles **25.** $x \approx \pm 3.9$ **27.** $x \approx \pm 1.4$ **29.** $x \approx \pm 5.9$ **31.** $x = 20$ **33.** $x = 9$

Lesson 2.3

1. 5 **3.** 8.1 **5.** 10.8 **7. a)** See student work.

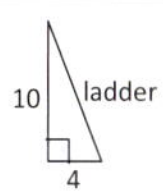

b) 10.8 feet **9. a)** 5 **b)** 25 **c)** 39 **11.** $x = 24$ **13.** $x = 24$ **15.** $x \approx 6.9$
17. See student work; ≈ 13.9 inches **19.** See student work; ≈ 12.7 inches **21. a)** 2.5 miles **b)**

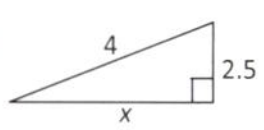

c) 3.1 miles
23. 4 and 5 **25.** 6 and 7
27. $x \approx \pm 7.4$

Lesson 2.4

1. Yes. See student work. **3.** Yes. See student work. **5.** No. See student work. **7. a)** True **b)** False; Pythagorean Triples are made of positive integers, not decimals. **c)** False; Pythagorean Triples are made of positive integers, not fractions. **9.** No **11.** Yes **13.** No **15. a)** 1.5, 2, 2.5 **b)** No, not all integers. **c)** Yes **17.** Yes; see student work. **19.** $x = 37$; $y = 10$ **21.** $x = 16$; $y \approx 23.0$ **23.** $x = 33$; $y \approx 8.4$

Lesson 2.5

1. Yes **3.** ≈ 228 miles **5.** ≈ 56.9 feet **7. a)** 15 meters **b)** 60 meters **c)** ≈ 21.2 meters **9.** ≈ 8.5 miles **11.** ≈ 5.4 feet **13. a)** 61.2 *cm* **b)** ≈ 60.1 *cm* **c)** 25 by 25 by 50 **15. a)** 6 faces **b)** 100 square meters **c)** 10 meters **d)** ≈ 17.3 meters **17.** 9 and 10 **19.** $x \approx 5.7$ **21.** $x \approx 3.2$ **23.** See student work. (Possible answer: 12, 16, 20.)

Lesson 2.6

1. $c = 5$ **3.** $c \approx 5.1$ **5.** ≈ 5.8 **7.** ≈ 6.7 **9.** ≈ 7.3 **11.** ≈ 10.6 **13.** ≈ 3.2 **15.** ≈ 13.8 units **17.** ≈ 23.4 units **19.** No **21.** Yes **23.** 3.2 kilometers **25.** ≈ 4.1 feet

Lesson 2.7

1. 5 **3.** ≈ 8.5 **5.** ≈ 4.2 **7.** 10 **9.** 5 **11. a)** 11.2 **b)** 10.8 **c)** 14.9 **d)** 36.9 **13. a)** 13 units **b)** 81.64 units **c)** 530.66 square units **15.** AB ≈ 6.3; BC ≈ 6.3; AC ≈ 5.7; $\overline{AB} \approx \overline{BC}$ so the triangle is isosceles. **17.** $x = 25.8$ **19.** $x = 24.4$ **21.** Yes, this triangle is possible. **23.** See student work; acute **25.** See student work; right

Block 2 Review

1. 5 **3.** 10 **5.** 8 **7.** 11 feet **9.** $x = \pm 9$ **11.** $x = \pm 2$ **13.** 8 and 9 **15.** Answers may vary (≈ 2.8). **17.** Answers may vary (≈ 7.2). **19.** ≈ 9.6 **21.** ≈ 110 feet **23.** $x = 10$ **25.** $x = 5$ **27.** $x \approx 6.7$ **29. a)** 5 meters **b)** ≈ 7.1 meters **31.** Yes. See student work. **33.** Yes **35.** No. All lengths are not integers. **37.** ≈ 5.8 inches

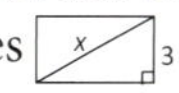

39. ≈ 277.3 miles

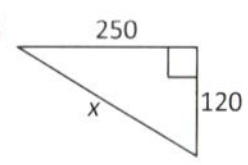

41. ≈ 11.2 meters

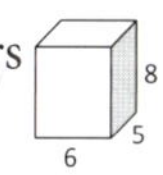

43. $c = 5$ **45.** 10 **47.** ≈ 8.9 **49.** ≈ 20.6 units **51.** ≈ 5.4 **53.** ≈ 6.4 **55. a)** 5 units **b)** 31.4 units **c)** 78.5 square units

BLOCK 3

Lesson 3.1

1. (5, −2)

3. a)

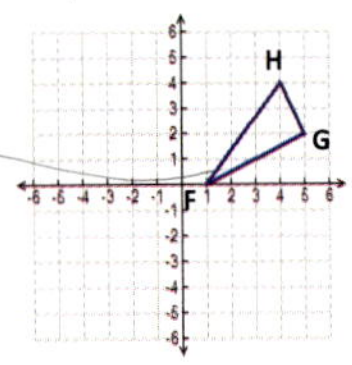

b) F'(−1, 0), G'(−5, 2), H'(−4,4)

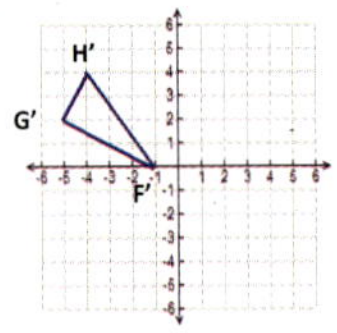

5. reflection over the y-axis; The x-coordinates of the image have the opposite sign of those in the pre-image.

7.

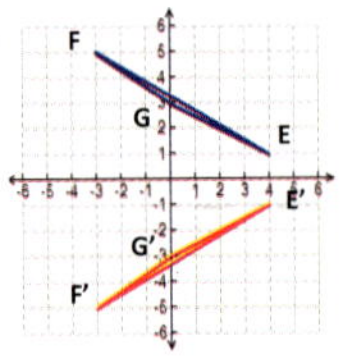

9.

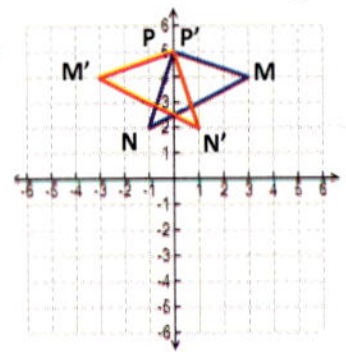

11. reflection over the y-axis

13. a)

x	y
−2	4
−1	1
0	0
1	1
2	4

b)

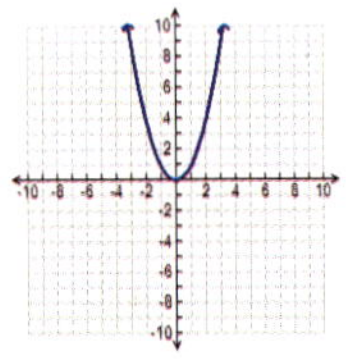

c) (−2, −4), (−1, −1), (0, 0), (1, −1), (2, −4)

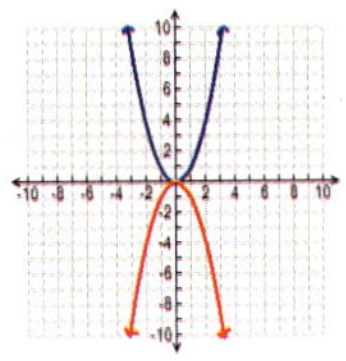

d) The x-coordinates would change to the opposite which means the image and pre-image would look the same.

15. B'(3, 0) **17.** No **19.** Yes **21.** Yes **23.** 90° **25.** 120°

Lesson 3.2

1. (2, 9) **3.** (−3, 10) **5.** (5, 20)

7. a)

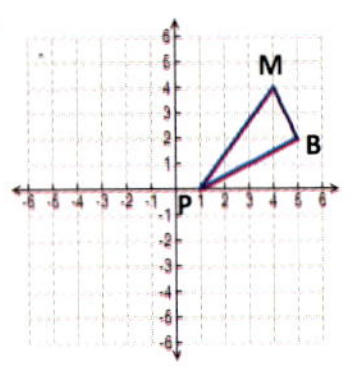

b) P'(−1, −3), B'(3, −1), M'(2, 1)

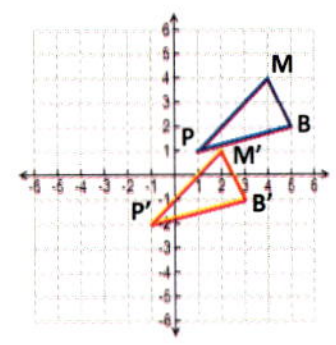

9. $(x, y) \to (x - 1, y - 2)$ **11.**

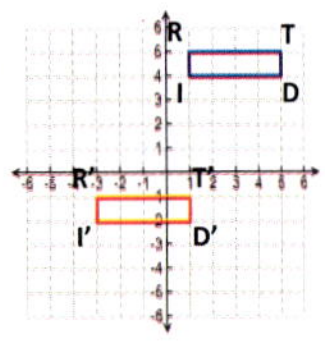

13. a) translation right 2 units and down 3 units **b)** $(x, y) \to (x + 2, y - 3)$

15. See student work.

17. $(x, y) \to (x + 4, y + 6)$ **19.** 8.9 feet

21. 9 inches and 12 inches

Lesson 3.3

1. K'(−1, −4) **3.** K'(−1, −4) **5.** #1 and #3 **7. a)**

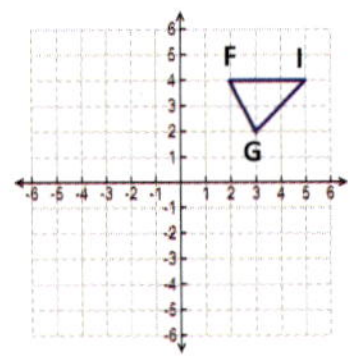

b)

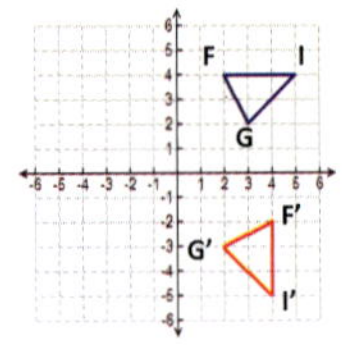

9. 180° about the origin **11.**

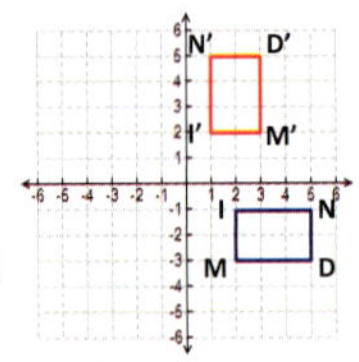

13. $(-y, x)$ **15.** No, the figures will be congruent; Answers may vary.

17. Agree; Answers may vary.

19. reflection; P'(−1, −3)

21. translation; P'(0, −5) **23.** translation; P'(5, 6) **25.** 8 and 9

27. 3 and 4 **29.** 5 and 6

Lesson 3.4

1. a)

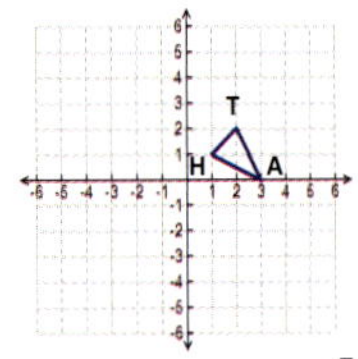

b)

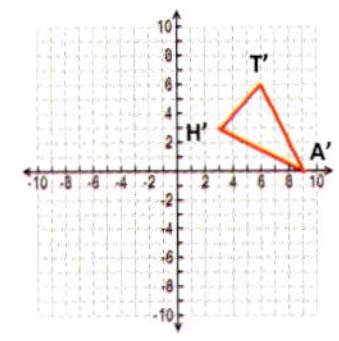

c) enlargement

3. 0.25; reduction **5.** $\frac{5}{2}$; enlargement **7.** $\frac{9}{10}$; reduction

9. $(x, y) \to (\frac{1}{3}x, \frac{1}{3}y)$ **11.**

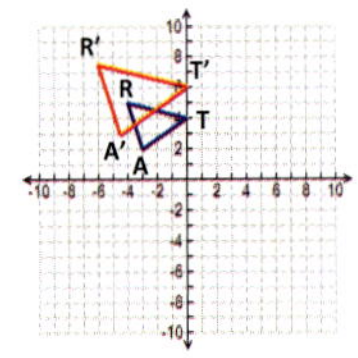

13. a) The perimeter of rectangle A is half the perimeter of rectangle B. **b)** The area of rectangle A is one-fourth the area of rectangle B. **15. a)** No. Only the y-coordinates were multiplied by the scale factor. **b)** R'(−4, 2), S'(4, 2), T'(0, −6) c) She only multiplied the y-coordinates by the scale factor.

17. $(x, y) \to (-x, -y)$ **19.** $(x, y) \to (x + 1, y + 5)$

21. $(x, y) \to (y, -x)$

Lesson 3.5

1. $(x, y) \to (-x, y)$ **3.** $(x, y) \to (3x, 3y)$ **5.** $(x, y) \to (x - 2, y - 8)$

7. dilation; similar **9.** reflection; congruent **11.** dilation

13. a) $(x, y) \to (0.5x, 0.5y)$ **b)** dilation **c)** similar

15. Answers may vary. **17.** Disagree. Triangles will be neither similar nor congruent. **19.** reflection; congruent

21. 13 **23.** 7.8 **25.** 12.1

Lesson 3.6

1. (0, −8) **3.** (4, 4) **5.** (10, −2) **7.** Answers may vary. **9.** Similar **11.** Similar **13.** Answers may vary. **15.** See student work. **17.** Answers may vary. **19.** Answers may vary. **21.** See student work. **23.** Vicki is correct. Evan's transformation series has an image of (−1, −5). **25.** $(x, y) \rightarrow (7x, 7y)$ **27.** $(x, y) \rightarrow (x + 1, y)$ **29.** $(x, y) \rightarrow (-y, x)$ **31.** $(x, y) \rightarrow (x, -y)$

Block 3 Review

1. (−1, −6)

3. a)

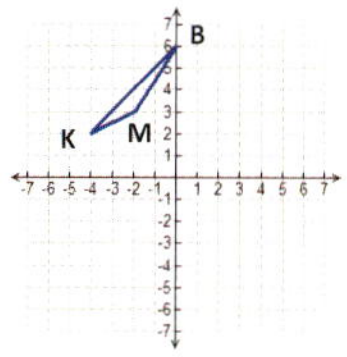

b) K'(−4, −2), B'(0, −6), M'(−2, −3)

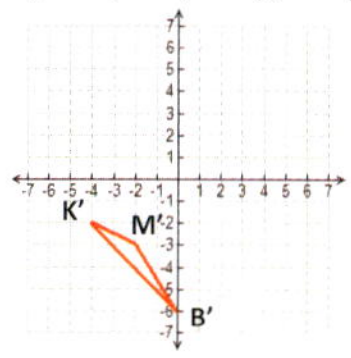

5. reflection over the x-axis **7.**

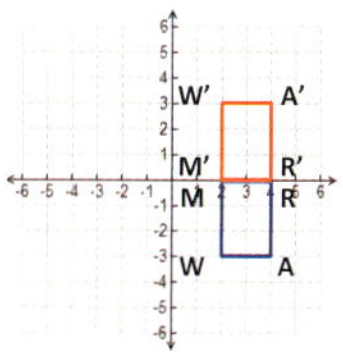

9. P'(5, −4) **11.** P'(8, −8)

13.

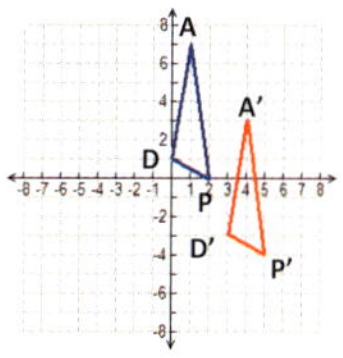

15. Q'(8, 3) **17.** 90° **19.** 90° **21.**

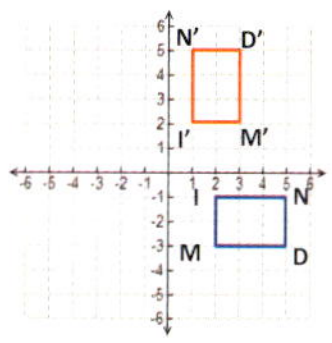

23. 0.3; reduction **25.** $\frac{4}{3}$; enlargement **27.** $(x, y) \rightarrow (4x, 4y)$

29.

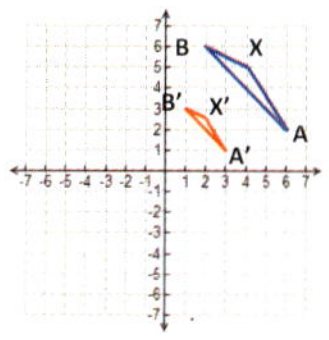

31. $(x, y) \rightarrow (x + 2, y + 5)$; congruent **33.** $(x, y) \rightarrow (-x, -y)$; similar **35.** rotation 90° clockwise about the origin; congruent **37.** dilation with a scale factor of 4; similar **39.** Answers may vary. **41.** V'(−1, 4) **43.** V'(−4, 7) **45.** congruent **47.** similar **49.** congruent; Answers may vary.

BLOCK 4

Lesson 4.1

1. x^5 **3.** p^5q^5 **5.** $w^7y^7z^7$ **7.** $25g^2h^4$ **9.** $0.125d^{27}f^6$ **11.** Answers may vary. **13.** Joe is right because the powers have different bases. **15.** $128w^4y^{22}$ **17.** $576x^{23}$ **19.** $1600a^{14}b^{13}$ **21.** $V = a^5b^6$ **23.** $A = 5x^6$ **25.**

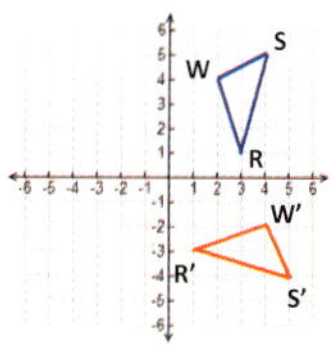

27.

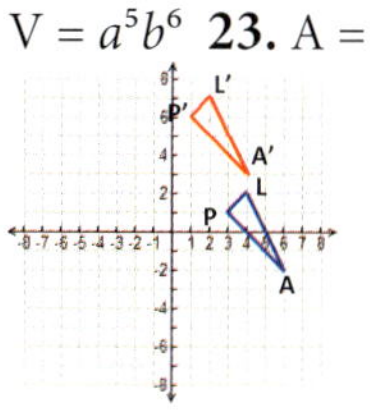

29. dilation

Lesson 4.2

1. 8^7 **3.** a^3b^4 **5.** $\frac{d^{10}}{g^{15}}$ **7.** 1 **9.** $\frac{1}{25}$ **11.** $\frac{m^2n^7}{k^3}$ **13.** $5x^3y^4$ **15.** 4^3 or 64 **17.** $\frac{6x^{10}}{y^6}$ **19.** $\frac{n^{15}}{r^6}$ **21.** $\frac{3q^2y^5}{2x^4w^3}$ **23.** $\frac{a^5m^5}{p^2}$ **25.** x^{15} **27.** $108x^{10}$ **29.** $2m^8$ **31.** $6p^3$

Lesson 4.3

1. 5.8×10^{-5} **3.** 2.0×10^9 **5.** 7.0×10^{-7} **7.** 5.6×10^2 **9.** -7.5×10^5 **11.** 0.0071 **13.** −0.0008 **15.** 431,700 **17.** 5,700,000 **19.** 9.9×10^{-3}, 0.06, 8×10^{-2}, 0.091 **21.** 140,000 kilometers **23.** 3 places to the left **25.** d^8f^3 **27.** $\frac{1}{4}a^2$ **29.** $\frac{5}{x^3y^4}$ **31.** $\frac{4}{y^4z}$ **33.** $16x^9y^4$

Lesson 4.4

1. 6×10^9; 6,000,000,000 **3.** 1.62×10^{11}; 162,000,000,000 **5.** 2.666×10^9; 2,666,000,000 **7.** Answers may vary. **9.** 7×10^8; 700,000,000 **11.** 8×10^{-5}; 0.00008 **13.** 2.5×10^5; 250,000 **15.** 2.5×10^3; 2,500 **17.** About 20,000 times **19.** 4.9275×10^{11} pounds or 492,750,000,000 lbs **21.** 25 inches **23.** 26 feet **25.** 50 *mm*

Lesson 4.5

1. a) B **b)** C **c)** D **d)** E **e)** A **3.** Distance cannot be negative. **5.** $x = \pm 7$ **7.** $x = -2$ **9.** $x = 3$ **11.** $x = 3$ **13.** $r = 3$ inches **15. a)** $b = 5$ **b)** $a = 20$ **c)** $c = 25$ **17.** $x = \pm 3.32$ **19.** $x = -2.80$ **21.** $x = \pm 27.71$ **23.** 8×10^6; 8,000,000 **25.** 3.36×10^8; 336,000,000 **27.** 3×10^8; 300,000,000 **29.** 600 times

Lesson 4.6

1. 552.64 cm^3 **3.** 904.32 m^3 **5.** 50.24 ft^3 **7. a)** 942 ft^3 **b)** Answers may vary. If the pool is filled to the top, when people get in it the water will overflow. **9. a)** 12 *in* **b)** 904.32 in^3 **c)** 40,694.4 in^3 **11.** 40 *in* **13. a)** 10 *cm* **b)** No. Physically, the cubes will not fit unless broken into granules. Space will remain between the side of the cylinder and the cubes. **15.** 3 *ft* **17.** 39.25 ft^3 **19. a)** 25,434 cm^3 **b)** 9 *cm* **21.** 135° **23.** $x = 3$ **25.** $x = 2$ **27.** $x = -7$

Lesson 4.7

1. 288 ft^3 **3.** 38 in^3 **5.** 6 m^3 **7.** See student work; 1,017.36 yd^3 **9.** 75.36 cm^3 **11.** 6 *in* **13.** 1 *in* **15. a)** length = 24 *in*, radius = 0.5 *in* **b)** See student work. **c)** 6.28 in^3 **17.** 3 *in* **19.** 268.21 ft^3 **21.** 5 units **23.** ≈ 9.9 units **25.** ≈ 401.92 cm^3 **27.** $\frac{25m^6}{4a^{10}}$

Lesson 4.8

1. 2,143.57 cm^3 **3.** 267.95 m^3 **5.** ≈ 65.42 in^3 **7. a)** ≈ 904.32 in^3 **b)** ≈ 10,851.84 in^3 of helium **c)** 10,852 in^3 of helium **9.** 6 *cm* **11.** ≈ 8 *in* **13.** ≈ 10 *yd* **15.** ≈ 2 *in* **17.** ≈ 5.76 *in* **19.** ≈ 12.6 *in* **21.** ≈ 2,744 in^3 **23.** $x = 3$ **25.** about 47 blocks **27.** ≈ 6.08 units

Block 4 Review

1. x^{12} **3.** $x^{12}y^6$ **5.** $m^2n^4w^9$ **7.** $128y^{11}$ **9.** $-675x^{18}$ **11.** x^9
13. $36y^7$ **15.** $3y$ **17.** $\frac{b^2c^5}{3}$ **19.** $\frac{25w^6}{4}$ **21.** $\frac{81}{d^{12}}$ **23.** $4ab$
25. 4.3×10^{-4} **27.** 8.5×10^5 **29.** 7.98×10^2 **31.** 80,000
33. 0.00195 **35.** −99,000,000 **37.** 5.76×10^4 **39.** Peter
41. 6×10^9; 6,000,000,000 **43.** 1.8×10^{-7}; 0.00000018
45. 9×10^5; 900,000 **47.** 5×10^1; 50 **49.** $x = 11$ **51.** $x = \pm 6$
53. $x = \pm 4$ **55.** $r \approx 9.8$ inches **57.** 7.35 **59.** $x \approx 1.59$
61. 100.48 m^3 **63.** **a)** 2,260.8 m^3 **b)** 25 fish **65.** $h = 12$ *cm*
67. The diameter of the glass is about 3 inches.
69. 50.24 *cm* 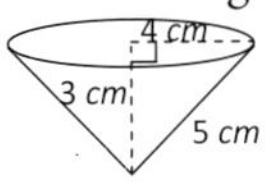

71. **a)** 37.68 in^3 **b)** 301.44 in^3

73. 2,411.52 cm^3 **75.** $r \approx 9$ *cm* **77.** 33.49 ft^3 **79.** 3,052.08 m^3
81. $d \approx 22$ *in*

INDEX

PROBLEM-SOLVING

UNDERSTAND THE SITUATION

- Read then re-read the problem.
- Identify what the problem is asking you to find.
- Locate the key information.

PLAN YOUR APPROACH

Choose a strategy to solve the problem:

- Guess, check and revise
- Use an equation
- Use a formula
- Draw a picture
- Draw a graph
- Make a table
- Make a chart
- Make a list
- Look for patterns
- Compute or simplify

SOLVE THE PROBLEM

- Use your strategy to solve the problem.
- Show all work.

ANSWER THE QUESTION

- State your answer in a complete sentence.
- Include the appropriate units.

STOP AND THINK

- Did you answer the question that was asked?
- Does your answer make sense?
- Does your answer have the correct units?
- Look back over your work and correct any mistakes.

DEFEND YOUR ANSWER

Show that your answer is correct by doing one of the following:

- Use a second strategy to get the same answer.
- Verify that your first calculations are accurate by repeating your process.

SYMBOLS

Algebra and Number Operations

SYMBOL	MEANING
$+$	Plus or positive
$-$	Minus or negative
$5 \times n, 5 \cdot n, 5n, 5(n)$	Times (multiplication)
$3 \div 4, 4\overline{)3}, \frac{3}{4}$	Divided by (division)
$=$	Is equal to
$\approx$	Is approximately
$<$	Is less than
$>$	Is greater than
%	Percent
$a : b$ or $\frac{a}{b}$	Ratio of a to b
$5.\overline{2}$	Repeating decimal (5.222…)
$\geq$	Is greater than or equal to
$\leq$	Is less than or equal to
x^n	The nth power of x
(a, b)	Ordered pair where a is the x-coordinate and b is the y-coordinate
$\pm$	Plus or minus
$\sqrt{x}$	Square root of x
$\neq$	Not equal to
$x = y$	Is x equal to y?
$\lvert x \rvert$	Absolute value of x
P(A)	Probability of event A

Geometry and Measurement

SYMBOL	MEANING
$\cong$	Is congruent to
$\sim$	Is similar to
$\angle$	Angle
$m\angle$	Measure of angle
ΔABC	Triangle ΛBC
$\overline{AB}$	Line segment AB
$\overrightarrow{AB}$	Ray AB
AB	Length of AB
π	Pi (approximately $\frac{22}{7}$ or 3.14)
$^\circ$	Degree